Introduction to
the Principles of
Ceramic Processing

Introduction to the Principles of Ceramic Processing

James S. Reed

New York State College of Ceramics
Alfred University
Alfred, New York

A WILEY-INTERSCIENCE PUBLICATION

JOHN WILEY & SONS

New York · Chichester · Brisbane · Toronto · Singapore

Library of Congress Cataloging-in-Publication Data:

Reed, James Stalford, 1938–
   Introduction to the principles of ceramic
processing.

   "A Wiley-Interscience publication."
   Includes bibliographies.
   1. Ceramics.   I. Title.
TP807.R36   1987   666   87-25310
ISBN 0-471-84554-X

Printed in the United States of America

10 9 8 7 6 5 4 3 2 1

Dedicated to Carol and Scott

Preface

Ceramic processing has traditionally been discussed in terms of the material formulations and industrial arts used in the production of classes of products. This empirical approach was understandably necessary, because the material systems were complex and not well characterized. But as the characterization of materials has improved and the scientific principles underlying ceramic processing have been elucidated, especially in the past 20 years, ceramic processing can now be explained in the context of general principles and the engineering processes involved. Ceramic processing is an increasingly important aspect of ceramic engineering, and advances in understanding the principles of ceramic processing will be needed to produce more advanced ceramic products and components of integrated products. Information on the principles of ceramic processing is widely scattered in research papers and in conference proceedings. The need for a book that provides an overview and introduces the principles of ceramic processing has been impressed on me in my teaching experience and in conversations with industrial scientists and engineers. This book has been written to be used as a reference book and as a textbook.

Part I of this book describes applications and compositions of modern ceramics, presents a perspective of science in ceramic processing, and reviews fundamentals of surface chemistry. Part II describes starting materials for ceramics and emphasizes the chemical preparation of inorganic chemicals and more advanced materials. The characterization and specification of particulate materials and the structure and functions of processing additives are explained in Parts III and IV, respectively. Part V presents the fundamentals of particle packing and discusses the rheology of processing systems; this very complex topic is approached by considering the contributions of the component phases and insights of particle mechanics from the field of soil science. Principles involved in enhancing the physical and chemical character of the process system, called beneficiation processes, are explained along with industrial practices in Part VI. These four chapters include crushing and milling processes; batching and mixing; particle separation, concentration, and washing processes; and spray granulation and spray drying. Part VII presents the principles of forming in the categories of pressing, plastic forming, which includes extrusion, plastic molding, jiggering, and injection molding, casting processes, which include slip casting, pressure casting, solid casting, and tape casting, and molecular polymerization forming, which includes sol-gel processing and chemical vapor deposition. Causes and the prevention of defects in formed products are explained. Part VIII discusses drying fundamentals; surface processing which includes surface finishing, film printing, and glazing; and microstructural changes and development during sintering.

I want to express my appreciation to those who have helped me in many ways. My greatest thanks go to my wife, Carol Reed, for her constant help and encouragement. Many people have kindled and supported my interest in ceramic engineering. In particular I would like to acknowledge the late Professor George W. Brindley, who first introduced me to the field of ceramics; Dean John F. McMahon, who encouraged my interest in graduate study and ceramic processing; and Professor James E. Young, who supported my interest in developing courses and teaching ceramic processing. Last but not least, I would like to acknowledge the numerous students whose questions and research theses over the past 20 years have helped to clarify insights and provide new understanding.

In the preparation of the book I am indebted to Mr. Frank Cerra for his early encouragement, Mrs. Sandi Congelli for typing the manuscript, Ms. Hollis Findeisen for drafting the majority of the line drawings, and Mr. Ward Votava for assistance with the photomicrographs.

JAMES S. REED

Alfred, New York
December 1986

Contents

VIII. DRYING, SURFACE PROCESSING, AND FIRING

Contents   xvii

Introduction to
the Principles of
Ceramic Processing

PART I

INTRODUCTION

Ceramic processing is an ancient art but a young applied science. The history of ceramic processing indicates periods of rapid development interspersed within long periods of rather slow development; the second half of the twentieth century is a period of intense development. The transition of ceramic processing to an applied science is the natural result of an increasing ability to refine, develop, and characterize ceramic materials, and systems of additives which aid in processing systems containing hard, brittle particles, improved equipment for processing these materials into products, and advances in understanding ceramic processing fundamentals. These topics are discussed in Chapter 1. Principles of physical chemistry that are a basis of ceramic processing science are reviewed in Chapter 2.

1

Ceramic Processing and Ceramic Products

This book is concerned primarily with understanding the scientific principles and technology involved in processing particulate ceramic materials into fabricated products. Our topic is commonly referred to as ceramic fabrication processes, ceramic processing technology, or simply ceramic processing. Ceramic processing technology is used to produce commercial products that are very diverse in size, shape, detail, complexity, material composition, structure, and cost. Several examples of both modern advanced ceramics and traditional ceramics are shown in Fig. 1.1. The applications of these products, as are indicated in Table 1.1, are also diverse and are dependent on the functions indicated in Table 1.2.

The functions of ceramic products are very dependent on their chemical composition and their atomic and microscale structure, which determines their properties. Compositions of ceramic products vary widely, and both oxide and nonoxide materials are used. Today the composition and structure of grains and grain boundary phases and the distribution and structure of pores is more carefully controlled to achieve greater product performance and reliability (Fig. 1.2). In the development and production of the more advanced ceramics, extraordinary control of the materials and processing operations is requisite to minimize microstructural defects. Recent successes in developing, producing, and applying advanced ceramics in high-tech applications have affected the consciousness of all engineers and heightened the interest in ceramic processing. An awareness that improvements in ceramic processing technology can improve manufacturing productivity and expand markets now pervades the industry.

1.1 A BRIEF HISTORY OF CERAMIC TECHNOLOGY

The history of ceramic processing technology is very interesting in that both simple processes developed in ancient times for natural materials, and

3

Fig. 1.1(a) Multilayer electronic packaging. The substrate used in the module is formed from sheets of unfired green ceramic. First, thousands of minute holes, or vias, are punched in each sheet. The wiring pattern that conducts the electrical signals is then formed by screening a metallic paste onto the sheet through a metal mask. The via holes are also filled with this paste to provide the electrical connections from one layer or sheet to another. The layers are then stacked and laminated together under heat and pressure, and the laminate is sintered in a process that shrinks it approximately 17%. (Photos courtesy of IBM, East Fishkill, NY.)

Fig. 1.1(b) Magnetic ceramic ferrites used in a wide variety of electrical power and electronic communications systems. (Photo courtesy of Magnetics Div., Spang and Co., Butler, PA.)

Fig. 1.1(c) Refractory "honeycomb" cordierite catalyst support used in automobile exhaust system. (Photo courtesy of Corning Glass Works, Inc., Corning, NY.)

Fig 1.1(d) Advanced alumina structural ceramics used in a wide variety of materials-processing technology applications (textile guides, valve seals, impellers, nozzle inserts, wear blocks, cutting tools, milling media, refractory insulation supports) and electrical applications (spark plug insulators, electronic substrates, electrical insulation). (Photo courtesy of Diamonite Products, Div. of W.R. Grace and Co., Shreve, OH.)

5

(e)

Fig. 1.1(e) Corrosion/erosion-resistant silicon carbide structural ceramic components used in the chemical processing industry. (Photo courtesy of Standard Oil Engineered Materials Co., Niagara Falls, NY.)

(f)

Fig. 1.1(f) Silicon nitride metal cutting tools (Photo courtesy of GTE Laboratories, Waltham, MA).

Fig. 1.1(g) Traditional silicate ceramic products include electrical porcelain, household and institutional porcelain products, refractories, and ceramic wall and floor tile. (Photo courtesy of R.T. Vanderbilt Co., Norwalk, CT).

Table 1.1 Products Produced by Ceramic Powder Processing

Electronics
    Substrates, chip carriers, electronic packaging
    Capacitors, inductors, resistors, electrical insulation
    Transducers, servisors, electrodes, igniters
    Motor magnets, spark plug insulators
Advanced structural materials
    Cutting tools, waer-resistant inserts
    Engine components
    Resistant coatings
    Dental and orthopedic prostheses
    High-efficiency lamps
Chemical processing components
    Ion exchange media
    Emission control components
    Catalyst supports
    Liquid and gas filters
Refractory structures
    Refractory lining in furnaces, thermal insulations, kiln furniture
    Recuperators, regenerators
    Crucibles, metal-processing materials, filters, molds
    Heating elements
Construction materials
    Tile, structural clay products
    Cement, concrete
Institutional and domestic products
    Cookware
    Hotel china and dinnerware
    Bathroom fixtures
    Decorative fixtures and household items

Table 1.2 Classification of Ceramics by Function

Function	Class	Nominal Composition[a]
Electrical	Insulation	α-Al_2O_3, MgO, porcelain
	Ferroelectrics	$BaTiO_3$, $SrTiO_3$
	Piezoelectric	$PbZr_{0.5}Ti_{0.5}O_3$
	Fast ion conduction	β-Al_2O_3, doped ZrO_2
	Superconductors	$Ba_2YCu_3O_{7-x}$
Magnetic	Soft ferrite	$Mn_{0.4}Zn_{0.6}Fe_2O_4$
	Hard ferrite	$BaFe_{12}O_{19}$, $SrFe_{12}O_{19}$
Nuclear	Fuel	UO_2, UO_2–PuO_2
	Cladding/shielding	SiC, B_4C
Optical	Transparent envelope	α-Al_2O_3, $MgAl_2O_4$
	Light memory	doped $PbZr_{0.5}Ti_{0.5}O_3$
	Colors	doped $ZrSiO_4$, doped ZrO_2, doped Al_2O_3
Mechanical	Structural refractory	α-Al_2O_3, MgO, SiC, Si_3N_4 $Al_6Si_2O_{13}$
	Wear resistance	α-Al_2O_3, ZrO_2, SiC, Si_3N_4, toughened Al_2O_3
	Cutting	α-Al_2O_3, ZrO_2, TiC, Si_3N_4, SIALON
	Abrasive	α-Al_2O_3, SiC, toughened Al_2O_3, SIALON
	Construction	Al_2O_3–SiO_2, CaO–Al_2O_3–SiO_2, porcelain
Thermal	Insulation	α-Al_2O_3, ZrO_2, $Al_6Si_2O_{13}$, SiO_2
	Radiator	ZrO_2, TiO_2
Chemical	Gas sensor	ZnO, ZrO_2, SnO_2, Fe_2O_3
	Catalyst carrier	$Mg_2Al_4Si_5O_{18}$, Al_2O_3
	Electrodes	TiO_2, TiB_2, SnO_2, ZnO
	Filters	SiO_2, α-Al_2O_3
	Coatings	NaO–CaO–Al_2O_3–SiO_2
Biological	Structural prostheses	α-Al_2O_3, porcelain
	Cements	$CaHPO_4 \cdot 2H_2O$
Aesthetic	Pottery, artware	Whiteware, porcelain
	Tile, concrete	Whiteware, CaO–SiO_2–H_2O

Source: Adapted from George B. Kenne and H. Kent Bowen, "High-Tech Ceramics in Japan: Current and Future Markets," *Am. Ceram. Soc. Bull.* **62**(5), 590–596 (1982).

[a]Whiteware is a family of porous-dense, fine-grained materials with a glassy matrix usually containing Al_2O_3, SiO_2, K_2O, and Na_2O. Porcelain is a type of whiteware that is nonporous, hard, and translucent. SIALON is a solid solution phase with the nominal composition $Si_4Al_2N_6O_2$. Toughened Al_2O_3 is a two-phase material containing a minor amount of doped or undoped ZrO_2.

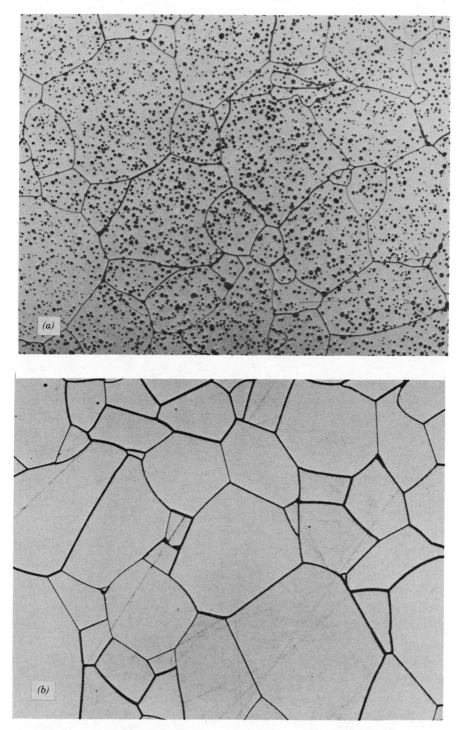

Fig. 1.2 Comparative microstructures in (a) a conventional dense alumina of 98% density, (b) an optically transparent alumina exceeding 99.9% density.

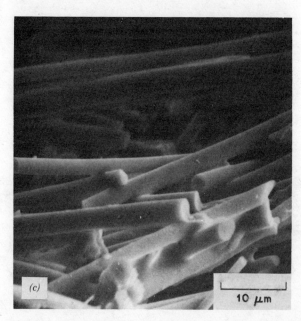

Fig. 1.2 A high-performance alumina thermal insulation of less than 10% density. (Photos a and b courtesy of General Electric Co., Richmond Heights, OH; photo c courtesy of Zircar Products Inc., Florida, NY.)

10

recently developed, relatively sophisticated processes dependent on synthetic materials are used extensively near the end of the 20th century.

Hand mixing, hand building, and scratch and slip decorating of earthenware date back to before 5000 BC. The first forming machine was probably the potter's wheel, which was used earlier than 3500 BC for throwing a plastic earthenware body and later for turning a somewhat dried, leather, hard body. Shaping by pressing material in fired molds and firing in a closed kiln were subsequent developments.

The most notable achievement early in the Christian era was the development in China of pure white porcelain of high translucency. Duplication in the West was frustrated until 1708, when a young German alchemist, Fredrich Bottger, under the direction of the celebrated physicist Count von Tschirnhaus, discovered that fine porcelain could be produced on firing a body containing a fire-resistant clay with fusible materials. Other inventions in the 18th century included the use of a template for forming, slip casting in porous molds, auger extrusion, transfer decoration, and firing in a tunnel kiln.

The introduction of steam power in the 19th century led to the mechanization of mixing, filter pressing, dry pressing, and pebble mill grinding. Near the end of that century, separate phases of silica were distinguished using optical microscopy, and silicon carbide was synthesized in an electric furnace. Pyrometric cones were developed by Seager to control firing.

The first half of the 20th century saw the rapid development of x-ray techniques for the analysis of the atomic structure of crystals and later electron microscopy for examining microstructure beyond the limit of the optical microscope. Material systems became more refined, and special compounds were developed, synthesized, and fabricated into products for refractory and electronic applications. Refined organic additives were purposefully introduced to improve the processing behavior. Industrial production became mechanized, and several stages of manufacturing were automated. Thermocouples were used routinely to monitor temperatures during firing.

The second half of the 20th century has witnessed major advances in the synthesis, characterization, and fabrication of ceramic products. Scanning electron microscopy is now used for routine microstructural analysis for quality control in manufacturing. Several different instrumented techniques have been developed for bulk chemical analysis at a concentration of less than a fraction of one part in a million and surface concentrations a few atomic layers in thickness. The particle size distributions of a material can be determined to below 0.1 μm in a few minutes. The flow behavior during forming is developed and controlled using a multicomponent system of additives. Testing apparatus and processing machinery are much more advanced. Computers are now used throughout the industry to monitor and/or control raw-material handling and preparation, fabrication, and firing.

1.2 INDUSTRIAL CERAMIC PROCESSING

The realization of a product depends on material factors and nonmaterial factors such as the economics of the marketplace, consumer response, dimensional and surface finish tolerances, apparent quality, and manufacturing productivity. The manufacture of ceramics is a complex interaction of raw materials, technological processes, people, and financial investment. Manufacturing managers are involved with all of these aspects. Production engineers, process design engineers, and scientists and engineers involved with ceramic materials and process research and development must be particularly knowledgeable of the applied science and industrial arts of ceramic processing.

Ceramic processing commonly begins with one or more ceramic materials, one or more liquids, and one or more special additives called processing aids. The starting materials or the batched system may be beneficiated chemically and physically using operations such as crushing, milling, washing, chemical dissolving, settling, flotation, magnetic separation, dispersion, mixing, classification, de-airing, filtration, and spray-drying. The forming technique used will depend on the consistency of the system (i.e., slurry, paste, plastic body, or a granular material) and will produce a particular unfired shape with a particular composition and microstructure. Drying removes some or all of the residual processing liquids. Additional operations may include green machining, surface grinding, surface smoothing and cleaning, and the application of surface coatings such as electronic materials or glaze. The finished material is then commonly heat-treated to produce a sintered microstructure. The sintered product may be a single component or a multicomponent composite structure. A general processing flow diagram indicating the sequences of operations used in forming a product is shown in Fig. 1.3.

1.3 SCIENCE IN CERAMIC PROCESSING

Up to about the first half of the 20th century, processing engineers depended on empirical correlations and practical intuition for innovating new process designs and process controls. Most new products were seen as inventions rather than the planned outcome of research and development. Material systems were typically complex, and laboratory tests and analyses were tedious and time-consuming. Wide margins existed for improving the performance and reliability of the products. Typically, a small amount of a new shipment of a raw material was processed through the plant to ascertain if it was usable or to suggest processing adjustments to minimize production losses.

Before the development of scientific insights of ceramic processing, the

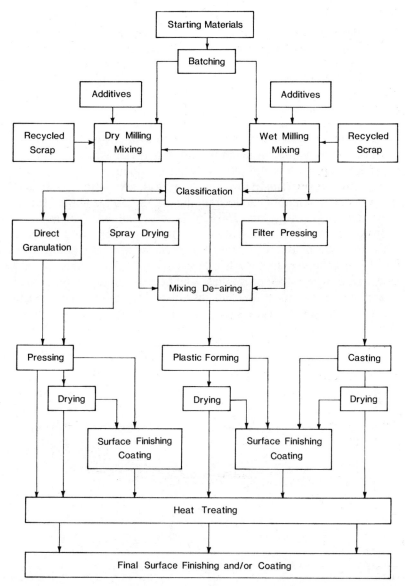

Fig. 1.3 A general processing flow diagram illustrating processing paths from raw materials to the fired product.

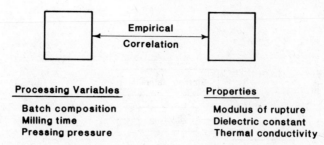

Processing Variables

Batch composition
Milling time
Pressing pressure

Properties

Modulus of rupture
Dielectric constant
Thermal conductivity

Fig. 1.4 Correlation between the final properties and processing variables may identify the more sensitive processing parameters but is empirical.

properties of the product were often correlated with changes in a processing operation to identify the more important superficial variables (Fig. 1.4). This empirical approach is still used and can be aided greatly using computers and statistical programs. However, empirical correlations such as these do not provide a scientific understanding of the fundamental causes of behavior during processing and forming. The probability that adjustments based on empirical correlations alone will produce significant advances is small, because the potential number of unsuccessful combinations of variables is always relatively large. Also, empirical correlations may be of little heuristic value when the processing engineer is faced with a lack of reproducibility in manufacturing, an insufficient reliability in the performance of nominally identical products, or the development of new products.

Viewed as a science, ceramic processing is the sequence of operations that purposefully and systematically changes the chemical and physical aspects of structure, which we call the characteristics of the system. The properties at each stage are a function of the characteristics of the system at each stage and the ambient pressure and temperature, as is shown in Fig. 1.5.

The objectives of the science of ceramic processing are to identify the important characteristics of the system and understand the effects of proces-

Ceramic Processing

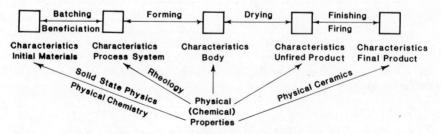

Fig. 1.5 Processing develops the characteristics of the system.

sing variables on the evolution of these characteristics. The objectives in process engineering should be to change these characteristics purposefully to improve product quality. Because of the key dependence on controlling characteristics, an understanding of techniques for characterizing the starting materials and the process system at each stage is an integral part of any discussion of ceramic processing.

Raw materials are now more beneficiated, more consistent, and often much simpler in composition. Modern instrumentation for analyzing ceramics materials and systems is more automated and precise, and with microcomputer accessories quantitative data are obtained quickly and presented in a convenient format. Computer-controlled, closed, raw material handling systems improve the precision, efficiency, health, and safety in batching and mixing particulate materials. The processing pressure and temperature are more precisely monitored and controlled. Wear-resistant materials are used in critical areas to minimize maintenance and contamination. In the factory, on-line monitoring and control of production processes is practiced in some stages of processing in some industries.

Principles of processing science can provide insights into fundamental causes of behavior, procedures for modifying and controlling materials and processes, and avenues for improving manufacturing productivity. The role of ceramic processing science in ceramic manufacturing will surely increase.

SUMMARY

Ceramic products are used in applications where the performance and reliability of the product must be predictable and assured and the product must be fabricated successfully in a productive manner. The manufacture of these products from a complex batch containing ceramic materials and processing additives into a finished product involves many operations. All of the materials and operations must be carefully controlled. Principles of science should be used in addition to empirical tests for understanding, improving, and controlling ceramic processing.

SUGGESTED READING

1. F. H. Norton, *Fine Ceramics*, Krieger, Malabar, FL, 1978.
2. Karl Schwartzwalder, "Processing Controls in Technical Ceramics," Chapter 2 in *Ceramic Processing before Firing*, George Y. Onoda Jr. and Larry L. Hench, Eds., Wiley-Interscience, New York, 1978.
3. Robert J. Charleston, *World Ceramics*, Chartwell, Secancus, NJ, 1977.
4. George B. Kenney and H. Kent Bowen, "High-Tech Ceramics in Japan: Current and Future Markets," *Am. Ceram. Soc. Bull.*, **62**(5), 590–596 (1983).

5. W. D. Kingery, "Social Needs and Ceramic Technology," *Am. Cer. Soc. Bull.*, **59**(6), 598–600 (1980); *"Needs and Opportunities for Ceramic Science and Technology,"* Proceedings of the 3d CIMTEC Meeting on Modern Ceramics Technologies, Elsevier, New York, 1980, pp. 11–16.

6. Haus Thurnauer, "Development and Use of Electronic Ceramics Prior to 1945," *Am. Ceram. Soc. Bull.*, **56**(2), 219–220, 224 (1977).

7. John F. McMahon, "Implications of Our Ceramic Heritage," *Am. Ceram. Soc. Bull.*, **56**(2), 221–224 (1977).

8. A. G. Pincus, "A Critical Compilation of Ceramic Forming Methods. 1. General Introduction," *Am. Ceram. Soc. Bull.* **43**(11), 827–828 (1964).

PROBLEMS

1.1 Explain why the correlation of a property such as apparent viscosity with a processing parameter such as mixing time is empirical in nature. What should you determine to explain the effect of the mixing time on the apparent viscosity?

1.2 Distinguish clearly among the performance, reproducibility, reliability, and quality of a product.

1.3 Productivity can be defined as the ability to make a product sooner, faster, better, at lower cost, and meeting technical and legal requirements. Explain why each factor is important.

1.4 What are the driving forces for further improvements in ceramic processing? What are the restraining forces? Tabulate your answers for each.

1.5 What is the meaning of "value added" in ceramic processing? What is the relative dependence of profits on raw-material costs when the value added is low or high? Explain.

1.6 How will the need for increased productivity and the need to retain manual skills be reconciled in future manufacturing?

1.7 How can a science of ceramic processing aid in improving the health and safety of the manufacturing workplace?

1.8 Ceramic processing is often said to be a "systems problem." Explain.

1.9 Do you agree with the statement, "Every processing scientist must also be a processing engineer"? Why or why not?

1.10 What was the beginning of "fine ceramics"? Is there a distinct beginning of engineered ceramics?

2

Surface Chemistry

Surface and interface phenomena play a particularly important role in ceramic processing because powder systems have a relatively high surface area/mass and the adsorption and distribution of additive phases on the surfaces may alter the microstructure and processing behavior quite markedly. In this chapter we will review several general principles of physical chemistry which are especially important for understanding the processing behavior of particle systems.

2.1 THE ATOMIC STRUCTURE OF THE SURFACE DIFFERS FROM THAT IN THE INTERIOR OF THE PARTICLE

Surfaces of liquids and solids have special properties, because they terminate the phase. An atom at a free surface is bonded to fewer neighboring atoms than an atom within the particle. Since bonding reduces the potential energy, a surface atom has extra energy called the surface energy. This extra energy can be partially reduced by slight adjustments in the composition, packing, and bonding between atoms in the surface. Nevertheless, surface atoms or ions are more active.

In terms of thermodynamics, the surface tension γ is defined as

$$\gamma = \left(\frac{\partial G}{\partial A}\right)_{P,T,N_i} \tag{2.1}$$

where G is the Gibb's free energy of the system. During the change in area A, the independent variables of pressure P, temperature T, and number of species N_i in the system remain constant. Surface tension is a property of a surface and should not be confused with elastic tension.

Consider a liquid film between two slidewires as shown in Fig. 2.1. The liquid can reduce its surface free energy when atoms move from the surface

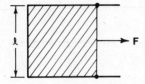

Fig. 2.1 A force is required to increase the surface area of a liquid film.

into the bulk, contracting the surface area. The work dW done by the force F on the slidewire of length l is

$$dW = F\,dx = 2\gamma l\,dx \tag{2.2}$$

and

$$\gamma = \frac{F}{2l} \tag{2.3}$$

The factor 2 appears because the film has two surfaces. For liquids and low-viscosity suspensions free of external forces, the surface tension will cause spherical droplets to form, because this shape has the lowest surface area per unit volume.

The surface tension of a crystal varies with the crystallographic direction of the faces. The more stable planes are usually those of lower surface free energy, which are usually more densely packed. Adsorbed atoms or ions can preferentially alter the surface tension of some planes and the crystal habit.

Table 2.1 Representative Surface Tensions of Liquids in Air

Liquid	Temperature (°C)	Surface Tension (mN/m)
Water	0	76
	20	73
	25	72
	50	68
	80	63
Methanol	20	23
Ethanol	20	22
Acetone	20	24
Ethylene glycol	20	48
Oleic acid	20	33
Mercury	25	474
Water: 7.5% methanol	20	61
Water: 5% ethylene glycol	25	58
Water: 0.001% dimethyl silicone	25	39
Nitrogen	−196	8.8

Source: Handbook of Chemistry and Physics, R.C. Weast Ed; CRC Press, Cleveland, 1983.

Surface tension values for ceramics are somewhat imprecise because of uncertainties concerning the purity of the surface and the accuracy of the technique. For solids, the surface tension is generally $< 100 \, \text{mN/m}$ for polymer materials, $100–2,000 \, \text{mN/m}$ for oxides, and $\geq 1,000 \, \text{mN/m}$ for metals and refractory carbides and nitrides. Representative values of the surface tension of several common liquids are listed in Table 2.1. The surface tension of most liquids decreases with temperature.

2.2 SURFACE ENERGY CAUSES A PRESSURE DIFFERENCE ACROSS A CURVED SURFACE

Consider the bubble of radius r in a liquid, as shown in Fig. 2.2. Surface tension will tend to contract the surface area and the internal volume, increasing the internal pressure by an increment ΔP. At equilibrium, the work of contraction $\Delta P \, dV$ is equal to the decrease in surface free energy $\gamma \, dA$, and

$$\Delta P \, 4\pi r^2 \, dr - \gamma 8\pi r \, dr = 0 \qquad (2.4)$$

and

$$\Delta P = \frac{2\gamma}{r} \qquad (2.5)$$

Equation 2.4 is a special case of the Laplace equation. For a surface with principal radii of curvature r_1 and r_2 at a point on the surface, the general Laplace equation is

$$\Delta P = \gamma \left(\frac{1}{r_1} + \frac{1}{r_2} \right) \qquad (2.6)$$

The radius r is negative when the curvature is concave.

This effect of surface curvature may cause the average chemical potential of atoms in microscopic particles to be greater than that in large particles. And atoms in a microscopic region of sharp positive curvature are of a higher chemical potential than atoms in a flat surface.

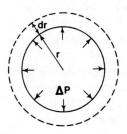

Fig. 2.2 Model for calculating the pressure caused by surface curvature.

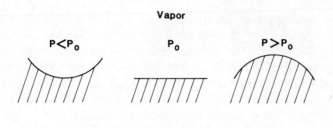

Fig. 2.3 The vapor pressure of a liquid varies with surface curvature.

The surfaces we have considered are really interfaces between a solid or liquid and its vapor. A consequence of the surface curvature effect is that the equilibrium vapor pressure P is a function of the surface energy and surface curvature, given by the Kelvin equation

$$\ln \frac{P}{P_0} = \frac{2\gamma V_{\text{mol}}}{rRT} \tag{2.7}$$

where V_{mol} is the molar volume of the condensed phase at temperature T and P/P_0 is the vapor pressure over a curved surface relative to that over a planar surface (Fig. 2.3).

Processing consequences of the surface curvature effect are the more rapid evaporation or dissolving of finer particles and in regions of sharp positive curvature (smoothing), preferential condensation in regions of sharp negative curvature, displacement of a curved phase boundary toward its center of curvature, and capillary phenomena.

2.3 WEAK VAN DER WAALS FORCES OR CHEMICAL BONDING CAUSES SOLID SURFACES TO ADSORB MOLECULES FROM GASES AND LIQUID SOLUTIONS

Primary chemical bonding between the surface and an adsorbed gas occurs in chemisorption, and the composition and structure of the surface are changed. Chemical adsorption is usually believed to provide monolayer molecular coverage. Chemisorption can occur above or below the critical temperature of the gas, and chemisorbed gas may be very difficult to remove. The oxidation of metals and the chemical hydration of oxides are examples of chemisorption.

General Van der Waals interactions cause a surface to physically adsorb a gas below its critical temperature. This physical adsorption may change the surface structure of a polymer solid but does not usually alter the structure of oxides and refractory metals. Physical adsorption is rapid and reversible;

i.e., it can be removed by lowering the pressure or increasing the temperature. Examples of physical adsorption in ceramic processing are the adsorption of polar molecules in solutions, and gases such as CO_2, N_2, and water vapor on fine oxide particles. Physical adsorption may provide multilayer coverage.

Langmuir considered the adsorption of gas molecules to be a dynamic process. After partial adsorption giving a fractional surface coverage V_a/V_m, only those molecules striking a portion of the surface not already covered $(1 - V_a/V_m)$ may be expected to be adsorbed. Thermal agitation may cause the desorption of some molecules. When the rates of adsorption and desorption are equal, an adsorption equilibrium is established and

$$\frac{V_a/V_m}{1 - V_a/V_m} = bP \tag{2.8}$$

where V_a is the amount of gas adsorbed on the surface of the adsorbent at a pressure P, V_m is the volume of gas required for monolayer coverage, and b is a constant for particular system and temperature. Equation 2.8 may also be written in the linear form

$$\frac{P}{V_a} = \frac{1}{bV_m} + \frac{P}{V_m} \tag{2.9}$$

Brunauer, Emmett, and Teller expanded Langmuir's dynamic model for monolayer adsorption to include multilayer adsorption before complete monolayer coverage. Their analysis for isothermal adsorption is expressed by the BET equation*

$$\frac{P/P_s}{V_a(1 - P/P_s)} = \frac{1}{V_m C} + \frac{(P/P_s)(C - 1)}{V_m C} \tag{2.10}$$

where P_s is the saturation vapor pressure at the temperature of the adsorbate-absorbent system and C is a constant related to the energy of adsorption.

Of the classic adsorption isotherms illustrated in Fig. 2.4, Type I is characteristic of chemical adsorption on surfaces. Type II is typical of multilayer physical adsorption, and this, coupled with capillary adsorption in pores, produces type IV adsorption. Langmiur's analysis is best used for type I isotherms and the BET equation for types II and IV. The BET equation can be used to determine V_m for cases of multilayer adsorption using data for $0.05 < P/P_s < 0.35$ when C is a constant. The surface area is calculated by multiplying V_m by a monolayer packing constant for the adsorbed gas.

*Arthur W. Adamson, *Physical Chemistry of Surfaces*, Wiley-Interscience, New York, 1976.

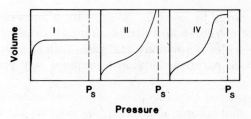

Fig. 2.4 Classical adsorption isotherms.

Above the critical temperature, adsorption is generally monomolecular unless adsorption occurs in pores a few multiples of the absorbate molecule size. In powder agglomerates or compacts, multilayer adsorption in pores is called capillary condensation. For a particular fractional pressure and temperature, Eq. 2.7 can be used to estimate the pore radius below which capillary condensation will occur.

In ceramic processing, adsorbed water vapor or other gases may promote the sticking of powders to surfaces and the agglomeration of ceramic powders, and soften binder phases. Heating above the boiling point is required to remove the adsorbed gas completely.

2.4   THE WETTING AND SPREADING OF A LIQUID ON A SOLID SURFACE DEPENDS ON SHORT-RANGE MOLECULAR FORCES THAT CAN BE SIGNIFICANTLY MODIFIED BY A MONOLAYER COATING

Consider the liquid in contact with a planar surface shown in Fig. 2.5. Spreading occurs when the contact angle θ measured through the liquid phase approaches zero. Wetting implies that the contact angle is less than 90°.

The interaction of the interfacial tensions at the liquid-vapor-solid juncture is described by the Young equation ($\theta > 0°$)

$$\cos \theta = \frac{\gamma_{SV} - \gamma_{SL}}{\gamma_{LV}} \tag{2.11}$$

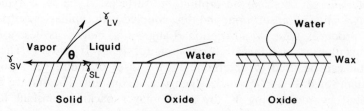

Fig. 2.5 Juncture of interfacial tensions for a liquid on a solid surface and cases of wetting and nonwetting.

where γ_{SV}, γ_{SL}, and γ_{LV} are the effective interfacial tensions as indicated in Fig. 2.5.

A nonvolatile surfactant that is adsorbed by the liquid and lowers γ_{LV} will reduce the contact angle. Similarly a film of oil, wax, or polymer of lower γ_{SV} on an oxide particle may cause a greater apparent contact angle. Reaction between the liquid and solid, reducing γ_{SL}, may also improve the wetting.

In ceramic processing, wetting and spreading phenomena affect the coating of particles with liquid, the dispersion of agglomerates, and the stability of air bubbles in suspensions.

2.5 WETTING MAY CAUSE COMPRESSIVE FORCES BETWEEN PARTICLES AND LIQUID MIGRATION

Liquid that wets the surfaces of two particles will spread over the surface and concentrate in the contact region, forming a neck (Fig. 2.6). If the particles are finer than about 100 μm, buoyancy forces can be neglected. The pressure difference across the curved meniscus is given by Eq. 2.6. Wetting situations for spherical and angular particles are shown in Fig. 2.6; note that the principal radii are of opposite signs. When the smaller of these is negative, ΔP is negative, and a compressive stress occurs in the contact region. Detailed analyses show that for spheres in contact, the compressive stress is greatest when $\gamma_{LV} \cos \theta$ is large and the volume of liquid is small. For angular particles in point contact, calculations* indicate that the compressive pressure increases as the volume of liquid increases; torques and shear forces can cause rotation and sliding, causing flat sides to come together.

The pressure difference across a curved meniscus can also cause the migration of liquid between pores of different sizes or the migration of

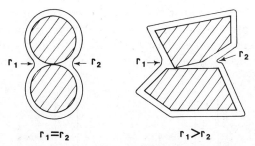

Fig. 2.6 Illustration of a liquid film of uniform thickness on two spherical particles and two angular particles.

*J. W. Cahn and R. B. Heady, *J. Am. Ceram. Soc.* **53**(7), 406–409 (1970).

liquid from a saturated region to a less saturated region. As shown in Fig. 2.7, liquid will rise in a capillary if it wets the surface, but it will be depressed if nonwetting occurs. For fine capillaries, the meniscus is approximately hemispherical, $r = R_C \cos \theta$, where r_C is the radius of the capillary, and from Eq. 2.6

$$\Delta P = \frac{2\gamma_{LV} \cos \theta}{R_C} \tag{2.12}$$

At equilibrium, this pressure difference will offset the hydrostatic pressure of a column of liquid of height H, which may rise above the meniscus external to the capillary, and

$$\Delta P = \frac{2\gamma_{LV} \cos \theta}{R_C} = D_L g H \tag{2.13}$$

or

$$H = \frac{2\gamma_{LV} \cos \theta}{D_L g R_C} \tag{2.14}$$

where D_L is the density of the liquid and g is the acceleration of gravity.

The average laminar flow velocity $\bar{v}$ of a column of liquid of length L and viscosity η in a horizontal cylindrical capillary is given by the Poiseville equation;

$$\bar{v} = \frac{\Delta P R_c^2}{8\eta L} \tag{2.15}$$

Combining Eq. 2.12 and Eq. 2.15,

$$\bar{v} = \frac{\gamma_{LV} \cos \theta}{\eta} \frac{R_C}{4L} \tag{2.16}$$

These equations indicate that the rate of penetration of liquid into an agglomerate or through a porous medium will be greater for a liquid of lower viscosity and higher surface tension and for pores of larger radius. Finer pores produce a greater suction. An increase in temperature may

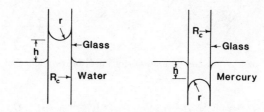

Fig. 2.7 Capillary rise and capillary depression.

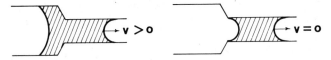

Fig. 2.8 An imbalance of the capillary forces may cause liquid migration.

reduce γ_{LV}/η and improve penetration. Surfactants that increase $\cos\theta$ also tend to reduce γ_{LV}.

Pore capillaries in packed particles will vary in radius along their length. As shown in Fig. 2.8, a wetting liquid will migrate into the smaller-diameter segment. Repellency occurs when the wetting angle is greater than 90°.

In ceramic processing, the capillary force provides a mechanism for the cohesion of wetted agglomerates, the migration of liquid in pores, and the rearrangement of particles during mixing. Capillary suction produces a driving force for the migration of liquid in ordinary slip casting and drying.

2.6 WEAK INTERMOLECULAR FORCES CAN CAUSE THE ADHERENCE OF MOLECULES ON A SURFACE AND THE AGGLOMERATION OF SMALL PARTICLES

The attractive Van der Waals forces have their origin in the attractive interaction between fluctuating electrical dipoles in one molecule and the induced dipole in a neighboring molecule and, if present, the attractive interaction between permanent dipoles of particular chemical species. The former are called London dispersion forces and are important in hydrogen bonding.

These forces, which can be summed for a particular geometry to obtain an estimate of the energy of interaction, are of the form of the product of an attraction constant and a geometrical term.

For two small spherical particles of diameter a, the potential energy of attraction U_A is

$$U_A = \frac{-Aa}{24h} \tag{2.17}$$

when $h \ll a$, where h is the separation between the surfaces of the two particles, and A is the Hamaker attraction constant. The attraction constant is a function of the composition and molecular structure of the particles and the medium between them. When the particles are dispersed in a liquid, the Hamaker constant is

$$A = (\sqrt{A_2} - \sqrt{A_1})^2 \tag{2.18}$$

where A_2 and A_1 are the Hamaker constants for the particles and dispersion medium, respectively. The attraction constant A is smaller when the particles and medium are more similar in composition.

In addition to the adsorption of molecules, the attractive Van der Waals forces are significant in ceramic processing, because they promote the agglomeration of dry powders and the coagulation of particles in liquid suspension. Mechanisms to counter these effects are discussed in Chapters 9 and 10.

SUMMARY

Atoms and molecules at surfaces may have unsatisfied chemical bonds that produce a surface tension that modifies the behavior of a material. The surface behavior is altered by surface curvature and material adsorbed on the surface. Surface wetting produces capillary phenomena in porous ceramic systems. Weak Van der Waals forces may produce agglomerates in powder systems when repulsion forces are insufficient to inhibit their formation.

SUGGESTED READING

1. Arthur W. Adamson, *Physical Chemistry of Surfaces*, Wiley-Interscience, New York, 1976.
2. W. D. Kingery, H. K. Bowen, and D. R. Uhlmann, *Introduction to Ceramics*, 2d Ed., Wiley-Interscience, New York, 1976.
3. Gordon M. Barrow, *Physical Chemistry*, 3d Ed., McGraw-Hill, New York, 1973.
4. S. J. Gregg and K. S. W. Sing, *Adsorption, Surface Area, and Porosity*, Academic Press, New York, 1967.

PROBLEMS

2.1 Compare the pressure differences ΔP and relative vapor pressure P/P_0 for water at 20°C when the concave surface radius is 0.1 μm and 10 μm.

2.2 The wetting angle of a drop of water on a glass surface is 16° when the temperature is 20°C. Would the wetting angle be expected to increase or decrease on heating to 60°C? Estimate the interfacial tension γ_{SL} if $\gamma_{SV} = 300$ mN/m at 20°C.

2.3 If the surface in problem 2.2 is first coated with a wax film for which $\gamma_{SV} = 30$ mN/m and $\gamma_{SL} = 30$ mN/m, is the wetting by water improved?

2.4 Estimate the capillary suction and potential height of capillary migration for water and ethyl alcohol in a porous solid with an effective capillary radius of 1 μm. Assume complete wetting and a temperature of 20°C.

2.5 Calculate the relative migration velocities of water in identical capillaries for temperatures of 20°C and 50°C. (Assume complete wetting.)

2.6 An agglomerated powder is charged into a mixing tank containing water. Assuming complete wetting and an effective pore diameter of 1 μm, estimate the depth of penetration of water into the agglomerates after 1 min when the water is 20°C. Assume that air is displaced out of the agglomerates.

2.7 Consider the neck of water between two spherical particles 10 μm in diameter (20°C, $\theta = 0$ deg). Derive an expression for the capillary pressure in the neck as a function of the x and r_n when $r_n \ll x$.

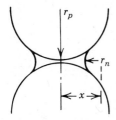

2.8 In problem 2.7, calculate the particle radius r_p for which the adhesion force is equal to the weight of the particle, when r_n is very small in relation to x. Assume that the liquid is water at 20°C and the spherical particle has a density of 4.0 g/cm^3.

2.9 Does the Van der Waals attraction force between two dry particles depend on the surface roughness?

PART II

CERAMIC
RAW MATERIALS

In studying ceramic processing it is necessary to be familiar with the types of raw materials available. Clay minerals, which provide plasticity when mixed with water; feldspar, which acts as a nonplastic filler on forming and a fluxing liquid on firing; and silica, which is a filler that resists fusion, have been the backbone of the traditional ceramic porcelains. Other silicate minerals are used in whitewares such as a ceramic tile, thermal shock-resistant cordierite products, and steatite electrical porcelains.

Silica, aluminosilicates, tabular aluminium oxide, magnesium oxide, calcium oxide, and mixtures of these minerals have long been used for structural refractories. Alumina, magnesia, and aluminosilicates are now used in some advanced structural ceramics. Silicon carbide and silicon nitride are used for refractory, abrasive, electrical, and structural ceramics. Finely ground alumina, titanates, and ferrites are the backbone of the electronic ceramics industry. Stabilized zirconias are used for advanced structural and electrical products and zircon, zirconia, and other oxides doped with transition and rare-earth metal oxides are widely used as ceramic pigments. These materials are commonly prepared by calcining particle mixtures, but some are now produced using special chemical techniques.

In Chapter 3, the more common ceramic materials produced in large tonnage and widely used in ceramics are considered. Special materials of exceptional purity and homogeneity which are being developed for research and some very advanced products are discussed in Chapter 4.

3

Common
Raw Materials

In this chapter we will briefly consider the nature of the starting materials, traditionally called raw materials, that can be purchased from a vendor and received at a manufacturing site. These materials can vary widely in nominal chemical and mineral composition, purity, physical and chemical structure, particle size, and price. Categories of raw materials include (1) nonuniform crude material from natural deposits, (2) refined industrial minerals that have been beneficiated to remove mineral impurities to significantly increase the mineral purity and physical consistency, and (3) high-tonnage industrial inorganic chemicals that have undergone significant chemical processing and refinement to significantly upgrade the chemical purity and improve the physical characteristics.

The choice of a raw material for a particular product will depend on material cost, market factors, vendor services, technical processing considerations, and the ultimate performance requirements and market price of the finished product. For products in which processing adds considerable dollar value, the cost of the starting material is a relatively small component of the production costs. Accordingly, a higher-quality and more expensive material may be acceptable for microelectronics, coatings, fibers, and some high-performance products. But the average cost of raw materials for building materials and traditional ceramics such as tile and porcelain must be relatively low. Cost-benefit considerations may suggest substitutions of materials of lower cost that do not impair the quality, or alternatively, a more expensive material, which may be more economically processed and/or which will increase the quality and performance of the product.

3.1 CRUDE MATERIALS

Many early ceramics industries were based near a natural deposit containing a combination of crude minerals that could be conveniently processed into usable products. Construction materials such as brick and tile and some pottery items are historical examples, and many are still identified by the regional name.

Some crude materials are of sufficient purity to be used in heavy refractories. Crude bauxite, a nonplastic ore containing hydrous alumina minerals, clay minerals, and mineral impurities such as quartz and ferric oxides, is used in producing some refractories. Today, however, most ceramics are produced from more refined minerals.

3.2 INDUSTRIAL MINERALS

Industrial minerals are used in large tonnages for producing construction materials, refractories, whitewares, and some electrical ceramics. They are used extensively as additives in glazes, glass, and raw materials for industrial chemicals. Common examples are listed in Table 3.1.

Clays are produced by the weathering of aluminosilicate rocks and sedimentation. Clay minerals are layer-type hydrous aluminosilicates which can be dispersed into fine particles (Fig. 3.1). Kaolin is a relatively pure, white firing clay composed principally of the mineral kaolinite $Al_2Si_2O_5(OH)_4$ but containing other clay minerals as indicated in Table 3.2 and a minor amount of impurity minerals such as quartz SiO_2, ilmenite $FeTiO_2$,

Table 3.1 Common Raw Materials

Category	Materials
Crude material	Shales, stoneware clay, tile clay, crude bauxite, crude kyanite, natural ball clay, bentonite
Industrial minerals	Ball clay, kaolin, bentonite, pyrophyllite, talc, feldspar, nepheline syenite, wollastonite, spodumene, glass sand, potter's flint, kyanite, bauxite, zircon, rutile, chrome ore, calcined kaolin, dolomite
Industrial inorganic chemicals	Calcined alumina (Bayer process), calcined magnesia (from brines, seawater), fused alumina, fused magnesia, silicon carbide (Atcheson process), soda ash, barium carbonate, titania, calcined titanates, iron oxide, calcined ferrites, calcined stabilized zirconia, zirconia pigments, calcined zircon pigments

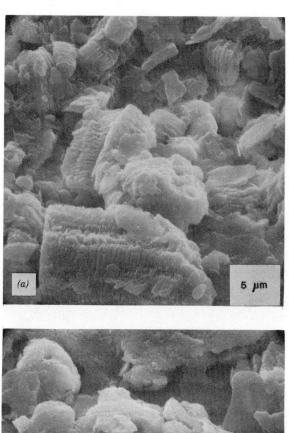

Fig. 3.1 Kaolin that is (a) aggregated and (b) dispersed.

Table 3.2 Clay Minerals

Mineral	Ideal Chemical Formula
Kaolinite	$Al_2(Si_2O_5)(OH)_4$
Halloysite	$Al_2(Si_2O_5)(OH)_4 \cdot 2H_2O$
Pyrophyllite	$Al_2(Si_2O_5)_2(OH)_2$
Montmorillonite	$(Al_{1.67}Na_{0.33}Mg_{0.33})(Si_2O_5)_2(OH)_2$
Mica	$Al_2K(Si_{1.5}Al_{0.5})_2(OH)_2$
Illite	$Al_{2-x}Mg_xK_{1-x-y}(Si_{1.5-y}Al_{0.5+y}O_5)_2(OH)_2$

After W. D. Kingery, *Introduction to Ceramics*,
Source: 1st Ed, Wiley-Interscience, New York, 1960.

rutile TiO_2, and hematite Fe_2O_3. Ball clay is a sedimentary clay of fine particle size containing complex organic matter ranging down to a submicron size. Bentonite is a complex clay containing a relatively high proportion of the clay mineral montmorillonite. Clays are used in whiteware formulations and aluminosilicate refractories to provide plasticity in forming and resistance to deformation when partial fusion occurs during firing.

Other layer-type hydrous silicates are talc $Mg_3Si_4O_{10}(OH)_2$ and pyrophyllite $Al_2Si_4O_{10}(OH)_2$ which are used extensively in compositions for ceramic tile, cordierite, and steatite porcelain. Commercial grades contain impurities such as calcite $CaCO_3$ or dolomite $(Mg,Ca)CO_3$ and other mineral impurities that depend on the source.

Crushed and milled quartz SiO_2 derived from relatively pure deposits of sandstone is a granular silicate mineral used extensively in whitewares, refractories, and glaze compositions (Fig. 3.2). Feldspars composed of the minerals albite $NaAlSi_3O_8$ and microcline or orthoclase $KAlSi_3O_8$ and nepheline syenite containing albite, microcline, and nephelite $K_{0.5}Na_{1.5}(Al, Si)_2O_8$ are the principal fluxes used in whitewares and silicate glazes. Wollastonite $CaSiO_3$ is used in some tile compositions and glazes. Petalite $LiAlSi_4O_{10}$ and spodumene $LiAlSi_2O_6$ are used as a secondary flux and to reduce the thermal expansion of the fired material.

Chrome ores composed principally of a complex solid solution of spinels $(Mg,Fe)(Al,Cr,Fe)_2O_4$ and impurities such as dolomite and magnesium silicates are used in combination with calcined magnesia MgO in basic refractories. Lime CaO produced by calcining limestone $CaCO_3$ and calcined dolomite $(Ca, Mg)O$ is bonded with tar and used for lining basic oxygen steel furnaces. Beneficiated kyanite $AlSiO_5$, bauxite, and zircon $SiZrO_4$ are also used in refractory compositions. Milled zircon is also used as an opacifier in glazes and in producing zircon pigments and is a precursor for zirconia ZrO_2. Calcined kaolin (Fig. 3.2) is used as a nonplastic filler in refractory mixes and mortars.

The beneficiation of industrial minerals begins with crushing and grinding to a small enough size to liberate undesired mineral phases. Further beneficiation may include settling and floatation to segregate minerals by

Fig. 3.2 Scanning electron micrographs of crushed (a) 140-mesh quartz and (b) calcined kaolin showing rough surfaces.

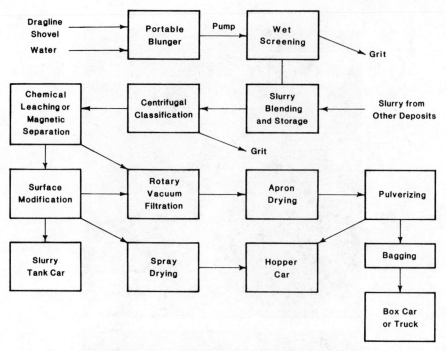

Fig. 3.3 Processing flow diagram for the beneficiation of kaolin.

density or size, the separation of magnetic minerals using powerful elec-
tromagnets, blending of different processing runs for consistency, and
perhaps particle size classification. Solids may be concentrated by filtration
or centrifugation, and a portion of the soluble impurities are eliminated with
the liquid. A typical flow diagram for refining kaolin is shown in Fig. 3.3.
Materials, especially clays, refined and ultimately used in suspension form,
may be purchased as an aged slurry of controlled viscosity that is shipped in
railroad tank cars.

Concentrated solids are usually dried using a rotary or belt dryer or by
spray drying. Some materials are calcined, and a hard aggregate is formed.
Dried cake or calcined materials may be pulverized or ground and then
sized or air-elutriated before bagging or loading in hopper cars. Many fine
materials are loaded and unloaded using pneumatic fluidization and are
stored at the plant site in large silos.

3.3 INDUSTRIAL INORGANIC CHEMICALS

Important industrial ceramic chemicals include tabular and calcined
aluminas, magnesium oxide, silicon carbide, silicon nitride, alkaline earth
titanates, soft and hard ferrites, stabilized zirconia, and inorganic pigments.

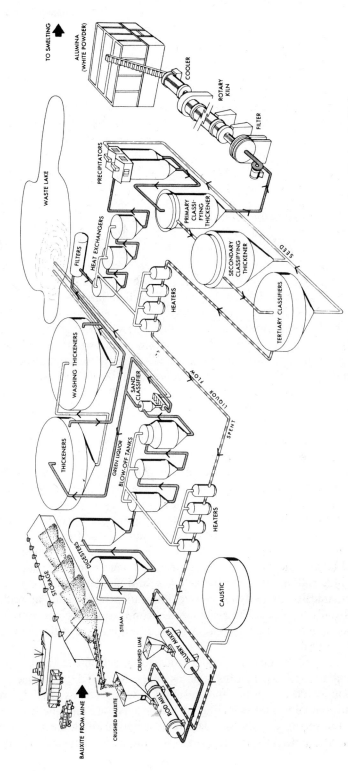

Fig. 3.4 The Bayer process for chemically refining bauxite into alumina. (Courtesy of Alcoa Inc., Pittsburgh, PA.)

Extensive chemical beneficiation reduces the content of accessory minerals and may increase the chemical purity up to about 99.5%. For many materials, the scale of operation is extremely large, which aids in lowering the unit processing costs and selling price

Alumina Al_2O_3 is the most widely used inorganic chemical for ceramics (Table 1.2) and is produced worldwide in tonnage quantities for the aluminum and ceramics industries using the Bayer process. The principal operations in the Bayer process are the physical beneficiation of the bauxite, digestion (in the presence of caustic soda NaOH at an elevated temperature and pressure), clarification, precipitation, and calcination, followed by crushing, milling, and sizing (see Fig. 3.4). During the digestion, most of the hydrated alumina goes into solution as sodium aluminate, i.e.

$$Al(OH)_3 + NaOH \rightarrow Na^+ + Al(OH)_4^- \qquad (3.1)$$

and insoluble compounds of iron, silicon, and titanium are removed by settling and filtration. After cooling, the filtered sodium aluminate solution is seeded with very fine gibbsite $Al(OH)_3$, and at the lower temperature the aluminum hydroxide reforms as the stable phase. The agitation time and temperature are carefully controlled to obtain a consistent gibbsite precipitate. The gibbsite is continuously classified, washed to reduce the sodium content, and then calcined. Material calcined at 1100–1200°C is crushed and ground to obtain a range of sizes (Fig. 3.5). Tabular aluminas are obtained by calcining to a higher temperature, about 1650°C.

Magnesium oxide MgO of greater than 98% purity is prepared by precipitating magnesium hydroxide in a basic mixture of treated dolomite and natural brines or seawater containing $MgCl_2$ and $MgSO_4$, followed by washing, filtration, drying, and calcination. Zirconia ZrO_2 of 99% purity is obtained by the caustic fusion of zircon $SiZrO_4$ and then chemically removing the silica.

Silicon carbide SiC is produced in large tonnages by reacting a batch consisting principally of high-purity sand and low-sulfur coke at 2200–2500°C in an electric arc furnace. The crystalline product is crushed, washed in acid and alkali, and then dried after iron has been removed magnetically. Granular material is used in refractories and bonded abrasives. Milled material chemically treated to remove impurities introduced in milling is used industrially for structural ceramics. Silicon nitride Si_3N_4 is prepared by reacting silicon metal powder with nitrogen or a mixture of silica and carbon powders with nitrogen.

The production of mixed metal oxides for electronic ceramics such as barium titanate $BaTiO_3$, ferrites such as $Mn_{0.5}Zn_{0.5}Fe_2O_4$ and $BaFe_{12}O_{19}$, mixed metal oxide resistors, and ceramic colors such as doped zirconia involves the batching and calcining of industrial inorganic chemicals, as is shown for the ferrite in Fig. 3.6. The concentration of chemical dopants is carefully controlled. Soluble material is sometimes removed by filtering

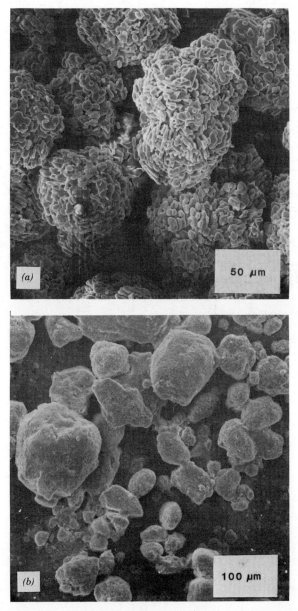

Fig. 3.5 Scanning electron micrograph of particles in calcined Bayer process alumina: (a) aggregates of grains before milling and (b) particle agglomerates of irregular shape in milled alumina powder.

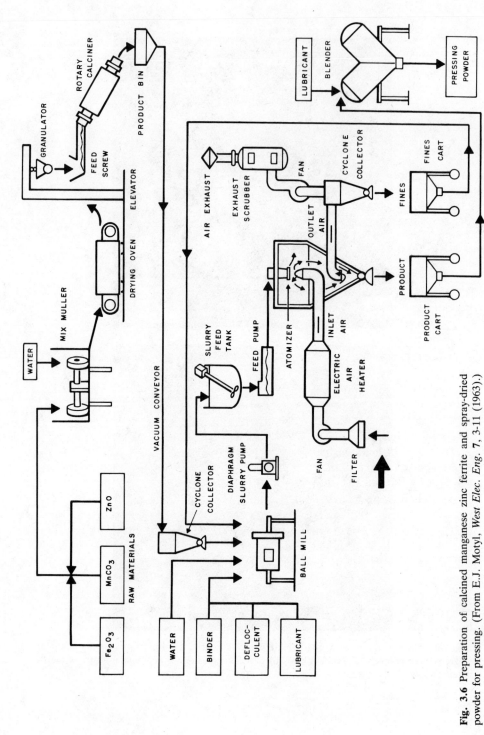

Fig. 3.6 Preparation of calcined manganese zinc ferrite and spray-dried powder for pressing. (From E.J. Motyl, *West Elec. Eng.* **7**, 3-11 (1963).)

before drying. Precursor industrial chemicals for these compounds are commonly powders finer than a few microns in size. Barium carbonate $BaCO_3$ and titania TiO_2 are commonly used for preparing the titanates, and manganese carbonate $MnCO_3$, zinc oxide ZnO, hematite Fe_2O_3, and barium carbonate for the ferrites.

Titania TiO_2 is produced by the sulfate or chloride process. In the sulfate process, ilmenite $FeTiO_3$ is treated with sulfuric acid at 150–180°C to form the soluble titanyl sulfate $TiOSO_4$, i.e.

$$FeTiO_3 + 2H_2SO_4 + 5H_2O \rightarrow FeSO_4 \cdot 7H_2O + TiOSO_4 \qquad (3.2)$$

After removing undissolved solids and then the iron sulfate precipitate, the titanyl sulfate is hydrolysed at 90°C to precipitate the hydroxide $TiO(OH)_2$, i.e.

$$TiOSO_4 + 2H_2O \rightarrow TiO(OH)_2 + H_2SO_4 \qquad (3.3)$$

The titanyl hydroxide is calcined at about 1000°C to produce titania TiO_2. In the chloride process, a high-grade titania ore is chlorinated in the presence of carbon at 900–1000°C, and the chloride $TiCl_4$ formed is subsequently oxidized to TiO_2.

Barium carbonate $BaCO_3$ is the primary source of barium oxide BaO for ceramics. Barite ore, nominally $BaSO_4$ is reduced at a high temperature to barium sulfide BaS, which is water-soluble. The reaction of an aqueous sulfide solution with sodium carbonate Na_2CO_3 or carbon dioxide CO_2 produces a barium carbonate precipitate which is then washed, dried, and ground.

The commercial iron oxide hematite α-Fe_2O_3 used for preparing ferrites is produced from the thermal decomposition of hydrated ferrous sulfate $FeSO_4 \cdot 7H_2O$ or by the precipitation of hematite and goethite α-$Fe_2O_3 \cdot H_2O$ from an oxygenated sulfate solution containing dispersed iron metal. The size and shape of the ultimate crystals of Fe_2O_3 are very dependent on the pH, temperature, time, and impurities during precipitation. Zinc oxide ZnO is produced by roasting a concentrate of the mineral sphalerite ZnS in air. Manganese carbonate $MnCO_3$ is derived from manganese sulfate $MnSO_4$.

When thermally reacting titanates and ferrites, the temperature, time, and atmosphere must be adequate to permit decomposition of the carbonate and promote interdiffusion of the reactants through the intermediate reaction product that may be several microns in thickness (Fig. 3.7). The time dependence of the relative amount x of reactant A of radius r_A transformed into reaction product is given by the Carter equation

$$[1 + (z - 1)x]^{2/3} + (z - 1)(1 - x)^{2/3} = z + 2(1 - z)\frac{Kt}{r_A^2} \qquad (3.4)$$

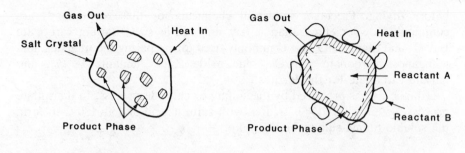

Salt Decomposition   Mixed Powder Reaction

Fig. 3.7 Formation of reaction product during salt decomposition and on reacting mixed powders.

where K is the apparent rate constant and z is the volume of product formed from a unit volume of reactant A.* The effect of temperature T is commonly expressed by the Arrhenius relation and

$$K = K_0 \exp \frac{-Q}{RT} \qquad (3.5)$$

where K_0 is the limiting rate constant that depends on the diffusion path length and Q is the apparent activation energy for diffusion. The reaction is a function of both time and temperature, but temperature has a greater influence on the rate. The time for total reaction eliminating the reactants varies directly with the maximum size of agglomerates or segregated material in the batch. When reacting micron-size oxide powders, thermal processing at a temperature in excess of 1200°C is commonly requisite, as is shown in Fig. 3.8 for the formation of spinel $MgAl_2O_4$.

Other important variables affecting solid-state reactions during calcining are the particle size distributions of the reactants, the mixedness of the reactants, the composition and flow of gases, the depth and turnover of material, and endothermic and exothermic effects. Sintering during calcination produces particle aggregates (Fig. 3.9), and considerable grinding of the hard aggregates is commonly required to produce a fine powder.

As indicated by the nominal solid-state reaction for the formation of barium titanate

$$BaCO_3 + TiO_2 \rightarrow BaTiO_3 + CO_{2(gas)} \qquad (3.6)$$

the partial pressure of CO_2 in the pores of the product influences the reaction kinetics. Also the nonequilibrium phase Ba_2TiO_4 initially forms between $BaTiO_3$ and unreacted $BaCO_3$ and is undersirable in the calcined product; this phase is minimized by dispersing agglomerates of titania and

---

* H. Schmalzried, *Solid State Reactions*, Academic Press, New York, 1974.

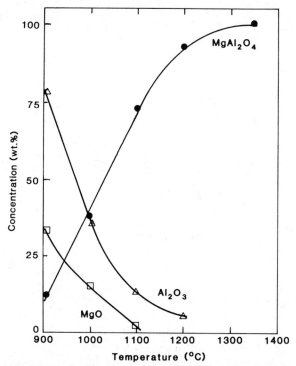

Fig. 3.8 The formation of spinel $MgAl_2O_4$ from the solid-state reaction of micron-size MgO and α-Al_2O_3 powder as a function of the reaction temperature for a constant reaction time of 8 h.

Fig. 3.9 Scanning electron micrograph of an aggregated, calcined barium titanate powder.

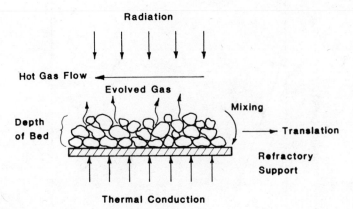

Fig. 3.10 Thermal transport, gas flow, and particle movement during rotary calcination.

mixing thoroughly to maximize the particle contacts and reduce the diffusion path between $BaCO_3$ and TiO_2. Calcination in a furnace, which provides a more uniform temperature in the material and mixing of material with air, as shown in Fig. 3.10, may produce a more uniform product. In calcining ferrites and pigments, the oxygen pressure of the air must be controlled to obtain the requisite oxidation states of the transition metal ions. The calcining temperatures and atmosphere must also be controlled to prevent the loss of nonrefractory chemical dopants.

SUMMARY

Few ceramics are produced today using crude raw materials. Industrial minerals are refined physically to reduce the concentration of undesirable mineral impurities and to produce a particular particle size distribution; water-soluble impurities are removed by washing. Industrial inorganic chemicals used to produce the majority of technical ceramics are chemically processed on a large scale to improve both the chemical and the mineral purity; the calcined product containing hard aggregates is commonly milled to disperse the aggregates and obtain a product of controlled size distribution. Mixed-oxide industrial chemicals are commonly produced by calcining a mixture of these industrial chemicals. The completeness of the reaction and uniformity of the product depend on the particle size and mixedness of the reactants and the time, temperature, and atmosphere and their uniformity during calcination. Different lots of processed materials are blended to maintain a higher level of uniformity.

SUGGESTED READING

1. *Process Mineralogy of Ceramic Materials*, Wolfgang Baumgart et al., eds, Elsevier, New York, 1984.

2. Kirk-Othmer, *Encyclopedia of Chemical Technology*, Wiley-Interscience, New York, 1983.

3. F. H. Norton, *Fine Ceramics*, Krieger, Malabar, FL, 1978.

4. W. D. Kingery et al., *Introduction to Ceramics*, 2d Ed, Wiley-Interscience, New York, 1976.

5. W. E. Worrall, *Clays and Ceramic Raw Materials*, Halsted Press Div., Wiley-Interscience, New York, 1975.

6. Rex W. Grimshaw, *The Chemistry and Physics of Clays and Other Ceramic Materials*, Wiley-Interscience, New York, 1971.

7. H. Okada et al., Effect of Physical Nature of Powders and Firing Atmosphere on $ZnAl_2O_4$ Formation, *J. Am. Ceram. Soc.* **68**(2), 58–63 (1985).

8. Betty L. Milliken, Color Control in a Pigment Manufacturing Plant, *Am. Ceram. Soc. Bull.* **62**(12), 1338–1340 (1983).

9. T. Nomura and T. Yamaguchi, TiO_2 Aggregation and Sintering of $BaTiO_3$ Ceramics, *Am. Ceram. Soc. Bull.* **59**(4), 453–455,458 (1980).

10. Materials of Advanced Ceramics and Traditional Ceramics, *Ceramic Industry Magazine*. Corcoran Publishers, Solon, OH, 1985.

11. *Annual Ceramic Industry Data Book*, Cahners, Boston MA.

12. *Ceramic Source*, American Ceramic Society, Columbus, OH.

PROBLEMS

3.1 Estimate the raw-material cost per unit weight for a porcelain body containing equal amounts of ball clay, kaolin, feldspar, and ground quartz, and compare this to the selling price of a porcelain product using the same weight basis.

3.2 Compare the product cost of a ceramic capacitor to the raw material cost using the same weight basis.

3.3 Construct a processing flow diagram for ceramic grade silica, and contrast this to the Bayer process for alumina.

3.4 When forming a compound oxide by the calcination of mixed powders, a small amount of a reactant may be present in the final powder. Does the completion of the reaction depend on the average particle size, the maximum size of a reactant particle, or the size of agglomerates? Explain.

3.5 State several reasons why a pigment calcined in an open crucible in a gas-fired furnace might differ in color from the same pigment batch fired in a closed sagger in an electric furnace.

3.6 Construct a processing flow diagram for the formation of a $BaTiO_3$ powder.

3.7   When may a multiphase product layer be expected during a solid state reaction? Does local equilibrium improve as the product grows?

3.8   Write the nominal reaction equation for the formation of a doped barium titanate $Ba_{1-x}M_x^{2+}TiO_3$ on heating a mixture of $BaCO_3$, MCO_3, and TiO_2 powders. Illustrate the diffusion paths.

3.9   If the large reactant particle in Fig. 3.7 is a porous aggregate, will the rate of formation of the product be expected to be higher or lower? Explain.

3.10 What is the implication of eq. 3.4 concerning the initial and final rate of a solid state reaction?

4

Special Inorganic Chemicals

New and potential uses of ceramics in what are called high-performance or high-technology applications have stimulated much interest in novel techniques for preparing special ceramic powders with special characteristics. Characteristics sought include a purity in excess of 99.9%, a precisely controlled, reproducible chemical composition including dopants, chemical homogeneity on an atomic scale, and a precisely controlled and consistent submicron particle size. In some applications, a special particle shape may be a goal. The variety of compounds prepared in the laboratory is extensive. Although ceramics have been produced for years from these special powders on a laboratory scale, relatively few of these special powders have been used in industrial processing. However, the successful commercial applications of special materials for products such as optical fibers, and thick film electronic ceramics and large potential markets for more advanced ceramics have increased the interest in and the evaluation of these techniques for industrial fabrication. In this chapter we will consider general technical aspects of different techniques and examples of materials that have been prepared.

4.1 POWDERS FROM CHEMICAL SOLUTION TECHNIQUES

Chemical solution techniques provide a relatively convenient means for achieving powders of high purity and fine size. First a suitable liquid solution containing the cations of interest is prepared and analyzed. A solid particulate phase may be formed by precipitation, solvent evaporation, or solvent extraction. Segregation is minimized by combining the ions in a precipitate or gel phase or by extracting the solvent in a few milliseconds from a microscopic drop. The solid phase is usually a salt that can be decomposed without melting by calcination at a relatively low temperature. A porous, friable calcine is ground relatively easily to a submicron size.

Precipitation Techniques

Chemical precipitation techniques of the type used for classical wet quantitative analyses can be used to prepare a wide variety of inorganic salts (Fig. 4.1). The addition of a chemical precipitant to the solution or a change in temperature or pressure may decrease the solubility limit and cause precipitation. Precipitation occurs by nucleation and growth. Impurity ions in solution that are adsorbed on particular surfaces of the particles may affect their growth rates. Relatively slow growth rates along particular crystallographic directions will cause the precipitate particles to have an anisometric shape. A higher degree of supersaturation may increase the nucleation rate and produce a smaller particle size, but if precipitation is extremely rapid, foreign ions tend to be occluded in the particle. The mixing rate and temperature must be controlled to obtain a controlled precipitate. When the cations in solution are of about the same size and chemically similar, the precipitation of a salt containing the cations in solid solution may occur; this is called coprecipitation. In heterogeneous precipitation, the concentration of an ion in the salt differs from that in the solution, and the composition of the coprecipitate may change as precipitation progresses. Less soluble isomorphs tend to concentrate in the salt.

The coprecipitation of ferrous oxalates for the preparation of soft ferrites has been studied in detail by Gallagher et al.* A nickel-iron solid solution oxalate is precipitated on admixing hot ammonium oxalate in a heated aqueous sulfate solution, i.e.:

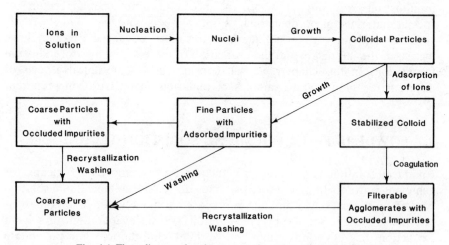

Fig. 4.1 Flow diagram for the preparation of a pure precipitate.

* P. K. Gallagher et al., *Am. Ceram. Soc. Bull.* **48**(1), 1053–1059 (1969).

$$NiSO_4 + 4FeSO_4 + 5(NH_4)_2C_2O_2 \cdot H_2O \xrightarrow{H_2O(60°C)}$$

$$5Ni_{0.2}Fe_{0.8}C_2O_4 \cdot 2H_2O + 10NH_4^+ + 5SO_4^{2-} \qquad (4.1)$$

The thermal decomposition of the oxalate salt in air at a temperature below 500°C produces a well-ordered nickel ferrite compound and is described by the reaction

$$3Ni_{0.2}Fe_{0.8}C_2O_4 \cdot 2H_2O_{(s)} + \frac{4+x}{2}O_2(g)$$

$$\rightarrow Ni_{0.6}Fe_{2.4}O_{4(s)} + xCO_{(g)} + (6-x)CO_{2(g)}$$

$$+ 6H_2O_{(g)} \qquad (4.2)$$

The atomic scale mixedness in the salt crystal enables the direct formation of the ferrite of fine particle size (see Fig. 3.7).

In some cases, it may be possible to precipitate a specific compound containing two cations, and the composition of the precipitate will be uniform regardless of the concentration of ions in solution. An example of the latter is $Ba(TiO(C_2O_4)_2) \cdot 4H_2O$ formed by the reaction[†]

$$BaCl_2 + TiCl_4 + 2H_2C_2O_4$$

$$\xrightarrow{H_2O(20°C)} BaTiO(C_2O_4)_2 \cdot 4H_2O + 6H_3O^+ + 6Cl^- \qquad (4.3)$$

The thermal decomposition of the barium titanyl oxalate salt produces barium titanate indirectly, but at a relatively low temperature (700°C), because of the fine particle size and mixedness, i.e.:

$$BaTiO(C_2O_4)_2 \cdot 4H_2O_{(s)}$$

$$\xrightarrow{(25-225°C)} BaTiO(C_2O_4)_{2(s)} + 4H_2O_{(g)} \qquad (4.4)$$

$$BaTiO(C_2O_4)_{2(s)} + \frac{1}{2}O_{2(g)}$$

$$\xrightarrow{(225-456°C)} BaCO_{3(s)} + TiO_{2(s)} + CO_{(g)} + 2CO_{2(g)} \qquad (4.5)$$

$$BaCO_{3(s)} + TiO_{2(s)} \xrightarrow{(465-700°C)} BaTiO_{3(s)} + CO_{2(g)} \qquad (4.6)$$

The hydrolysis of mixed alkoxides has been used to produce a variety of oxides with particle sizes finer than 20 nm. When water is admixed into alcohol solutions of the alkoxides barium isopropoxide $Ba(OC_3H_7)_2$ and

[†] M. D. Rigterink, *J. Can. Ceram. Soc.* 37, LVI–LX (1968).

titanium amyloxide $Ti(OC_5H_{11})_4$, the simultaneous hydrolytic decomposition produces barium titanate according to the reaction,*:

$$Ba(OC_3H_7)_{2(soln)} + Ti(OC_5H_{11})_{4(soln)}$$

$$\xrightarrow{H_2O(20°C)} BaTiO_{3(s)} + 2C_3H_7OH_{(1)} + 4C_5H_{11}OH_{(1)} \qquad (4.7)$$

An oxide containing several different but chemically similar ions may be prepared by admixing a solution of the ion in an excess of the precipitant. The very high degree of supersaturation causes the rapid precipitation of all ions. The precipitation system and conditions must be well controlled to obtain a reproducible precipitate; important variables include solution concentrations, pH, mixing and stirring rates, and temperature. Precipitates may be purified by digestion, washing, and, in some cases, reprecipitation prior to filtration (Fig. 4.1). Digestion is growth of the larger precipitate particles at the expense of the finer particles while the precipitate is in the solution; surface-adsorbed impurities decrease as the specific area decreases. Washing will improve the purity if surface-adsorbed impurities are removed

2 μm

Fig. 4.2 Scanning electron micrograph of an aggregate of α-Al_2O_3 produced from the alum process.

---

* K. S. Mazdiyasni et al., *J. Am. Ceram. Soc.* **52**(10), 523–526 (1969).

without precipitating other ions in solution films on particles. Dissolving and reprecipitation in a fresh solution may reduce the concentration of minor impurities. Alum $NH_4Al(SO_4)_2 \cdot 12H_2O$ dissolves in a hot aqueous solution. The reprecipitated alum formed on cooling has a lower concentration of alkali and transition metal impurities:

$$NH_4Al(SO_4)_2 \cdot 12H_2O_{(impure)} \xrightarrow{H_2O \text{ (heat)}} \text{solution}$$

$$\text{solution} \xrightarrow{\text{(cool)}} NH_4Al(SO_4)_2 \cdot 12H_2O_{(purified)}$$

$$+ H_2O_{(impure)} \tag{4.8}$$

This technique is used commercially to produce alumina with a purity exceeding 99.995% (Fig. 4.2).

Precipitation techniques have been widely investigated for preparing submicron-size, high-purity oxide powders. Particle sizes as small as 2 nm have been produced for some systems, and these have been used as commercial catalysts. Precipitation techniques are currently being considered for the industrial production of more advanced ferrites.

Solvent Evaporization and Extraction Techniques

An alternative procedure used to prepare special powders is to disperse the solution containing the ions of interest into microscopic volumes and then remove the solvent as a vapor, forming a salt. Maintenance of atomic-scale homogeneity will be possible for multicomponent systems only when the components are of about equal solubility or when the salt forms extremely rapidly.

Spray drying has been used to produce and dry salt particles 10–20 μm in diameter. The fast drying that occurs in a few milliseconds reduces segregation. This technique has been reported to give good results for the preparation of ferrites from mixed sulfates and Mg-stabilized beta alumina from mixed nitrates.[*] Porous calcined agglomerates are reported to be easily comminuted into the component particles. A variation of this general procedure is the atomization of the solutions in a hot furnace to combine the drying and calcination in one step; because the heating rate is several hundred degrees per second, complete decomposition occurs only if the salt decomposes at a relatively low temperature. Unaggregated submicron particles of magnesia were produced from the spray pyrolysis of an acetate solution[†]:

[*] J. G. M. de Lau, *Am. Ceram. Soc. Bull.* **49**(6), 570–574 (1970).
[†] T. Gardner and G. Messing, *Am. Ceram. Soc. Bull.* **63**(12). 1498–1501 (1984).

$$Mg(C_2H_3O_2)_2 \cdot 4H_2O_{(s)} \xrightarrow{\text{H}_2\text{O}(20°\text{C})} \text{solution}$$

$$\xrightarrow{\text{air}(500°\text{C})} MgO_{(s)} + CO_{2(g)} + \overset{\cdot}{C}O_{(g)} + H_2O_{(g)} \qquad (4.9)$$

Another approach used has been to adsorb the chemical solution into a microporous organic material such as cellulose. The loaded fiber is first pyrolysed and then calcined in a controlled atmosphere. The porous agglomerates are easily comminuted. This technique has been used to produce

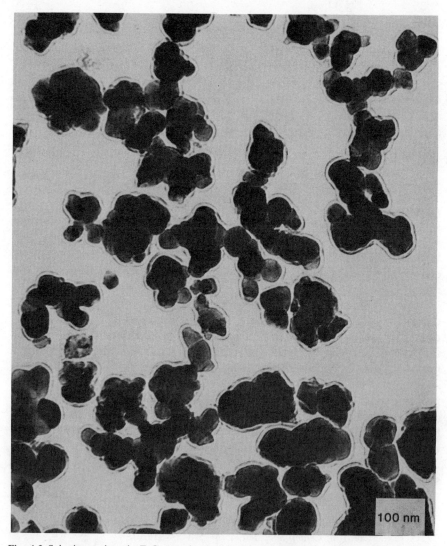

100 nm

Fig. 4.3 Submicron zirconia ZrO_2 powder prepared by adsorption, pyrolysis, and calcination. (Sample dispersed using ammonium polyacrylate; transmission electron micrograph.)

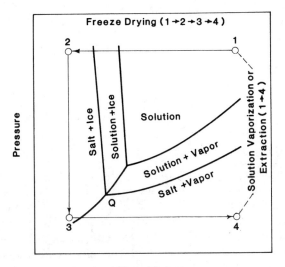

Fig. 4.4 A pressure-temperature phase diagram for an aqueous salt solution and the processing path for freeze-drying. (After M. Rigterink, *Am. Ceram. Soc. Bull.* **51**(2), 161 (1972).)

colloidal sized particles or fibers of a wide variety of oxides, carbides, and metals (Fig. 4.3). Other approaches for extracting the solvent have included spraying of the solution into a immiscible desiccating liquid (liquid/liquid drying) or spraying into a hot immiscible liquid to produce vaporization.

In freeze-drying, a salt solution is sprayed into a cold liquid such as hexane (60°C) and quickly frozen into a salt and ice. The solvent is sublimated at a pressure below the triple point by slowly raising the pressure and temperature, as indicated in Fig. 4.4. Calcination of the homogeneous salt yields porous aggregates which are milled to a micron-size powder with good sinterability.[*]

Sol-Gel Techniques

Sol-gel processing has attracted much interest both for the preparation of special powders and for forming thin coatings and cast or extruded shapes. In sol-gel processing, colloidal particles or molecules in a suspension, a sol, are mixed with a liquid, which causes them to join together into a continuous network, called a gel. Polymerization greatly restricts chemical diffusion and segregation. The gel is dried, calcined, and milled to form a powder.

Silica gel is produced by the hydrolysis and polymerization of the alkoxide tetraethyl orthosilicate in a solution, as indicated by the following reactions;

[*] M. D. Rigterink, *Am. Ceram. Soc. Bull.* **51**(2), 158–161 (1972).

$$(OC_2H_5)_4Si + H_2O \xrightarrow{\text{hydrolysis}} (OC_2O_5)_3SiOH + C_2H_5OH \qquad (4.10)$$

$$(OC_2H_5)_3SiOH + OH^- \xrightarrow{\text{dehydration}} (OC_2H_5)_3SiO^- + H_2O \qquad (4.11)$$

$$(OC_2H_5)_3SiO^- + (OC_2H_5)_3SiOH \xrightarrow{\text{gelling}} (OC_2H_5)_3SiOSi(H_5C_2O)_3$$
$$+ OH^+ \qquad (4.12)$$

Polymerized gels may also be formed by the hydrolysis of other simple or mixed alkoxides such as titanium and zirconium esters. Gels may also be formed when an aqueous sol of a hydrated aluminum or iron oxide is mixed with a dehydrating agent; a hydrogen-bounded alumina gel is produced by mixing an aqueous boehmite AlO(OH) sol with acetone. This technique has also been used to produce titania and zirconium oxide powders. In the citrate process, citric acid is added to a prepared salt solution; after partial alcohol dehydration to a viscous liquid, the solution is further dehydrated and decomposed by spraying into a furnace.

4.2 POWDERS FROM VAPOR PHASE REACTIONS

Vapor phase reactions have been used to produce special oxide and nonoxide powders (Fig. 4.5). Oxides can be formed by reacting a metal chloride with water vapor at a high temperature, as is indicated for the formation of titania by the reaction

$$TiCl_{4(g)} + 2H_2O_{(g)} \rightarrow TiO_{2(s)} + 4HCl_{(g)} \qquad (4.13)$$

Salts dissolved in alcohol have been pyrolysed in an atomizing burner to produce compound oxide powders. Fine oxide powders have also been produced by oxidizing evaporated metals.

Silicon nitride powder may be formed by reacting silicon tetrachloride and ammonia in a plasma below 1000°C, i.e.:

$$3SiCl_{4(g)} + 4NH_{3(g)} \rightarrow Si_3N_{4(s)} + 12HCl_{(g)} \qquad (4.14)$$

The thermal decomposition of $(CH_3)_2SiCl_2$ and CH_3SiH_5 vapor has also been used to produce high-purity silicon carbide SiC.

Vapor phase techniques produce submicron-size, well-dispersed particles entrained in large volumes of gas. Large, complex collection systems are required to remove the powder.

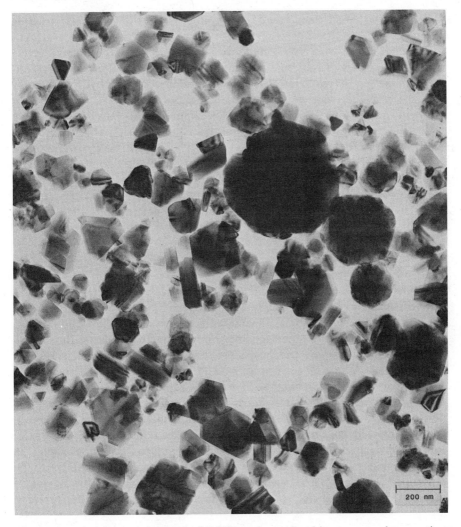

Fig. 4.5 Scanning electron micrograph of β-SiC powder produced from a vapor phase reaction.

4.3 OTHER TECHNIQUES

Techniques somewhat intermediate between the solid-state reaction of mixed-oxide particles and the special chemical techniques may offer a compromise between added expense and improved homogeneity. Mixtures of oxide powders or gels may form a crystalline mixed hydroxide compound when heated to 180–700°C in the presence of steam having a pressure of 1–100 MPa. A mixed-oxide compound of fine grain size is formed on calcining to a relatively low temperature. This technique, called hydrothermal synthesis, has been used to produce a variety of mixed oxide compounds.*

Decomposition of a salt commonly produces a relatively fine, reactive product. Mixtures of salts that decompose at a similar temperature, such as an alkaline earth carbonate and an aluminum or iron hydroxide, may produce a mixed-oxide compound of fine particle size at a relatively low temperature. Silicon nitride and silicon aluminum oxynitride have been produced by heating discrete oxide particles in a nitrogen atmosphere at a high temperature.

A special chemical technique may be used to dope a fine powder or gel. After dispersion of the powder in a chemical solution, a precipitation or gelation reaction may precipitate or immobilize the dopant on the surface. The doped powder produced on drying and then heating to a relatively low temperature may be finer and require less comminution to disperse the aggregates.

SUMMARY

Powders and colloidal materials that have special or very precisely controlled characteristics have been developed for research studies and for the industrial production of some advanced ceramics. In preparing the special materials, relatively pure precursors are first mixed on an atomic scale in a liquid or gas. A salt or gel containing the ions of interest is formed by a reaction producing a precipitate or by very rapidly removing the solvent. Chemical segregation which reduces the homogeneity must be prevented during the preparation process. Decomposition should occur at a temperature below that causing sintering of the product into a hard aggregate. Powders prepared using special chemical techniques may range widely in chemical and physical form. Special chemical techniques continue to be developed and evaluated for the production of powders for advanced ceramics. To be beneficial, they must enable production of a better product.

SUGGESTED READING

1. *Better Ceramics through Chemistry*, Vol. 32 *of Materials Research Society Proceedings*, C. J. Brinker, D. E. Clark and D. R. Ulrich (eds), North Holland, New York, 1984.

2. D. W. Johnson Jr., Powder Preparation–Ceramics, Chapter 2 in *Advances in Powder Technology*, Gilbert Y. Chin (ed), Am. Soc. Metals, Metals Park, OH. 1982.

* S. Komarneni et al., *Adv. Ceram. Mater.* **1**(1), 87–92 (1986).

3. P. E. D. Morgan, Chemical Processing for Ceramics, in *Processing of Crystalline Ceramics*, *Materials Science Research*, Vol. 11, H. Palmour, R. E. Davies, and T. M. Hare (eds), Plenum, New York, 1978, pp.67–77.

4. D. W. Johnson Jr. and P. K. Gallagher, Reactive Powders from Solution, Chapter 12 in *Ceramic Processing before Firing*, George Y. Onoda Jr. and Larry L. Hench (eds), Wiley-Interscience, New York, 1978.

5. David W. Johnson Jr., Nonconventional Powder Preparation Techniques, *Am. Ceram. Soc. Bull.* **60**(2), 221–224, 243 (1981).

6. P. K. Gallagher et al., Preparation of a Nickel Ferrite from Coprecipitated $Ni_{0.2}Fe_{0.8}C_2O_4 \cdot 2H_2O$, *Am. Ceram. Soc. Bull.* **48**(1), 1053–1059. (1969)

PROBLEMS

4.1 Considering the nature of the chemical diffusion fluxes during a solid-state reaction, can complete atomic-scale homogeneity be expected in a compound formed by the reaction of mixed powders?

4.2 Explain why the choice of the salt phase is critical for both the precipitation stage and the calcination stage in preparing a complex oxide by coprecipitation and decomposition.

4.3 Contrast the structure of the salt particles from the coprecipitation of two ions in a solid solution phase and from the simultaneous precipitation of each ion in separate phases.

4.4 What are the technical and market factors that will favor the production of powders using special chemical methods rather than the solid-state reaction of powders?

4.5 Design a process for producing a iron oxide powder that is doped with cobalt only in the near surface region.

4.6 Draw a binary phase diagram for a salt that is soluble in water. Comment on the progression of phases as a drop of this solution is cooled to a temperature at which salt and ice are formed.

4.7 State three possible mechanisms that may cause chemical heterogeneity in a powder prepared by calcining a homogeneous salt.

4.8 For heterogeneous precipitation, the distribution of an impurity I and host H in the surface layer is

$$\frac{dI_{ppt}}{dH_{ppt}} = \lambda \frac{I_o - I_{ppt}}{H_o - H_{ppt}}$$

How does the amount of impurity I precipitated vary with the amount of host precipitated when the distribution coefficient λ is 1 and 0.4?

PART III

MATERIAL CHARACTERIZATION

The properties of a fired product are very dependent on the characteristics of the starting material and their subsequent modification during processing. In order to select and control these materials, it is necessary to have a knowledge of their more important characteristics. Some of this information is supplied in the material specifications provided by the raw-material supplier. However, the material system will have to be characterized at different processing stages, and the processing engineer must be familiar with the techniques used and the information obtained. The general characteristics of a material system and common specifications of several commercial ceramic materials are presented in Chapter 5. Chemical and microstructure analyses are discussed in Chapter 6. Chapters 7 and 8 describe principles and techniques for characterizing the size, shape, density, surface area, and porosity of ceramic particle systems.

5

Characteristics and Specifications of Ceramic Materials

The characteristics of a material are those parameters that specify the chemical and physical aspects of its composition and structure. "Composition" denotes the proportions of chemically and physically different constituents. "Structure" refers to the spatial distribution, orientation, and association of these constituents.

The properties of a material are its responses to changes in the physical or chemical environment. Every particle system will have particular properties—e.g., a particular thermal conductivity, elastic modulus, and dielectric constant. Flow and deformation properties are commonly referred to as rheological properties. Responses to the chemical environment such as adorption or dissolution are chemical properties. Porous particle systems may have special properties such as capillarity, permeability, and electroosmotic flow. Dispersed systems have special properties such as settling rate, electrophoretic mobility, and optical scattering. A system is said to be anisotropic with respect to a particular property if the property varies with direction in the material.

This chapter will consider the general characteristics of particle systems and specifications provided for ceramic raw materials. The student should keep in mind the distinct difference between the properties and characteristics of a material system.

5.1 PARTICLES, POWDERS, COLLOIDS, AND AGGLOMERATES

A particle is a discrete, solid unit of material and may be single or multiphase in composition. Groups of particles that are weakly bonded together may behave as a fragile, larger pseudoparticle called an agglomerate (see Fig. 3.5). If strongly bonded together, the larger particle is not easily dispersed and is referred to as an aggregate or a hard agglomerate. Bonds in hard agglomerates are generally primary chemical bonds formed by a chemical reaction or sintering. In soft agglomerates, the relatively weak bonds may be of electrostatic, magnetic, Van der Waals, or capillary adhesion type.

The magnitude of the inertial force of a particle relative to surface forces has a major effect on particle behavior. A particle system is said to be granular if the gravitational force is predominant (the material is free-flowing), a powder if the surface force is of the same order as the gravitational force (naturally agglomerates), and colloidal if the particles are so fine that the inertial force of a particle is insignificant and the surface forces dominate the behavior. The surface forces are dependent on the environment of the particles. But for practical purposes, particles larger than 44 μm (opening in a 325-mesh sieve) can generally be considered to be granular and particles smaller than 1 μm as colloidal. Colloids dispersed in a low-viscosity liquid typically exhibit Brownian motion at 20°C. The behavior of powders and colloids can be markedly altered by adsorbed surfactants that modify the surface forces.

5.2 RAW-MATERIAL SPECIFICATIONS

The general characteristics of one particle and a system of particles are listed in Table 5.1. Complete characterization is an impossible task, and for each

Table 5.1 Characteristics of a Particle System

Single Particle	Particle System
1. Primary chemical composition	1. Distribution of chemical composition
2. Impurity composition, distribution, and partitioning	2. Distribution of impurities
3. Phase composition	3. Distribution of phase composition
4. Point and line defects, domains, etc.	4. Distribution of crystal defects
5. Structure of phases, boundaries	5. Porosity and pore structure
6. Porosity and pore structure	6. Particle structure distribution
7. Size	7. Particle size distribution
8. Shape	8. Particle shape distribution
9. Density	9. Particle density distribution
10. Specific surface area	10. Bulk density
	11. Specific surface area

material and application we must consider what characterization is necessary and sufficient. Some of this information may be supplied by the raw-material vendor on a specification sheet for the material. Table 5.2 lists typical specifications for special very high purity, very fine alumina materials. The impurity composition is quite complete, and the crystalline phase is identified. The specific surface area and nominal information about agglomeration and crystal size are presented. Since these are relatively expensive materials, the customer will certainly determine additional characteristics of each lot and perhaps process and fabricate a small amount of the material in the laboratory or factory to verify that it is satisfactory.

Typical specifications of three calcined Bayer process aluminas are listed in Table 5.3. The parameters specified are similar to those for the purer

Table 5.2 Specification of Special High-Purity Aluminas

Characteristic	Calcined[a]	Calcined[a]	Calcined[b]
Crystal phase	>90% gamma	85% alpha	alpha
Purity (%)	99.99	99.99	99.99
Impurity analysis of ceramic grade (ppm)			
Na	20	20	
Pb	4	4	
Si	18	18	
Cr	4	4	
Fe	10	10	
Ga	15	15	
Ca	10	10	
Mg	5	5	
Zn	4	4	
Ti	5	5	
Mn	3	3	
V	3	3	
Cu	2	2	
SiO_2			<50
Fe_2O_3			<20
CaO			<10
Na_2O			<10
Ga_2O_3			<10
Others			<10
Untimate particle size (μm)	0.01	0.15	<0.5
Specific surface area (m^2/g)	115	10	5–50
Agglomerate size (μm) mean	2	0.6	0.5
Crystal density (Mg/m^3)	3.67	3.98	3.98
Apparent bulk density (Mg/m^3)	0.12	0.51	

[a]Products of Baikowski International Corp., Charlotte, NC.
[b]Product of Aluminum Company of America, Pittsburgh, PA.

Table 5.3 Specifications of Three Bayer Process Aluminas

Characteristic	Calcined Intermediate Soda	Reactive Low Soda	Tabular (−325 Mesh)
Chemical analysis (%)			
Al_2O_3	99.4	99.7	99
SiO_2	0.02	0.02	0.2
Na_2O	0.25	0.08	0.10
Fe_2O_3	0.04	0.01	0.3
CaO	0.04	0.01	0.07
LOI (1100°C)	0.2		
Total water[a]	0.3		
α Alumina phase (%)	90+	~100	~100
Ultimate crystal	<5.0	>0.5	
Size (μm)			
Particle size distribution			
Sieve analysis (wt %)			
+100 mesh			
+200 mesh			
+325 mesh			
−325 mesh			>95
Sedimentation analysis[b] (μm)			
90%<	40	1.5	
50%<	12	0.5	
10%<	3	0.2	
Specific surface area[c] (m^2/g)	1.0	3–6	
Specific gravity	3.8	3.98	>3.4
Bulk density (Mg/m^3)	1.0		

Source: Products of Aluminum Company of America, Pittsburgh.
[a] 1100°C ignition loss after adsorption at 44% relative humidity.
[b] Gravity settling.
[c] Nitrogen adsorption.

alumina, but we can readily see from the specifications that these aluminas are quite different in chemical purity and particle size (see Figs. 3.5 and 4.2). A plant engineer examines the preshipment specifications supplied by the vendor for the particular lot of material. Depending on past experience and the particular application, a set of characteristics may be determined for a small sample using standardized test procedures, in the purchaser's laboratory, before authorizing or rejecting a shipment.

Specifications for three different barium titanate powders used for electronic ceramics are listed in Table 5.4. These materials are prepared with different chemical and particle size characteristics, as indicated in the specifications. The concentration of CO_2 and SO_3 indicates the incomplete

Table 5.4 Typical Specifications of Calcined Barium Titanates

Character	Capacitor	MLC	Piezoelectric
Chemical analysis (wt %)			
SiO_2	0.10	0.12	0.15
Al_2O_3	0.10	0.14	0.16
TiO_2	34.64	33.95	33.38
SrO	0.90	0.78	0.91
BaO	63.59	64.28	64.18
Na_2O	0.10	0.17	0.15
SO_3	0.15	0.14	0.18
CO_2	0.09	0.15	0.43
LOI	0.17	0.32	0.57
Size analysis $(\mu m)^a$			
90%<	4.5	5.5	5.4
50%<	1.6	2.3	2.0
10%	0.8	1.0	0.8
+325 Mesh (%)	0.02	0.02	0.02
Bulk density (Mg/m^3)	1.8	2.4	2.0
Electrical Property Analyses (body contains 10% calcium zirconate and 1% magnesium zirconate):			
Dielectric constant (25°C)	5250	4000	4400
Dissipation factor (% at 25°C)	1.18	0.83	0.67
Δ Dielectric constant (100°C)	−52.9	−48.9	−54.0
Δ Dielectric constant (−10°C)	−33.7	−2.4	−4.9
Fired density (Mg/m^3)	5.30	5.54	5.60

^{a}Products of TAM Ceramics Inc., Niagara Falls, NY.

decomposition of reactants during calcination. Differences in the electrical properties of a fired body reflect the variations in the chemical stoichiometry and physical characteristics of the raw materials and microstructure developed during firing.

Specifications for the commerical kaolins listed in Table 5.5 include the basic chemical and particle size characteristics. The composition of clay mineral types and mineral impurities such as free quartz are not listed. The MBI index is a relative indication of the specific surface area determined by adsorption of methylene blue dye. Clay bodies are usually processed as suspensions, and the pH index of a suspension may suggest the compatibility or change in pH if one clay is substituted for another. The pyrometric cone equivalent (PCE) indicates the relative resistance of a material to vitrification and creep on heating. The MOR is the flexural strength of dried bars formed by extrusion. The pH, PCE, and MOR are not characteristics; rather, they are indices that indicate something about effects of soluble chemical impurities, impurity phases, and the particle size distribution on the chemical, thermal, and mechanical behavior, respectively.

Table 5.5 Typical Specifications of Ceramic-Grade Kaolins

Characteristic	NC[a]	GA-P[b]	GA-C[b]
Chemical analysis (%)			
SiO_2	47.72	45.36	45.74
Al_2O_3	37.53	38.26	38.25
Fe_2O_3	1.16	0.36	0.41
TiO_2	0.08	1.52	1.55
CaO		0.47	0.06
MgO		0.04	0.12
K_2O	1.17	0.21	0.06
Na_2O	0.15	0.11	0.14
LOI	14.04	13.47	13.66
Total	99.85	99.80	99.99
Particle size analysis (cumulative mass percent finer)			
20 (μm)	97.5	98.0	97.0
10	89.0	93.5	88.0
5	75.0	83.0	74.5
2	53.0	65.0	54.0
1	35.0	48.5	38.0
0.5		32.0	21.5
0.2		15.0	11.0
MBI (meq/100 g)	—[c]	7.8	2.0
pH	5	7.2	4.3
PCE	33–34	34–35	34–35
Dry MOR (MPa)	1.2	3.4	0.9

[a]North Carolina kaolin, Harris Mining Co. Inc., Spruce Pine, NC.
[b]Georgia kaolin, Cyprus Industrial Minerals, Inc., Sandersville, GA. (GK-P for plastic forming; GK-C for casting.)
[c]Contains halloysite with tubular particle shape.

SUMMARY

The characteristics of a material are the parameters necessary for its identification or description. Specifications provided by suppliers of materials provide some of these characteristics. More complete specifications or tighter specifications of the lot-to-lot reproducibility of a material commonly increase the material cost. Processing alters the characteristics of the particle system. Materials processors should determine the materials characteristics that are requisite for control of the processing and the properties of their products.

SUGGESTED READING

1. F. H. Norton, *Fine Ceramics*, Krieger, Malabar, FL, 1978.
2. W. M. Flock, Characterization and Process Interactions, Chapter 4 in *Ceramic Processing before Firing*, George Y. Onoda Jr. and Larry L. Hench (eds), Wiley-Interscience, New York, 1978.
3. G. Y. Onoda Jr. and L. L. Hench, Physical Characterization Terminology, Chapter 5 in *Ceramic Processing before Firing*, George Y. Onoda Jr. and Larry L. Hench (eds), Wiley-Interscience, New York, 1978.
4. Y. S. Kim, Effects of Powder Characteristics, in *Treatise on Materials Science and Technology*, Vol. 9. *Ceramic Fabrication Processes*, Franklin F. Y. Wang (ed), Academic Press, New York, 1976, pp. 51–67.

PROBLEMS

5.1 Compare the impurity analyses of the two alpha aluminas in Table 5.2 using both an elemental and an elementary oxide basis. Which has the lower content of Ca?

5.2 Compare the agglomerate size to the ultimate particle size (size of crystallites) for the three alumina powders in Table 5.2. Which has the largest and which the smallest agglomerate/crystallite size ratio?

5.3 Calculate the bulk density as a percent for the gamma and alpha alumina powders in Table 5.2 and the calcined Bayer alumina in Table 5.3. Correlate the results in terms of other physical characteristics of the materials.

5.4 Calculate the ratio of moles $(BaO + SrO)$/moles TiO_2 for each of the barium titanate powders in Table 5.4. A coarse-grained sintered material is generally produced when the ratio is less than 1 if no grain growth inhibitor is added. In which of these powders would exaggerated grains be expected on firing?

5.5 Calculate the concentration of SO_3 and CO_2 in parts per million for the barium titanate powders in Table 5.4. What is the original source of these impurities?

5.6 The chemical analyses (wt %) of three production lots of calcined capacitor grade barium titanates are as follows.

SrO	0.95	0.82	0.84
BaO	63.71	64.00	63.94
TiO_2	34.67	34.61	34.59

What is the reproducibility of the molar ratio of $(BaO + SrO)/TiO_2$?

5.7 Explain the tendency for agglomeration of a dry powder in terms of the Van der Waals force per unit weight. How does the tendency vary with particle size?

6

Chemical and Phase Composition

Wet chemical techniques have been used routinely for the analyses of major elements in ceramic materials. However, the analysis of impurities and many major elements is now commonly performed using instrumented techniques, which are faster and more accurate. Microscale chemical analyses are determined using electron beam techniques. Surface analysis techniques coupled with ion beam machining may provide an analyses of surface material. Structural techniques such as x-ray diffraction, infrared spectroscopy, and light and electron microscopy are used to determine the identity of the phases, their structure, and the microstructure. Thermal analysis techniques are used to infer a change in the composition and structure of the system from the effects of a change in temperature or atmosphere on the chemical and physical properties of the material system.

6.1 BULK CHEMICAL ANALYSIS

The conventional qualitative elemental analysis of a ceramic material does not usually pose problems. Most industrial ceramic minerals contain at least 30 detectable elements, but fewer than 10 are commonly present at a level greater than 0.01–0.05%. Care must be taken to provide a representative sample.

Small samples are removed systematically from the bulk to obtain a representative sample for analysis. The samples may be analyzed individually or pooled and split down using a riffler or by cone and quartering to obtain a statistically random sample for analyses. When individual samples are analyzed, the data may indicate variations in the bulk and, when pooled, the mean analysis. Airborne samples may be collected in a filter using a suction device. The sample for analysis is often less than 1 g.

In classical wet chemical techniques, the material is fused by heating in the presence of a flux, dissolved in acid, and then analyzed using a standard precipitation, titration, or colorimetric technique. Wet chemical techniques are still widely used for the routine analyses of major elements in a wide variety of materials, and the sensitivity may sometimes exceed 1 ppm using colorimetric analysis.

Instrumented spectroscopic techniques listed in Table 6.1 are used for qualitative survey analyses, quantitative impurity analyses, and for the quantitative analyses of some major elements in systems that are not easily or accurately analyzed using wet chemistry techniques. The energy of free atoms, ions, and molecules is quantized, and each species has a set of characteristic energy levels. The absorption of radiant energy may cause a transition from a lower to a higher energy level; the return to a lower energy state causes the emission of radiation that is characteristic of the particular species. Electron transitions between higher energy levels produce radiation ranging from the near infrared to the ultraviolet. Transitions of electrons nearer the nucleus produce x-rays. In solids, liquids, and gases, transitions

Table 6.1 Spectroscopy Techniques

Method	Principle	Detect
Emission spectroscopy (powder)	Thermal stimulation of electrons in atom	Emitted line spectra (visible–UV)
Flame emission spectroscopy (liquid solution)	Thermal stimulation of electrons in atom	Emitted line spectra (visible-UV)
Atomic absorption spectroscopy (liquid solution)	Thermal stimulation of electrons in atom	Emitted line spectra (visible-UV)
X-ray fluorescence (powder)	X-ray stimulation of electrons in atom	Emitted line spectra (x-rays)
Mass spectroscopy (gas from solid, liquid, gas)	Ions deflected in a magnetic field	Mass/charge of ion
Infrared spectroscopy (solid, liquid, gas)	Molecular vibrations with a change in dipole moment absorb IR radiation	Absorption spectra (infrared)

can occur between energy levels associated with the vibration and rotation
of molecules and cause adsorption in the infrared.

In emission spectrographic analysis, the sample of compacted powder is
excited in an electric arc or by a laser flash. Chemical information is
provided by the wavelength and intensities of the characteristic line spectra
emitted in the visible to ultraviolet region of the spectrum (Fig. 6.1).
Emission spectroscopy is often used to quickly obtain the qualitative "survey" analysis of an unfamiliar material.

Alkalies emit in the visible range; samples containing alkali are commonly dissolved in a liquid, and the alkali is analyzed using flame emission
spectrometry. Atomic absorption spectrometry may be used to analyze as
many as 30–40 elements present at a concentration of <0.1%. It has
become the standard technique for the industrial analyses of impurities and
is now used for analyzing major elements in some systems. The sample in
solution form is sprayed into a flame to dissociate it into its elements.
Radiation from a cathode lamp containing the element of interest is also
passed through the flame. Dissociated atoms in the flame absorb the
spectrum emitted by the lamp, which reduces the transmitted intensity.
Concentrations down to the parts-per-million level are determined by
comparing the absorbence of standards and the sample solution.

X-ray fluorescence in which x-rays are used to stimulate secondary
characteristic x-radiation may be used for the quantitative analyses of major
and minor elements with an atomic number greater than that of sodium.
Characteristic x-rays are diffracted by an analyzing crystal, and the diffraction angle and intensity provide information about the element and its
concentration. This technique is accurate and fast for routine analyses.

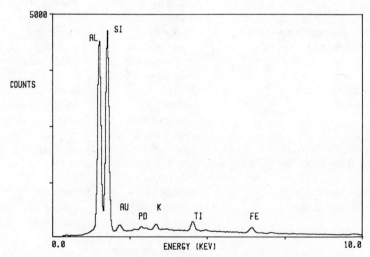

Fig. 6.1 Line spectrum for a mullite refractory tile determined using energy-dispersive spectroscopy. (Courtesy of W. Votava, Alfred University, Alfred, NY. Au and Pd from specimen
preparation.)

The components of a gas in pores, or a gas produced by heating or sputtering, may be determined using gas chromatography or mass spectroscopy. The partitioning of the unknown gas between the carrier gas and a sorbent as it flows through an absorption column may provide a characteristic adsorption spectrum enabling identification of the compounds in the gas phase. Mass spectrometry may be used to determine the identity and concentration of inorganic or organic ions present at concentration of less than 10 ppm in a gas produced by heating or sputtering, because their deflection in a magnetic field is proportional to the charge divided by the mass of the ion.

Infrared spectrometry determines the adsorption of infrared radiation due to characteristic vibrations and rotations of atoms in molecules and solid compounds. It is used to determine molecular structure and the presence of trace molecular anion impurities in calcined materials (Fig. 6.2).

Table 6.2 Instrumental Analysis Techniques

Bulk Techniques	Comments
Emission spectrography (ES)	Elemental analyses to the ppm level, frequently used for qualitative survey analyses, 5 mg powder sample
Flame emission spectroscopy (FES)	Quantitative analyses of alkali and Ba to the ppm level, ppb detectability for some elements, solution sample
Atomic absorption spectroscopy (AAS)	Industry standard for quantitative elemental impurity analyses; detectability to ppm level, solution sample
X-ray fluorescence (XRF)	Elemental analyses, detectability to 10 ppm, $Z > 11$, solid/liquid samples
Gas chromatography/mass spectrometry (GC/MS)	Identification of compounds and analysis of vapors and gases
Infrared spectroscopy (IRS)	Identification and structure of organic and inorganic compounds, mg dispersed power in transparent liquid or solid or thin-film sample
X-ray diffraction (XRD)	Identification and structure of crystalline phases, quantitative analysis to 1%, mg powder sample
Nuclear magnetic resonance (NMR)	Identification and structure of organic (NMR) compounds, sample to 5 mg for H and 50 mg for C

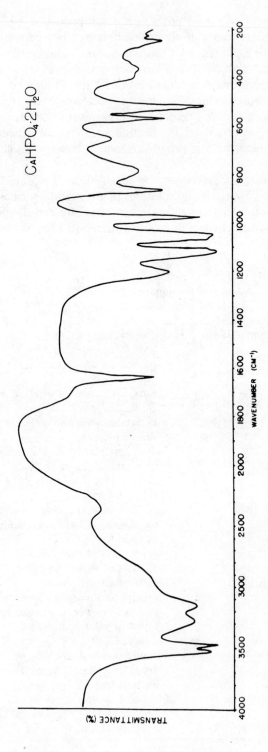

Fig. 6.2 Transmission infrared spectrum of brushite CaHPO$_4 \cdot$ 2H$_2$O indicates H$_2$O adsorption bands at about 3500 and 3200 cm^{-1} due to stretching mode and a band at 1650 cm^{-1} due to a bending mode. (From F. Casciani and R. Condrate, *Spec. Lett.* **12**(10), 704 (1979).)

Neutron activation analysis is used to determine quantitatively the concentration of radioactive isotopes in a material. A neutron source may be used to convert susceptible elements to a radioactive isotope. Capabilities of instrumented analysis techniques are summarized in Table 6.2.

6.2 PHASE ANALYSIS

Crystalline phases diffract x-rays according to the Bragg law,

$$n\lambda = 2d \sin \theta \qquad (6.1)$$

where θ is the diffraction angle for a lattice spacing d, λ is the wavelength of the x-rays, and n is an integer. Powder or polished polycrystalline specimens are used, and the diffraction 2θ angles are recorded. The identification of a phase is accomplished by comparing the d spacings and relative intensities of the sample material with reference data for known materials (Fig. 6.3). Quantitative phase analysis to about 1% is possible when phases are randomly oriented and diffraction lines of different phases are clearly distinguished.

Optical microscopy has long been used to identify phases in thin sections of polycrystalline and partially vitrified systems. Optical microscopy is also used routinely to examine surface topography and the microstructure of polished and etched specimens down to about 0.2 μm. Scanning electron microscopy (Fig. 6.4), having a greater resolution, is now widely used for microstructure analyses, because it is convenient and versatile and because

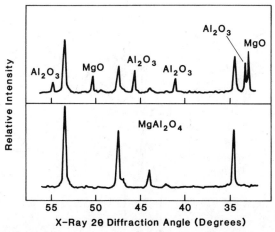

Fig. 6.3 X-ray diffraction pattern for (top) an incompletely reacted 1:1 mixture of periclase MgO and corundum Al_2O_3 containing spinel $MgAl_2O_4$ and (bottom) a completely reacted mixture.

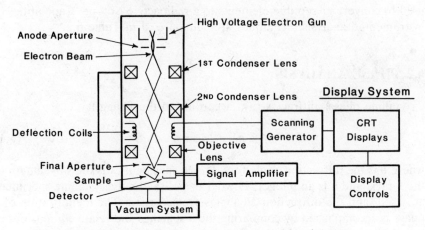

Fig. 6.4 Schematic diagram of a scanning electron microscope.

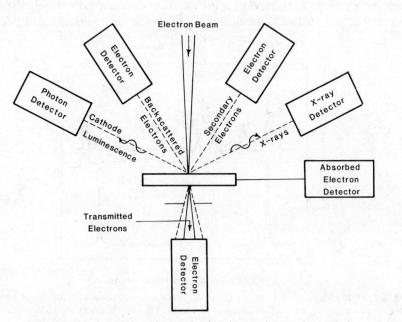

Fig. 6.5 Signals and detection modes for electron beam microanalysis.

Table 6.3 Microscopic Characterization Techniques

Technique	Comments
Light microscopy (LM)	Microstructure of etched, polished, or thin sections; surface topography; phase analyses; resolution to 0.2 μm
Scanning electron microscopy (SEM) with dispersive x-ray spectroscopy (EDS)	Microstructure of fracture, polished, or etched surface; resolution 10 nm; qualitative and semiquantitative analyses with 2 μm resolution using EDS; 0.1% detectability $Z > 11$, microstructure of dispersed organic in backscatter mode
Transmission electron microscopy (TEM)	Microstructure of thin sections 20–200 nm thick, resolution 1 nm, identification of crystal structure by electron diffraction, qualitative and semiquantitative analyses using scanning TEM with EDS with 30- to 50-nm resolution

micrographs with a large depth of focus can be obtained quickly to document the microstructure. Detection modes provided by the interaction of the electron beam with the sample are showed in Fig. 6.5. With an energy-dispersive spectroscopy attachment, microscale qualitative chemical analyses can be obtained rather easily, and this chemical information can aid greatly in the interpretation of microstructures. Although less convenient to use, TEM can provide analysis down to a resolution of about 1 nm; the structure of defect phases and grain boundaries can be determined using the electron diffraction or scanning modes with EDS analyses (Table 6.3).

6.3 SURFACE ANALYSIS

The analysis of surface and near-surface material, which may vary from that of the bulk, has been aided greatly by recent improvements in electron and ion beam instrumentation (Table 6.4). In Auger electron spectroscopy, a scanning beam of electrons excites the surface of the specimen, and the energy of emitted "Auger electrons" provides information about the atomic numbers of the elements present. Bombarding the surface with ions, called ion milling, can remove atomic layers of material for depth profiling. Subsequent analyses provide information about near-surface concentration gradients. In electron microprobe analyses, the characteristic x-rays emitted

Table 6.4 Surface Analysis Techniques

Surface Technique	Comments
Auger electron spectroscopy (AES)	Elemental analyses to a resolution of 3 μm lateral and 5 nm thick, detection to 0.1%, depth profile by ion milling
Electron microprobe analyzer (EMA)	Qualitative analyses $Z > 5$, quantitative analyses of polished sections to <0.1%, lateral and depth resolution of 2 μm
Secondary ion mass spectrometry (SIMS)	Identification of elements of all Z by mass spectrometry of sputtered ions, depth profiles by ion milling, depth resolution to 10 nm, detectability to ppm level
X-ray photoelectron spectroscopy (XPS)	Elemental analyses of layers 5 nm thick, lateral resolution 1 nm, depth profiles by ion milling, chemical bonding information
Fourier transform infrared spectroscopy (FTIR)	Identification of structure of adsorbed molecules and coating

when electrons scan a microscopic region of the surface are detected and used to identify and quantify chemical elements present.

If the surface is excited using monochromatic x-rays, the photoelectrons emitted from the surface contain information about the type of atoms and their oxidation state and structure in the surface. This technique is called x-ray photoelectron spectroscopy and is used for chemical analyses. Bombarding the surface with monoenergetic low-energy ions will remove surface ions by sputtering, and these can be analyzed using a mass spectrograph. This technique is called secondary ion mass spectroscopy. Because of the importance of surfaces in ceramic processing, these surface techniques are becoming more available, and their use is increasing in the development of more advanced ceramics.

6.4 THERMOCHEMICAL AND THERMOPHYSICAL ANALYSES

Thermochemical techniques are used to determine thermodynamic changes in individual materials and reactions between materials in a batch or between a material and the atmosphere, usually when the temperature is changed. For thermogravimetric analysis (TGA), material is suspended from a balance, and the weight is monitored during controlled heating or cooling or under isothermal conditions (Fig. 6.6). In differential thermal analysis DTA, thermocouples in contact with a powder specimen and a reference powder, respectively, indicate the test temperature and any differential temperature due to an endothermic or exothermic transition or reaction in the sample as a function of temperature or time (Fig. 6.7). When the enthalpy change is determined, the technique is called differential scanning calorimetry (DSC). Important parameters that must be controlled include the rate of heating or cooling, the thermal conductivity of the container and packed powder sample and reference, the particle size of the

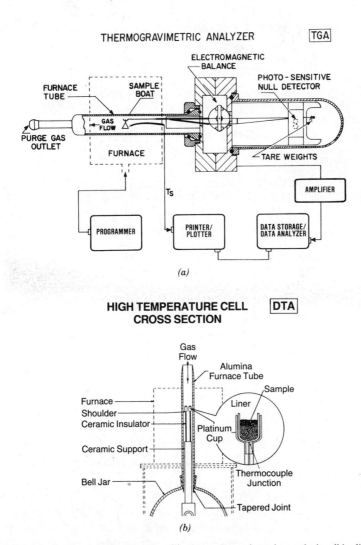

Fig. 6.6 Experimental configurations for (a) thermal gravimetric analysis, (b) differential thermal analysis.

DSC CELL CROSS-SECTION [DSC]

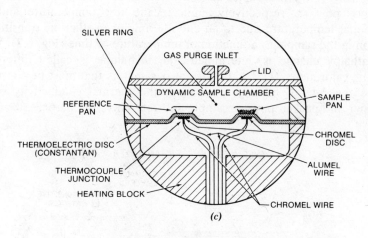

(c)

THERMOMECHANICAL ANALYZER [TMA]

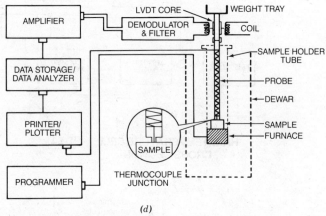

(d)

Fig. 6.6 (c) Differential scanning calorimetry, and (d) thermal mechanical analysis. (Figures courtesy of Du Pont Inc., Wilmington, DE.)

sample, and the composition and flow of the atmosphere. Thermochemical information supplemented with chemical, phase, and microstructural analyses of heated material is used to identify the changes such as the elimination of liquids, the oxidation and vaporization of organic additives, transitions in materials, and reactions between materials (Fig. 6.8) or between materials and the gaseous environment, vitrification, and recrystallization.

Thermophysical analyses include the monitoring of expansion or shrinkage during heating or cooling and the resistance to mechanical penetration

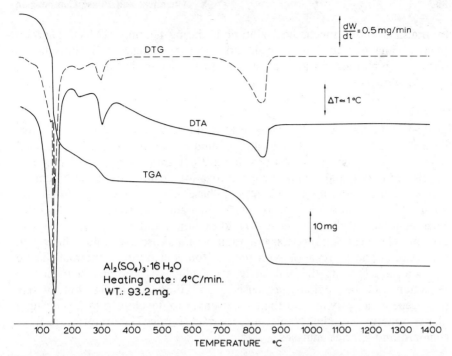

Fig. 6.7 Thermal gravimetric analysis (TGA), differential thermal gravimetric analysis (DTG), and differential thermal analysis (DTA) for the decomposition of alum $Al_2(SO_4)_3$ $16H_2O$. Decomposition yields water up to 300°C and SO_3 between 600 and 875°C. (Courtesy of Mettler Instrument Corp., Hightstown, NJ.)

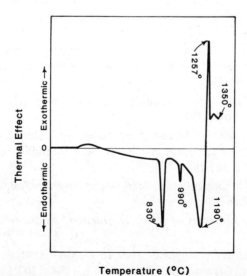

Fig. 6.8 Differential thermal analyses for a mixture of $BaCO_3$ and TiO_2 heated to 1350°C in air. Endothermic peaks at 830 and 990°C correspond to structural transitions in $BaCO_3$; the peaks at 1190° and 1257° correspond to the decomposition of $BaCO_3$ and formation of Ba_2TiO_4 and $BaTiO_3$, which is dependent on the pressure of CO_2. (After L. Templeton and J. Pask, *J. Am. Ceram. Soc.* **42**(5), 213 (1959).)

or transmission of mechanical vibrations during heating (TMA). Thermophysical analyses are used to identify phase changes and sintering of inorganic materials during calcination and changes in the properties of organic binders.

SUMMARY

Ceramic materials must be judiciously characterized to control and reproduce the material. The analyses and techniques selected should be planned wisely so that those characteristics that are determined supply the essential information. Nominal chemical characterization includes the major chemical composition and the impurity analysis. Phase analysis provides information about the type and amount of crystalline and amorphous constituents. Microanalytical techniques provide information about the surface composition and chemical and phase segregation on a microscopic scale. Changes in the material and its reaction with the environment during a thermal change are determined using thermal analysis techniques. Basic materials characterization can be tedious, expensive, and time-consuming, but, wisely practiced, it can provide information essential to developing or producing a higher-quality ceramic. Numerous contract service laboratories provide standard and special analyses.

SUGGESTED READING

1. Franklin F. Y. Wang, Powder Characterization, Chapter 3 in *Advances in Powder Technology*, Gilbert Y. Chin (ed), American Society for Metals, Metals Park, OH, 1982.

2. R. Nathan Katz, Characterization of Ceramic Powders, in *Treatise on Materials Science and Technology*, Vol. 9: *Ceramic Fabrication Processes*, Academic Press, New York, 1976.

3. L. L. Hench and R. W. Gould, *Characterization of Ceramics*, Marcel Dekker, New York, 1971.

4. Robert A. Condrate, Infra-red Spectroscopy, Chapter 6 in *Analytical Methods for Material Investigation*, G. A. Kirkendale (ed), Gordon and Breach, New York, 1971.

5. P. D. Garn, *Thermoanalytical Methods of Investigation*, Academic Press, New York, 1965.

6. B. D. Cullity, *Elements of X-ray Diffraction*, Addison-Wesley, Reading, MA, 1956.

7. Carlo G. Pantano, Surface and In-Depth Analysis of Glass and Ceramics, *Am. Ceram. Soc. Bull.* **60**(11), 1154–1163,1167 (1981).

8. Norman C. Kenny, Raw Material Acceptance Testing for Dielectric Applications, *Am. Ceram. Soc. Bull.* **59**(2), 241 (1980).

9. H. Bennett, Trends in Ceramic Analyses II, *Ceramic Age* **5**, 9–10 (1974).

PROBLEMS

6.1 The impurity analysis (ppm) of a material is reported as follows: SiO_2 50–100, Fe_2O_3 50–100, MgO < 10, CaO < 10, $TiO_2 < 10$, others <10. What is the minimum purity of the material in weight percent?

6.2 Two specimens of zinc aluminate spinel nominally $ZnO \cdot Al_2O_3$ are prepared by reacting zinc oxide and aluminum oxide. The weights of zinc oxide and alumina are 80 and 100 g in batch 1 and 81.38 and 101.94 g in batch 2. Calculate the stoichiometry of each spinel assuming the components react completely.

6.3 What technique would you use for a survey of impurities in zinc manganese ferrite?

6.4 What technique would you use for a quantitative analysis of lithium, sodium, and strontium impurity in barium titanate?

6.5 How are the concentrations of Fe^{2+} and Fe^{3+} in a ferrite determined?

6.6 How would you determine quantitatively the reaction temperature and amount of unreacted zinc oxide for the reaction in problem 2?

6.7 How would you determine if zinc oxide volatilizes during the reaction in problem 2?

6.8 How would you analyze for the presence of carbonate impurity in barium titanate made by reacting barium carbonate and titania? How would you determine the amount present?

6.9 What technique would you use to analyze the concentration of impurities in grain boundaries of sintered alumina of 10-μm grain size?

6.10 How would you determine the amount of monoclinic and tetragonal zirconia on the surface of a zirconia wear plate?

6.11 How can the distribution and concentration of silicate phases in an electrical porcelain be analyzed?

6.12 A clay has the following chemical analysis: SiO_2 (48.00), Al_2O_3 (36.0), Fe_2O_3 (0.44), TiO_2 (0.54), CaO (0.26), MgO (0.18), K_2O (1.70), Na_2O (0.36), and ignition (12.06 wt%). Estimate the composition of kaolinite, mica, montmorillonite, calcite, quartz, and free H_2O in the sample. Assume that the K_2O and Na_2O are present in mica and that the MgO is present in montmorillonite.

7

Particle Size and Shape

The particle sizes of most conventional materials fall in the range of 50 nm to 1.0 cm. However, sizes as large as 10 cm are used in some refractory castables and concrete and particles as small as 5 nm are observed in some chemically prepared materials. This range of sizes extends from about the size of coarse gravel to the size of viruses, as is shown in Fig. 7.1.

Two or more size analysis techniques may be required to cover the size range of interest. Fortunately, recent advances in instrumentation have

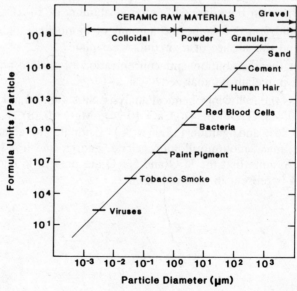

Fig. 7.1 Variation of number of Al_2O_3 formula units per particle as a function of particle diameter. (Size ranges of ceramic materials and several biological substances are shown for reference.)

greatly improved the precision and reduced the time and tedium involved in obtaining an analysis, and we can now use particle size results in a more timely and effective way. The accuracy of the particle size data depends somewhat on the sample preparation, the particle shape, and the technique used for the analyses. Because the particle size distribution is one of the most important characteristics of a particle system, it is important that we understand principles involved in instrumental techniques used and factors that can affect the data and their interpretation.

7.1 ANALYSIS TECHNIQUES

Just as for chemical analyses, it is important to obtain a representative sample of the appropriate amount. In particle size analysis, we must recognize that sampling and handling may change the physical state of the sample. Dispersing the material in a liquid may reduce the concentration and sizes of agglomerates. Conversely, agglomeration may occur in a well-dispersed suspension if a slow analysis technique is used. We should not forget the intended use of the size information when selecting an instrument and specifying the sample preparation procedure.

Microscopy Techniques

Microscopy can quickly show the nominal size and shape of the particles in a sample, and obtaining a representative micrograph of the sample is often the first step (Fig. 4.5). Additional micrographs are often taken at a higher magnification to observe atypical particles and finer details of the shape and surface characteristics. The full range of sizes can be examined using optical and electron microscopes. Often a scanning electron microscope will cover the size range of interest. Dry powder can be mounted on the sample stub by inserting a stub coated with an adhesive into the powder. Alternatively, a dilute suspension of particles can be placed on the stub, which will form a deposit of particle on evaporation of the liquid. Extreme dilution, chemical dispersion, fast drying, or freeze-drying may be used to minimize agglomeration.

Quantitative image analysis requires a photographic enlargement or an electronic display. For statistical accuracy, it is necessary to measure and count at least 700 particles lying in a plane. It will be difficult to distinguish and measure fine particles in a micrograph when the particles are poorly dispersed and when coarse particles are more than 1 order of magniude larger than the fines are present (Fig. 7.2). Agglomerates in powders will be apparent, if present, and with care it will be possible to determine if these are hard or soft agglomerates (Fig. 7.3).

Particles may vary widely in shape (Fig. 3.1, 3.2, 3.5). Crushed minerals and calcined aggregates are generally angular and have rough surfaces.

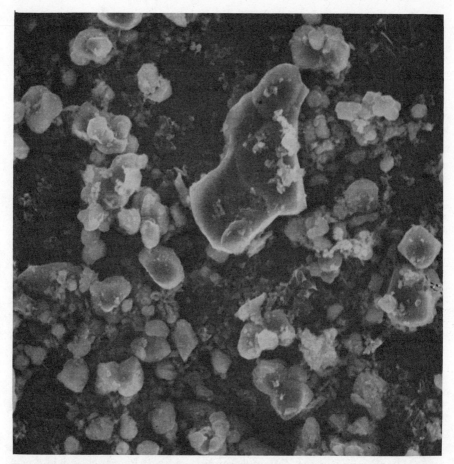

Fig. 7.2 Scanning electron micrograph of a calcined alumina with a wide size distribution (20–0.1 μm). (*Note*; small particles are indistinct.)

Milled materials are commonly more rounded, but aggregates may have a rough surface. Coarser kaolin particles may be in the form of lineated or fairly isometric stacks of platelike particles; fine kaolin contains both small stacks and platelike particles.

The characteristic size of particles that are spherical or cubical may be defined in terms of the diameter or edge length. For most real particles, the definition of the characteristic size is less precise. The maximum chord length in a particular reference direction between two tangents on opposite sides of the image is often used as the characteristic size of the particle, as shown by the parameter a in Fig. 7.4. Another characteristic size can be defined as the length of the chord that bisects the area of the particle. The diameter of the circle with an equal projected area is sometimes used as the size. Particle dimensions can be measured manually on a photographic

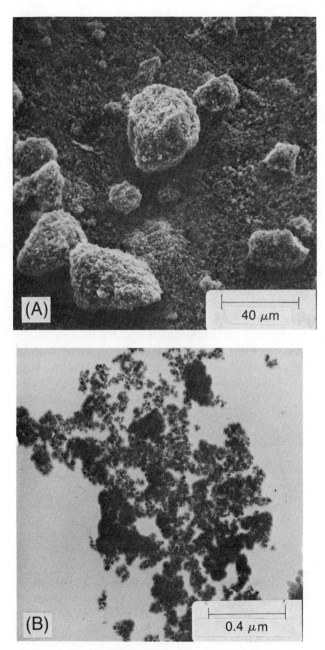

Fig. 7.3 As received, ultrafine stabilized zirconia powder contains (A) soft agglomerates that are easily dispersed into (B) smaller aggregates.

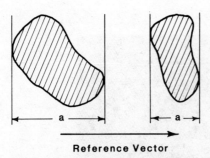

Fig. 7.4 The distance between two normals tangent to the image of the particle may be used as the "characteristic length" of the particle.

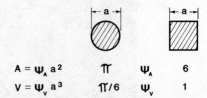

Fig. 7.5 Figure showing the definition of the area shape factor ψ_A and volume shape factor ψ_V. The size of the particle is a, and the area and volume are A and V, respectively.

enlargement using vernier calipers or a piezoelectric matrix pad. Images from light and electron microscopes can also be displayed on the screen of an image-analyzing computer which can rapidly and automatically scan the image and determine a preselected characteristic dimension of each region of differing optical density. A light-pen accessory enables the operator to select or omit particular features (such as agglomerates) in the analysis.

The aspect ratio of a particle is defined as the longest dimension divided by the shortest dimension, and a particle with an aspect ratio greater than 1 is said to be anisometric. The aspect ratio of the particles may be estimated by measuring the major and minor axes of particles suitably oriented in a photomicrograph. The constant of proportionality between the particle size and its area A or volume V can be considered a numerical shape factor. Examples are given in Fig. 7.5. The ratio ψ_A/ψ_V relative to that for a sphere is sometimes used as an index of angularity.

Error in microscopy techniques can occur owing to an inaccurate magnification, a nonrandom arrangement of particles, and insufficient sampling. The magnification of the microscope must be calibrated. Anisometric particles will tend to rest with their longer dimension parallel to the plane of the image, and this introduces a bias in the data. Extreme fines are often incompletely analyzed because of human bias or agglomeration in the specimen.

Sieving Techniques

Sieving is the classification of particles in terms of their ability or inability to pass through an aperture of controlled size. Particles are introduced onto a stack of sieves with successively finer apertures below, and the particles are agitated to induce translation until blocked by an aperture smaller than the particle size. Sieves with openings greater than 37 μm are constructed using wire mesh and are identified in terms of mesh size and the aperture size; the relationship between mesh size and aperture size for U.S. Standard sieves is listed in Appendix A. The aperture size can be calibrated using an optical micrometer. Monolithic electrodeposited metal sieves are now commercially available with uniform apertures as small as 2 μm.

Sieving is the most widely used technique for sizes down to 44 μm. A set of sieves often follows a $\sqrt{2}$ progression of sizes. Agglomeration becomes a problem below about 44 μm and can introduce error in the analysis. In some cases, dry sieving analyses can be extended to 20 μm by admixing a desiccant of known size into a very dry powder sample. Alternatively, the sample can be dispersed in a liquid and sieved wet.

For an equal weight of sample, the number of particles increases as (size)$^{-3}$, and apertures of the finer sieves become blocked ("blinded") if the sample size is too large and the sieving mechanism is inefficient. Older sieving apparatus depended on the low-frequency mechanical shaking for particle motivation. Modern apparatus that imparts a relatively high frequency air pulse produces a more efficient sieving of the small sizes. The combination of a low-frequency mechanical pulse to translate coarse particles and the higher-frequency air pulse can provide precise dry sieving analyses for a wide range of sizes if the particles are regular or only slightly anisometric in shape. An elongated particle does not have a single "characteristic size," and the precision of sieve analyses depends somewhat on the aspect ratio of the particle.

Wet sieving with "micromesh" sieves finer than 37 μm requires suction to pull liquid and fine entrained particles through the apertures. Particle motivation can be assisted by supporting the sieve in an ultrasonic tank or using a mechanical vibrator. Wet sieving on a 44- or 37-μm sieve is often used when analyzing the "subsieve" portion of a clay. Wet sieving to 2 μm for routine control analysis has been reported.

Sedimentation Techniques

A spherical particle of density D_p and diameter a released in a viscous fluid of viscosity η_L of lower-density D_L will momentarily accelerate and then fall at a constant terminal velocity v. Laminar liquid flow occurs when the particle Reynolds number Re is less than 0.2. The Reynolds number is calculated from the equation Re $= (vaD_L)/\eta_L$. For ceramic particles, the upper size limit is about 50 μm. The terminal velocity is related to the particle diameter according to the Stokes equation (Fig. 7.6);

$$v = a^2(D_p - D_L)g/18\eta_L \qquad (7.1)$$

where g is the acceleration due to gravity or centrifuging. The time t for settling a height H is

$$t = \frac{18H\eta_L}{a^2(D_p - D_L)g} \qquad (7.2)$$

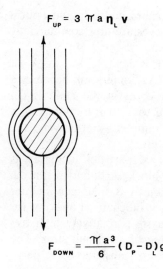

$$F_{UP} = 3\pi a \eta_L v$$

$$F_{DOWN} = \frac{\pi a^3}{6}(D_p - D_L)g$$

Fig. 7.6 Equilibrium of forces during the settling of a particle in a Newtonian fluid with laminar flow.

For particles of alumina in water, the time for gravity settling 1 cm is about 1 min for a 10-μm particle but about 2 h for a 1-μm particle.

Commercial apparatus utilizes either a layer technique or a homogeneous technique. In the layer technique, a very dilute suspension is carefully placed on a column of settling liquid, and the fraction of particles that have settled to a depth H is measured with time. Settling times are reduced by centrifuging, for which the Stokes equation is

$$t = \frac{18\eta_L \ln(r_t/r_0)}{a^2(D_p - D_L)\omega^2} \tag{7.3}$$

where ω is the angular velocity of the centrifuge and r_0 and r_t are the radial positions of the particles before and after centrifuging.

In the homogeneous technique, the fraction of particles in a plane at some depth H in a dilute (<4 vol %) homogeneous suspension is determined. Particles of all sizes may settle across the plane, but only particles of a particular size and smaller, in accordance with Stokes' law, will remain at or above the plane after some particular settling time. The analysis time for sizes finer than 1 μm is reduced considerably by centrifuging. The concentration in suspension is normally determined from the relative intensity I/I_0 of transmitted light or x-rays;

$$-\ln(I/I_0) = k\Sigma N_i a_i^2 \tag{7.4}$$

where k is a constant and N_i is the number of particles of size a_i, and

$$\Sigma \text{ particle volume} = -a_i \ln(I/I_0) \tag{7.5}$$

The analysis time varies with the particle density and fineness, but analysis times are typically in excess of 10 min and may range up to several hours for a gravitational settling to $0.02\ \mu$m using commercial instrumentation.

Possible sources of error in sedimentatin size analyses include hindered settling due to particle interactions, the tendency of fine particles to be pulled along behind larger ones, and agglomeration caused by Brownian motion. An adsorbed hull of liquid molecules and/or deflocculant can increase the size but reduce the average density of submicron particles (Fig. 4.3). Incomplete dispersion and agglomeration during settling may also affect the size data. Ultrasonic dispersion and an admixed deflocculant are commonly used for an ultimate size analysis.

Settling techniques cannot be used for samples containing particles of different density. For nonspherical particles, the diameter of a sphere that settles at the same rate, the equivalent spherical diameter, is obtained from the analyses. Theoretical models indicate that the Stokes' drag force is insignificantly affected by small distortions from sphericity.

Electrical Sensing Techniques

The resistance of an electrolyte current path through a narrow orifice between two electrodes increases when a ceramic particle passes through the orifice. The pulse in resistance is proportional to the volume of electrolyte displaced. In electrical sensing techniques, resistance pulses for a stream of dispersed particle passing through the orifice are converted into voltage pulses, amplified, scaled, and counted electronically (Fig. 7.7).

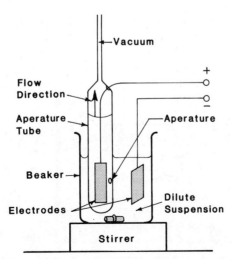

Fig. 7.7 Schematic of the electrical sensing zone technique.

Monosize samples of known size are used for calibration. Using one orifice tube, the analysis time for a predispersed sample is less than 1 min with very high statistical confidence. The working range is typically 2–40% of the aperture size, or 0.3–400 μm using several aperture tubes. This technique can be used for samples containing particles of different density and an aspect ratio up to about 10. Extreme dilution to less than a few parts per million is required to minimize coincident counting. Samples must be dispersed in special electrolytes.

Laser Diffraction Techniques

Dispersed particles momentarily traversing a collimated light beam will cause Fraunhofer diffraction of light outside of the cross section of the beam when the particles are larger than the wavelength of the light. The intensity of the forward-diffracted light is proportional to the particle size squared, but the diffraction angle varies inversely with particle size. A He–Ne laser is commonly used for the light source. The combination of an optical filter, lens, and photodetector or a lens and multielement detector coupled with a microcomputer enables computation of the particle size distribution from the diffraction data (Fig. 7.8). The sample may be either a liquid or gas suspension of particles or droplets of about 0.1 vol % concentration. A typical analysis time is about 2 min, which permits on-line analysis. The possible size range capability of this technique is about 1–1800 μm. However, the working size range of a particular instrument varies with the model, and some selection is possible. This technique is independent of the density of the particulates and is factory calibrated. Particles finer than the detection limit are not included in the data.

Submicron sizing analyses may be obtained by using filtered and polarized light from a W-halogen light source and analyzing the wide-angle Mie scattering for light of different wavelengths. The popularity of light diffraction instruments for particle size analysis has grown rapidly in recent years

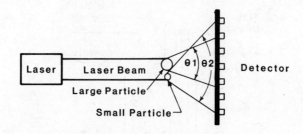

Diffraction Angle θ1 For Large Particles.
Diffraction Angle θ2 For Small Particles.

Fig. 7.8 Particles larger than the wavelength of the laser beam produce Fraunhofer diffraction.

because of their precision, speed, ease of operation, versatility, and ease of maintenance.

Light Intensity Fluctuation Techniques

Dispersed particles finer than 5 μm constantly diffuse randomly through the liquid and can scatter light. The frequency of the fluctuations may be detected using a photomultiplier, and the particle size is calculated from the frequency using the Stokes-Einstein equation

$$a = (k_B T)/(3\pi\eta_L D_T) \tag{7.6}$$

where D_T is the translational diffusion constant at a temperature T. An average size and polydispersity index is rapidly obtained using a digital correlation computer. The working size range is from about 5 to 0.005 μm.

Other Techniques

X-ray line-broadening techniques may be used to determine the average crystallite size in a powder finer than about 0.5 μm and small-angle x-ray scattering below 0.1 μm. Calibration is required, and particles should not be polycrystalline.

Air permeability techniques have also been used to determine an index of average particle size in the range 50–0.2 μm. Air permeability is a function of the number and geometry of capillaries in a bed of powder. The geometry of the capillaries depends on particle packing characteristics and is not simply a function of some nominal particle size. However, the air permeability technique is quick and is sometimes used for quality control purpose.

An average particle size may also be calculated from the specific surface area of the powder. This technique is discussed in Chapter 8.

7.2 SELECTING A TECHNIQUE

The specification of an instrument and sample preparation procedure should be based on experience as well as principles. Instrument cost, flexibility, size range capability, specimen preparation, and analysis time are all important factors. Ease of operation and maintenance and the direct readout of results in printed or graphical form is preferable. But most important, the result should be useful. Capabilities of the techniques discussed above are summarized in Table 7.1.

Comparative size analyses for representative materials using different size

Table 7.1 Summary of Particle Size Analysis Techniques

Method	Medium	Size Capability (μm)	Sample Size (g)	Analysis Time[a]
Microscopy				
Optical	Liquid/gas	400–0.2	< 1	S–L
Electron	Vacuum	20–0.002	< 1	S–L
Sieving				
	Air	8000–37	50	M
	Air	5000–37	5–20	M
	Liquid	5000–5	5	L
	Inert gas	5000–20	5	M
Sedimentation				
(Gravity)	Liquid	100–0.2	< 5	M–L
(Centrifuge)	Liquid	100–0.02	< 1	M
Electrical sensing zone				
	Liquid	400–0.3	< 1	S–M
Light scattering				
Fraunhofer	Liquid/gas	1800–1	< 5	S
Mie	Liquid	1–0.1	< 5	S
Intensity fluctuation				
	Liquid	5–0.005	< 1	S

[a] S = short (< 20 min); M = moderate (20–60 min); L = long(> 60 min).

analysis techniques and optimum procedures are quite limited. Zwicker[*] has reported very good agreement in the size analysis for finely ground alumina using micromesh sieving, centrifugal sedimentation, and electronic counting. Kahn[†] showed that control of agglomeration was critical to obtain a precise analysis for barium titanate finer than a few microns. Croft[‡] reported that sedimentation analyses were more precise and more sensitive to the submicron size distribution of clay suspensions than an electrical sensing zone technique. Frock and Plantz[§] reported reproducible differences in the size analyses of a common alumina powder when using different techniques, as indicated in Fig. 7.9. For a sample of columnar-shaped particles of wollastonite, the size distribution determined by gravity settling was reproducibly finer than that obtained by light diffraction, as shown in Fig. 7.10.

ASTM studies for centrifugal sedimentation and electrical sensing zone analyses have indicated that a precision of ± 1% (15% confidence level) at each size and ±3% between laboratories can be expected for each method using very careful procedures.

[*] J. D. Zwicker, *Am. Ceram. Soc. Bull.* **46**(3) 303–306 (1967).
[†] M. Kahn, *Am. Ceram. Soc. Bull.* **57**(4), 448–451 (1978).
[‡] T. Croft, Chapter 1 in *Processing Consequences of Raw Material Variables*, J. Varner et al. (eds.), Alfred University Press, Alfred, NY, 1985.
[§] H. Frock and P. Plantz, *ibid*.

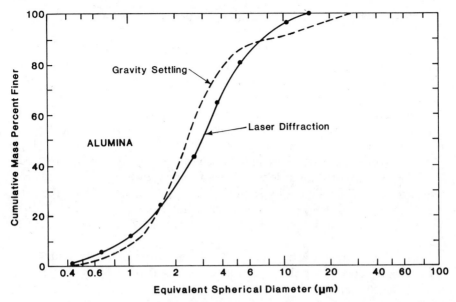

Fig. 7.9 Comparative size data for a calcined alumina obtained by gravity settling and laser light diffraction. (After H. Frock and P. Plantz, *Processing Consequences of Raw Material Variables*, Alfred University Press, Alfred, NY, 1985.)

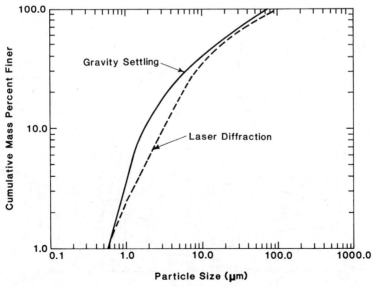

Fig. 7.10 Comparative size data for well-dispersed, milled $CaSiO_3$ particles having an aspect ratio ~4.

93

7.3 PRESENTATION OF PARTICLE SIZE DATA

Particle size data are usually presented in tabular or graphical form for analyzing and comparing the mean size and distribution of sizes. A description of the data in terms of a particular mathematical distribution function is also desirable but not always possible.

Tabulated size data were presented in the specifications of materials in Chapter 5. A histogram may be constructed from data classified according to the number, mass, or volume fraction for particular intervals of size (Fig. 7.11). Data determined on a volume basis are equal to data on a mass basis when the density of the particles does not vary with size. The cumulative size distribution is obtained by summing the fractions of particles that are finer (CNPF or CMPF) or alternatively larger (CNPL or CMPL) than specific sizes between the smallest and largest sizes of the distribution (Table 7.2).

When the number fraction ΔN of particles in each size interval Δa is plotted as a function of size, the fractional distribution function $f_N(a)$ is obtained in the limit as Δa approaches 0, i.e.:

$$f_N(a) = \lim_{\Delta a \to 0} \frac{\Delta N}{\Delta a} = \frac{dN}{da} \qquad (7.7A)$$

Similarly, the fractional distribution by mass $f_M(a)$ is

$$f_M(a) = \lim_{\Delta a \to 0} \frac{\Delta M}{\Delta a} = \frac{dM}{da} \qquad (7.7B)$$

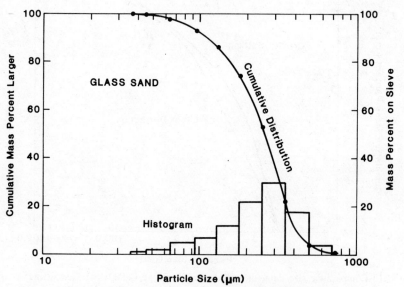

Fig. 7.11 Histogram and cumulative size distribution CMPL of glass sand.

Table 7.2 Particle Size Distribution of Ball Milled Petalite LiAlSi$_4$O$_{10}$

Sieve Size (μm)	Mass on Sieve		CMPF (%)	CMPL (%)
	(g)	(%)		
1000	0	0.0	100.0	0.0
500	0.18	0.9	99.1	0.9
355	0.50	2.5	96.6	3.4
250	1.45	7.3	89.3	10.7
180	3.00	15.0	74.3	25.7
125	3.96	19.8	54.5	45.5
90	3.26	16.3	38.2	61.8
63	3.18	15.9	22.3	77.7
44	1.94	9.7	12.6	87.4
0	2.52	12.6	0.0	100.0
Total	20.00	100.0		

The fractional distribution functions are related to the cumulative size distribution $F(a)$ by the equations

$$f_N(a) = \frac{dF_N(a)}{da} \tag{7.8A}$$

and

$$f_M(a) = \frac{dF_M(a)}{da} \tag{7.8B}$$

where $F(a)$ is the cumulative fraction finer than size a and the subscripts N and M indicate number and mass bases, respectively. For distributions extending over a wide range of size, the fractional distribution with the logarithm of size is used;

$$f_N(\ln a) = \frac{dF_N(a)}{d(\ln a)} = \frac{a(dF_N(a)}{da} \tag{7.9}$$

Particle sizes corresponding to $F_M(a)$ of 90, 50, and 10% are now often listed in the specifications of a material, as is indicated in Tables 5.3 and 5.4. These values provide an approximate basis for comparing the apparent mean size and range of sizes.

It is common practice to fit a mathematical function to the cumulative data. The fractional distribution is then obtained by taking the derivative of the cumulative function. Mathematical functions used to describe common particle size data are discussed in Section 7.5.

7.4 CALCULATING A MEAN PARTICLE SIZE

Different mean particle sizes may be defined for a material in terms of the fractional distribution function $f_N(a)$ and an appropriate shape factor. The total number of particles N is

$$N = \int_{a_{min}}^{a_{max}} f_N(a)\,da \qquad (7.10)$$

The sum of the lengths L of all of the particles is

$$L = \int_{a_{min}}^{a_{max}} af_N(a)\,da \qquad (7.11)$$

and the mean length $\bar{a}_L$ is

$$\bar{a}_L = L/N = \frac{\displaystyle\int_{a_{min}}^{a_{max}} af_N(a)\,da}{N} \qquad (7.12)$$

The total area A of the particles is

$$A = \int_{a_{min}}^{a_{max}} \psi_A a^2 f_N(a)\,da \qquad (7.13)$$

If $\bar{a}_A$ is the size of the particle having the average surface area,

$$A = N\psi_A \bar{a}_A^2 \qquad (7.14)$$

When ψ_A is a constant

$$\bar{a}_A = \frac{\displaystyle\int_{a_{min}}^{a_{max}} a^2 f_N(a)\,da}{N} \qquad (7.15)$$

Definitions of other mean particle sizes are given in Table 7.3. When the mathematical distribution function is not known, the average size can be calculated using a finite summation (Table 7.3). Using finite summation, the number of particles N is

$$N = \sum_{i=1}^{i=N_c} f_N(\bar{a}_i) \qquad (7.16)$$

where $f_N(\bar{a}_i)$ is the fraction of particles in an interval Δa_i having a mean size $\bar{a}_i$, and N_c is the number of size classes.

Table 7.3 Definitions of Mean Particle Size

Mean	Equation	Finite Summation
Length ($\bar{a}_L$)	$\bar{a}_L = \dfrac{\displaystyle\int_{a_{min}}^{a_{max}} a f_N(a)\, da}{N}$	$\bar{a}_L = \dfrac{\displaystyle\sum_{i=1}^{i=N_c} \bar{a}_i f_N(\bar{a}_i)}{N}$
Surface ($\bar{a}_A$)	$\bar{a}_A = \sqrt{\dfrac{\displaystyle\int_{a_{min}}^{a_{max}} a^2 f_N(a)\, da}{N}}$	$\bar{a}_A = \sqrt{\dfrac{\displaystyle\sum_{i=1}^{i=N_c} \bar{a}_i^2 f_N(\bar{a}_i)}{N}}$
Volume ($\bar{a}_V$)	$\bar{a}_V = \sqrt{\dfrac{\displaystyle\int_{a_{min}}^{a_{max}} a^3 f_N(a)\, da}{N}}$	$\bar{a}_V = \sqrt{\dfrac{\displaystyle\sum_{i=1}^{i=N_c} \bar{a}_i^3 f_N(\bar{a}_i)}{N}}$
Volume/surface ($\bar{a}_{V/A}$)	$\bar{a}_{V/A} = \dfrac{\displaystyle\int_{a_{min}}^{a_{max}} a^3 f_N(a)\, da}{\displaystyle\int_{a_{min}}^{a_{max}} a^2 f_N(a)\, da}$	$\bar{a}_{V/A} = \dfrac{\displaystyle\sum_{i=1}^{i=N_c} \bar{a}_i^3 f_N(\bar{a}_i)}{\displaystyle\sum_{i=1}^{i=N_c} \bar{a}_i^2 f_N(\bar{a}_i)}$

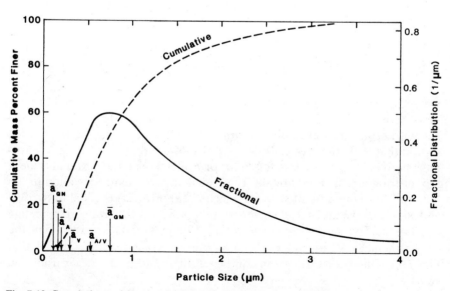

Fig. 7.12 Cumulative and fractional ($f_M(\ln a)$) distributions of a superground calcined alumina. The distribution is log-normal; calculated mean sizes are indicated on the size scale.

If the size data are on a mass basis, $f_N(a)$ may be calculated from $f_M(a)$ for substitution into the equations in Table 7.3, i.e.:

$$f_M(a) = (D_p \psi_v a^3) f_N(a) \qquad (7.17A)$$

or

$$f_M(\bar{a}_i) = (D_p \psi_v \bar{a}_i^3) f_N(\bar{a}_i) \qquad (7.17B)$$

Because of the different dependences on particle size, the rank of the different mean sizes is $\bar{a}_L < \bar{a}_A < \bar{\alpha}_V$, as is shown in Fig. 7.12. The mode for the distribution by number will appear at a smaller size than the mode corresponding to the mass distribution because of the greater number of particles in a unit mass of finer particles.

7.5 PARTICLE SIZE DISTRIBUTION FUNCTIONS

When a particular mathematical function can be used to satisfactorily approximate the size data for a material, the characteristic parameters of the distribution and the mathematical form may be used for purposes of describing and comparing materials. The characteristic parameters are a characteristic mean size or maximum size and a distribution modulus indicating the range of sizes.

The size distribution of many granulated powders produced by spray-drying and many powders produced by milling fined-grained minerals and calcined aggregate is often approximated by the log-normal LN distribution function

$$f(\ln a) = \frac{1}{\sqrt{2\pi} \ln \sigma_g} \exp - \frac{(\ln a_i - \ln \bar{a}_g)^2}{2(\ln \sigma_g)^2} \qquad (7.18)$$

where the frequency function $f(\ln a)$ may be either a number, volume, or mass basis. The parameter $\bar{a}_g$ is the geometric mean size, and σ_g is the geometric standard deviation indicating the range of sizes. The log-normal distribution is characteristically skewed as shown in Fig. 7.12. Using logarithmic probability paper, data for $F_N(a)$ or $F_M(a)$ will plot as a straight line if the distribution is log-normal. The geometric standard deviation σ_g is obtained from the ratio of sizes for which $F_N(a)$ or $F_M(a)$ is equal to 84.13% and 50.00%. The geometric mean size ($\bar{a}_{g_N}$ or $\bar{a}_{g_M}$) is the size corresponding to 50.00%. The geometric mean size $\bar{a}_{g_M}$ for mass data is larger than the geometric mean $\bar{a}_{g_N}$ for number data for the same material, as is indicated in Fig. 7.13. One geometric mean can be calculated from the other using the equation

$$\ln \bar{a}_{g_M} = \ln \bar{a}_{g_N} + 3(\ln \sigma_g)^2 \qquad (7.19)$$

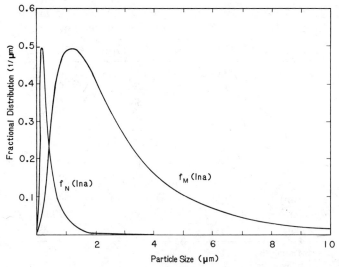

Fig. 7.13 Fractional number and mass size distributions for a milled zircon powder with a log-normal size distribution ($\bar{a}_{g_M} = 1.2\ \mu$m, $\sigma_g = 2.25$).

Linear, area, and volume mean sizes may be calculated for a LN size distribution using the equations in Table 7.4. Size data for an industrial spray-dried ferrite powder, obtained using two different techniques, and the characteristic log normal parameters are listed in Table 7.5. And size data for ball-milled zircon and barium titanate powders that are log-normal in distribution are listed in Table 7.6.

Another empirical distribution function that has been observed to describe the particle size data of some milled coarse-grained materials is the Rosin-Rammler (RR) equation;

$$F_M(a) = 1 - \exp - (a/a_{RR})^n \tag{7.20}$$

where a_{RR} is a characteristic size parameter for which $F_M(a) = 0.37$ and n^{-1}

Table 7.4 Equations for the Log-Normal Distribution

$$\ln \bar{a}_L = \ln \bar{a}_{g_N} + 0.5(\ln \sigma_g)^2$$

$$\ln \bar{a}_A = \ln \bar{a}_{g_N} + 1(\ln \sigma_g)^2$$

$$\ln \bar{a}_V = \ln \bar{a}_{g_N} + 1.5(\ln \sigma_g)^2$$

$$\ln \bar{a}_{V/A} = \ln \bar{a}_{g_N} + 2.5(\ln \sigma_g)^2$$

Table 7.5 Size Data and Log-Normal Parameters for an Industrial Spray-Dried Manganese Zinc Ferrite Powder

Size (μm)	CNPF (%)[a]	Sieve Size (μm)	CMPF (%)[b]
120	100.0	177	100
90	94.2	125	91.4
70	82.0	88	60.2
60	65.2	62	28.0
50	50.9	44	4.0
40	22.6		
30	11.0		

LN Parameters

$\bar{a}_{g_N}$ (μm)	50.5		–
$\bar{a}_{g_M}$ (μm)	–		79.0
σ_g (μm)	1.43		1.39

[a] SEM analysis.
[b] Seive analysis.

Table 7.6 Size Data and Log-Normal Parameters for Finely Milled Zircon and Barium Titanate Powders

Milled Barium Titanate		Milled Zircon
Size (μm)	CNPF (%)[a]	CMPF (%)[b]
10.0		100
6.0		97.5
4.0	100	93.0
3.0	99[c]	86.5
2.0	98	71.3
1.5	92	60.0
1.0	80	40.5
0.6	58	19.4
0.4	36	9.0
0.2	10	1.4
0.1	2	

LN Parameters

$\bar{a}_{g_N}$ (μm)	0.52	–
$\bar{a}_{g_M}$ (μm)	–	1.2
σ_g (μm)	2.1	2.25

[a] Scanning electron microscopy.
[b] Stokes sedimentation.
[c] Aggregates.

**Table 7.7 Size Data and Characteristic Parameters for
Two Milled Materials**

Calcined Alumina		−200-Mesh Quartz	
Size (μm)	CMPF (%)	Size (μm)	CMPF (%)
50	79.5	50	92.0
40	67.0	40	81.6
30	53.5	30	69.0
20	35.0	20	50.0
15	24.5	15	39.2
10	14.0	10	27.5
6	6.0	6	17.5
4	2.5	4	12.3
RR parameters		*GS parameters*	
a_{RR} (μm)	36	a_{max} (μm)	47
n	1.5	n	0.85

is an index indicating the dispersion of size. For the RR distribution, a plot of $-\ln(1 - F_M(a))$ vs. size on log-log paper is linear; n corresponds to the slope, and a_{RR} is the size for which $(-\ln(1 - F_M(a)) = 1$. The mean size $\bar{a}_{V/A} = a_{RR}/\Gamma(1 - 1/n)$ where Γ is the gamma function. The RR distribution is generally more skewed than the log-normal distribution. Size data and the calculated RR parameters are listed in Table 7.7 for a ball-milled alumina initially of coarse grain size.

The size distribution of rather coarse industrial minerals produced by crushing and grinding can sometimes be approximated by the Gaudin-Schuhmann (GS) equation;

$$F_M(a) = (a/a_{max})^n \qquad (7.21)$$

where a_{max} is the apparent maximum particle size obtained by extrapolation and n is the constant indicating the range of sizes. The size distribution of milled quartz in Table 7.7 is approximated by the GS distribution and is nearly linear when plotted on log-log graph paper.

The size data of real-particle systems depend on the type of fracture, milling, blending of sizes, and other processing. Particles size data are illustrated and discussed in several chapters presenting information on particle packing, milling, spray-drying, forming, and sintering processes.

SUMMARY

The particle size distribution of a particle system may be analyzed using several different techniques. Microscopy techniques provide information about both the size and shape of the particles and the presence of particle agglomerates; the potential range of sizes that may be analyzed is extremely large, and the analysis time is greatly reduced by using a computerized image analyzer. Sieving is convenient and widely used for the size analysis of particles larger than 44 μm. Laser diffraction techniques are very fast and convenient and have become very popular for size analysis from about 100 to 0.4 μm. Sedimentation techniques are versatile and enable analyses from about 44 to less than 0.1 μm in size in one analysis. Electrical sensing techniques are fast and convenient when the range of sizes is moderate. Light intensity fluctuation analysis permits size analysis to below 0.01 μm. The physical principle of each technique is the basis on which the size is defined, and precise size data for the same powder may vary somewhat using different techniques.

Particle size data are commonly presented in tabular form or graphically for puropses of analysis and comparison. Several different mean particle sizes and the range of sizes may be calculated to summarize the size data. The size distribution of powders produced by milling calcined aggregates and spray-dried powders may often be approximated by a regular mathematical equation such as the log-normal distribution function. When the distribution function is known, the size data are summarized in terms of the type and characteristic parameters of the distribution.

SUGGESTED READING

1. *Modern Methods of Particle Size Analyses*, H. Barth (ed), Wiley-Interscience, New York, 1985.

2. Terrance Allen, *Particle Size Measurement*, Wiley-Interscience, New York, 1981.

3. Bruce Weiner, Introduction to Particle Sizing Using Photon Correlation Spectroscopy, in *Processing Consequences of Raw Material Variables*, J. Varner, J. Funk, P. Johnson, and J. Reed (eds), Alfred University Press, Alfred, NY, 1985.

4. Harold N. Frock and P. E. Plantz, Rapid Analyses of Fine Particles Using Light Scattering, *ibid*.

5. Harold N. Frock and Richard L. Walton, Particle Size Instrumentation for the Ceramic Industry, *Am. Ceram. Soc. Bull.* **59**(6), 650–651 (1980).

6. T. Allen and M. G. Baudet, The Limits of Gravitational Sedimentation, *Powder Tech.* **18**, 131–138 (1977).

7. J. P. Oliver, G. K. Hickin, and Cylde Orr Jr., Rapid, Automatic Particle Size Analyses in the Subsieve Range, *Powder Tech.* **4**, 257–263 (1970/71)

8. J. D. Zwicker, Comparison of Particle Size Analysis Methods, *Am. Ceram. Soc. Bull.* **4**(3), 303–306 (1967).

9. John Elvans Smith and Myra Lee Jordan, Mathematical and Graphical Interpretation of the Log-Normal Law for Particle Size Distribution Analyses, *J. Colloid Sci.* **19**, 549–559 (1964).

PROBLEMS

7.1 How is the characteristic particle size a specified for each of the techniques listed in Table 7.1?

7.2 Plot the data in Table 7.2 in the form of a histogram. Draw the curves for CMPF and CMPL.

7.3 Using the CNPF data in Table 7.5, calculate the mean sizes $\bar{a}_L$, $\bar{a}_A$, $\bar{a}_{V'}$, and $\bar{a}_{V/A}$.

7.4 Using the CMPF data in Table 7.5, calculate $\bar{a}_{g_N}$ and $\bar{a}_{V/A}$.

7.5 Plot the CMPF data in Table 7.6 on log probability paper, indicating a $\pm 2\%$ uncertainty in the mass, and determine the geometric mean $\bar{a}_{g_N}$ and geometric standard deviation σ_g.

7.6 Calculate $\bar{a}_{g_N}$ and σ_g from the CNPF data in Table 7.6 using the fundamental definitions

$$\ln \bar{a}_{g_N} = \frac{\Sigma(n_i \ln \bar{a}_i)}{\Sigma n_i}, \qquad \ln \sigma_g = \sqrt{\frac{\Sigma n_i (\ln \bar{a}_i - \ln \bar{a}_{g_N})^2}{\Sigma n_i}}$$

7.7 Are the size data in Table 7.2 represented by the LN, RR, or GS distribution equations? (Assume uncertainty is $\pm 2\%$).

7.8 What are the largest and the smallest aspect ratios for the set of new coins: penny, nickel, dime, quarter.?

7.9 Calculate ψ_A, ψ_V, and ψ_A/ψ_V for a nickel. Assume that the characteristic size was determined by sieving.

7.10 Compare ψ_A/ψ_V for a sphere and a cube.

7.11 Using the method of finite summation, calculate $\bar{a}_{g_M}$, $\bar{a}_{g_N}$, and $\bar{a}_{V/A}$ for milled quartz. Assume reasonable values for the particle density and shape factor and $a_{max} = 60 \ \mu m$.

7.12 Calculate the maximum size of a spherical alumina particle settling in water at 20°C, for which $Re < 0.2$.

7.13 What is the gravitational settling velocity for nonporous alumina particles of 10- and 1-μm diameter in water at 20°C (assume $D_p = 3.98 \ g \ cm^{-3}$).

7.14 The diameters of six spherical particles are 2, 4, 8, 16, 32, and 64 μm. Calculate $\bar{a}_L$, $\bar{a}_A$, and $\bar{a}_V$.

7.15 Contrast the settling times for alumina particles of 0.1-μm diameter for a settling height of 1 cm under gravitational conditions and in an 1800-rpm centrifuge with an initial radial distance of 12 cm.

7.16 Estimate the fluctuation time constant t_c for the Brownian motion of an alumina particle of 0.1-μm diameter in water at 20°C, given $(t_c)^{-1} = 2D_T K^2$. Here $K = (4\pi n/\lambda_0) \sin(\theta/2)$, where n is the index of refraction of the liquid, $\lambda = 633$ nm for a He-Ne laser, and $\theta = 90°$.

7.17 Calculate the mass and number of formula units for alumina particles having diameters of 10^{-2}, 10^0, and 10^{+2} μm.

7.18 The CMPF size data for a powder are 27 (90), 20 (70), 13 (50), 10 (43), 5 (24), 3 (17), and 2 μm (13 wt %). Are these data approximated by a regular distribution function?

7.19 The LN parameters for a milled alumina powder are $\bar{a}_{g_M} = 1.3$ μm and $\sigma_g = 2.4$. Draw curves of $fm(a)$ and $fm(\ln a)$ on the same graph using a linear size scale.

8

Density, Pore Structure, and Specific Surface Area

Knowledge of the densities of ceramic phases, particles, processing additives, and bulk systems is important in ceramic processing for several reasons. The raw-material storage capacities are directly dependent on the bulk density of the material. We need to know the particle density of raw materials and the phase density of liquids and additives to calculate volumetric proportions from the gravimetric batch composition and to ascertain settling tendencies. A change in a very precisely determined particle density may indicate a change in the phase structure, chemical composition, or porosity of the material.

Pore size analysis is used to obtain information about the sizes and structure of pores in calcined aggregates, soft agglomerates and granules, and unfired and fired materials.

The specific surface area of a powder is very dependent on the particle size and shape and the content of submicron size particles and finer pores or fissures in the surfaces of particles. Specific surface area analysis may detect the presence of colloidal material that is not detected in some size analysis techniques. The specific surface area of a powder containing nonporous particles may be used as a relative index of absorptivity for approximating the amount of surface-modifying processing aid required in a particular processing operation.

In this chapter we will examine principles involved in determining the density, specific surface area, and pore structure of different ceramic materials.

8.1 DENSITY

The bulk density of a particle system is the mass per unit volume of the particles and interstices. The density of a particle is the mass/volume ratio of the particle. In a system of particles, the particle density refers to the mean density of all the different size particles. When the particles are nonporous, the particle density is referred to as the ultimate particle density and is the mean density of the solid phases constituting the particles. The density calculated from the chemical formula weight and the volume of the unit cell is termed the x-ray density. Often the ultimate density and x-ray density are nearly identical, but heavy elemental impurities may cause a measurable difference if these impurities are neglected in calculating the x-ray density.

For porous particles, the particle density must be defined more particularly. Particles dispersed in a fluid may adsorb molecules of the fluid into surface-accessible pores. The accessible porosity may be all or only a fraction of the porosity of the particles. Penetration of the fluid depends on the size of the molecules in the fluid relative to the pore size, wetting, and absence of another fluid in the pores. The ultimate volume is the volume of the solid phases. For a porous particle dispersed in a fluid, the apparent volume is the volume of the solid and the inaccessible pores (volume of fluid displaced). The particle bulk volume is the volume of solid and all porosity within the particles. Each of these particular volumes may be used to define the ultimate (D_u), apparent (D_a), and bulk (D_b) particle density (Fig. 8.1). The specific gravity is the weight of the particle relative to the weight of an equal volume of water.

Liquid pycnometry, as shown in Fig. 8.2, is commonly used for determining the mean apparent particle density of coarser powders to a precision of about $\pm 0.005 \, \mathrm{Mg/m^3}$. A calibrated pycnometer bottle containing powder is weighed, liquid of known density is added, and the bottle is reweighed. Accuracy depends on wetting the particles completely and eliminating air bubbles. A drop of wetting agent (antifoaming agent), which reduces the surface tension of water, aids in wetting and minimizes the floating of micron-size particles. Nonaqueous liquids of low vapor pressure such as kerosene are also used. A calibrated pycnometer bottle may be used to determine the density of another liquid.

A gas pycnometer may be used for measuring the density of powders and is more convenient when a significant fraction is finer than 10 μm in size.

Porous Fragment

19.65 g
10.0 cm³ Total
0.5 cm³ Closed Porosity
2.0 cm³ Open Porosity
D_u = 2.62 Mg/m³
D_a = 2.46
D_b = 1.96

Fig. 8.1 Illustration of a porous fragment and bulk, apparent, and ultimate density values.

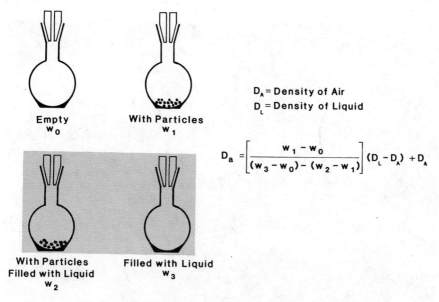

D_A = Density of Air
D_L = Density of Liquid

$$D_a = \left[\frac{w_1 - w_0}{(w_3 - w_0) - (w_2 - w_1)} \right] (D_L - D_A) + D_A$$

Fig. 8.2 Liquid pycnometer technique.

The powder density determined using gas pycnometry may be slightly higher than that obtained using liquid pycnometry when a small gas molecule such as He is able to penetrate into very small pores. In some materials, adsorption and condensation of gas molecules at room temperature in pores of a nanometer size can cause an erroneously high density, as indicated in Table 8.1.

The crushing and grinding of porous particles increases the relative fraction of surface pores (Fig. 8.3), and the apparent density of the finer particles may be higher (as indicated in Fig. 8.4) for tabular alumina. In many calcined powders that have been milled, the extremely coarse particles may be porous aggregates of slightly lower density, although most of the

Table 8.1 Apparent Particle Densities Determined by Liquid and Gas Pycnometry at 20°C

Material	Nominal Size (μm)	Fluid	Density (Mg/m^3)
Tabular alumina	590–2000	Water	3.65
(Al_2O_3)		Nitrogen	3.65
Rutile	44–250	Water	4.25
(TiO_2)		Nitrogen	4.35
Bentonite[a]	0.3–20	Water	2.22
(clay minerals/		Kerosene	2.53
montmorillonite)		Nitrogen	4.86[a]

[a] Significant error due to adsorption and condensation in pores.

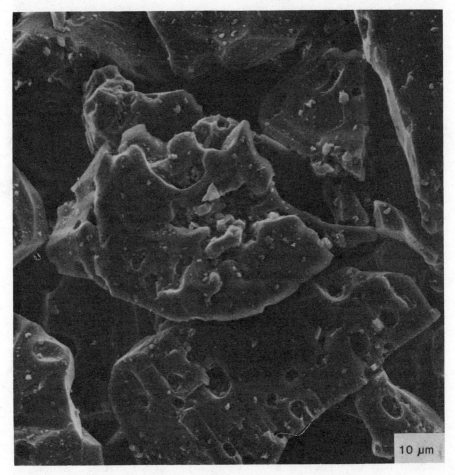

10 μm

Fig. 8.3 Pores in −325-mesh tabular alumina particles.

particles in the distribution are nonporous. The apparent particle density is equal to the ultimate particle density when all pores in particles are penetrated by the fluid and the volume of fluid displaced is equal to the volume of solid phases.

The density of a particle larger than about 1 mm can be determined from its buoyancy in a calibrated heavy liquid in the range of 1–5 Mg/m^3. Small calibrated divers may be used to calibrate the density of a uniform liquid or the density gradient in a liquid solution to an accuracy of 0.001 Mg/m^3 (Fig. 8.5). The liquid should completely wet the particle.

An Archimedes' balanced may be used to determine the density of a particle larger than about 1 cm^3 to a precision of about ±0.0002 Mg/m^3. Liquid wetting the filament suspending the particle produces a force F_γ and viscous drag,

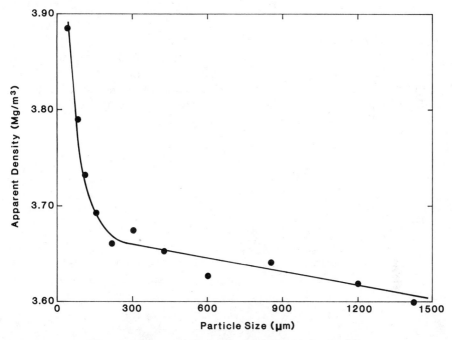

Fig. 8.4 Apparent density of sized tabular alumina particles.

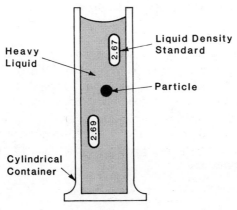

Fig. 8.5 Sink-float behavior of a particle and density standards in a heavy liquid.

$$F_\gamma = 2\pi r_f \gamma_{LV} \cos \theta \tag{8.1}$$

where r_f is the radius of the filament.

The drag force is reduced by using an extremely thin filament and a liquid system that minimizes $\gamma_{LV} \cos \theta$. When the filament is very thin or the meniscus of the liquid intersects the filament at the same position during each weighing, the buoyancy of the wire does not need to be considered, and the particle volume (V_p) and mass (M_p) are

$$V_p = \frac{F_\gamma/g + M_A - M_S}{D_L - D_A} \tag{8.2}$$

and

$$M_p = M_A + D_A V_p - M_f \tag{8.3}$$

where M_A is the mass suspended in air, M_S is the mass suspended in the liquid, and M_f is the mass of the filament. The apparent density D_a of the particle is

$$D_a = M_p/V_p \tag{8.4}$$

8.2 POROSITY AND PORE STRUCTURE

Many natural mineral particles, calcined aggregates, and soft agglomerates contain pore interstices, microfissures, and pores. These pore defects range widely in size and structure and may or may not be surface-accessible. General details of surface and internal pores can be observed by microscopically examining external surfaces and fracture or polished surfaces of internal structure. The approximate range of pore sizes is readily obtained by image analysis. However, the quantitative, volumetric description of the pore size distribution and structure from two-dimensional images is a very complex problem.

Pore size information determined microscopically may be augmented using mercury intrusion porosimetry (Fig. 8.6). Mercury does not wet most ceramic materials, and pressure must be applied to force it into an evacuated surface pore. A special sample holder is partially filled with dried sample, evacuated, and then filled with mercury. Pressure applied to the mercury causes penetration of pores larger than a particular size. The intruded radius R at any applied pressure P is calculated from the Washburn equation for a cylindrical pore;

$$R = \frac{-2\gamma_{LV} \cos \theta}{P} \tag{8.5}$$

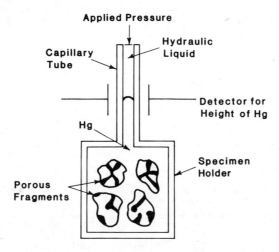

Fig. 8.6 Mercury penetration porosimetry technique.

The contact angle of mercury on most oxides is about 130–140° and in the range of 110–140°. Pressures up to 200 MPa are commonly used, and instruments with a 420-MPa capability are available. This range of pressures enables penetration of pores ranging from about 200 μm to 2 nm, as indicated in Table 8.2.

Mercury intrusion results are commonly presented in the form of cumulative volume penetrated per mass of specimen as a function of pore radius, as shown in Fig. 8.7. A rise in the curve indicates penetration of pores of a particular entry size. For unfired bodies containing packed particles or a bulk sample of particles or granules, the initial penetration normally corresponds to penetration into interstices among particles. Penetration into smaller pores in agglomerates or aggregates occurs at a higher pressure. The total penetration volume at the highest pressure is used to calculate the apparent volume. When two regions of penetration are clearly resolved, the relative fraction of pores in each size range may be estimated, and both

Table 8.2 Dependence of Intruded Radius on Pressure ($\gamma_{Hg} = 474$ mN/m)

Radius (μm)	$\theta = 130°$ Pressure (MPa)	$\theta = 140°$ Pressure (MPa)
100	0.006	0.007
10	0.06	0.07
1	0.6	0.7
0.1	6.0	7.0
0.01	60	70
0.001	600	700

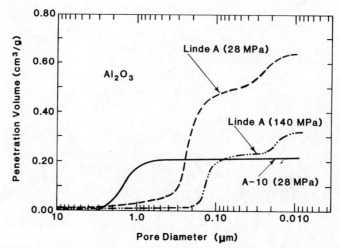

Fig. 8.7 Cumulative mercury penetration behavior for a nonporous calcined alumina powder and porous aggregates of alum-derived alumina powder compacted at the pressures indicated.

the bulk and apparent particle density can be determined from a porosimetry analysis.

The fractional pore size distribution $f(R) = dV/dR$, where V is the volume of penetration, may be derived directly from the cumulative pore size distribution or from the cumulative penetration with pressure. Differentiating Eq. 8.5,

$$PdR + RdP = 0 \tag{8.6}$$

and

$$f(R) = \frac{P\ dV}{R\ dP} \tag{8.7}$$

The relation between the cumulative and fractional pore size distribution curves is shown in Fig. 8.8. The fractional curve more clearly illustrates small changes in pore size.

The entry radius for pores that are constricted at the surface will be smaller than the effective radius of the cavity being filled with mercury. Pores of this shape produce a hysteresis in the penetration volume, and the penetration is dependent on the size of the specimen. Hysteresis and residual mercury in the pores after decompression indicate the presence of these "ink well" pores. This effect is minimal in highly porous agglomerates but can be quite pronounced in relatively dense, sintered aggregates. The migration of binders and segregation of fine particles at the surface can also reduce the size of surface pores.

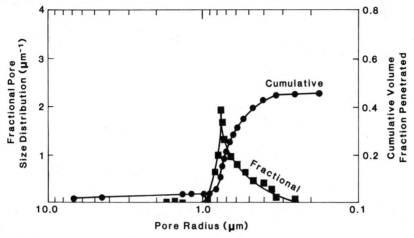

Fig. 8.8 Cumulative and fractional pore size distribution of a compacted calcined alumina powder.

8.3 SPECIFIC SURFACE AREA

The specific surface area is the surface area of the particles per unit mass or volume of material. It is commonly determined by the physical adsorption of a gas or by the chemical adsorption of a dye such as methylene blue. For a porous material, the surface area determined experimentally depends on the size of the adsorbed molecule relative to the size of the pores. Small gas molecules may penetrate pores smaller than 2 nm, whereas a bulky molecule is excluded.

Physical adsorption of a gas at a cryogenic temperature may be used to determine the specific surface area of a ceramic powder. For type II or type IV adsorption isotherms (Fig. 2.4), the volume of gas providing monolayer coverage on adsorption V_m may be calculated using Eq. 2.10. The specific surface area per unit mass S_M is given by the equation

$$S_M = \frac{N_A V_m A_m}{V_{mol} M_s} \tag{8.8}$$

where N_A is Avogadro's number, A_m is the area occupied by one adsorbate molecule (16.2×10^{-20} m^2 for N$_2$ and 19.5×10^{-20} m^2 for Kr), V_{mol} is the volume of ^{1} mole of gas at the standard temperature and pressure of V_m, and M_s is the mass of the sample.

Prior to adsorption, the surface of the powder sample is degassed by vacuum drying or by flushing an inert gas through the mass of heated powder. The adsorption of N$_2$ at the boiling point of liquid nitrogen (77.4°K) occurs quickly and is used routinely when S_M of the powder exceeds 1 m^2/g. Desorption on heating is commonly used for the analysis (Fig. 8.9).

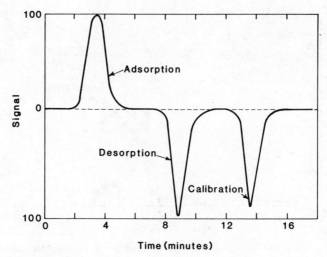

Fig. 8.9 Typical adsorption and desorption of N_2 gas from an oxide powder during cooling and heating. (Courtesy of Quantachrome Corp., Greenvale, NY.)

The specific surface area of a powder consisting of nonporous particles can be calculated from the mean particle size $\bar{a}_{V/A}$ using the equation

$$S_M = \frac{\psi_A/\psi_V}{\bar{a}_{V/A}D_a} \tag{8.9}$$

A shape factor ratio ψ_A/ψ_V of 6 is usually assumed when the specific shape factors of the particles are unknown. The specific surface area is sensitive to variations in the shape and size of particles finer than 1 μm (Figs. 8.10, 8.11) or a change in the concentration of a minor phase of much higher surface area. It is not normally sensitive to agglomeration if chemical additives have not been introduced.

Type II isotherms (Fig. 2.4) are characterized by an initial portion with a diminishing slope, an intermediate portion (often linear), and a final upturn at a P/P_s of about 0.7–0.9. Behavior in these three regions is interpreted as monolayer adsorption, multilayer adsorption, and capillary condensation, in pores between particles respectively. If the particles contain pores a few multiples of the size of the adsorbate gas, the capillary condensation may occur at a much lower relative pressure.

The chemical adsorption of bulky molecules from a solution in contact with the powder may be used to determine a quantitative index of the external surface area of clays, bentonites, and powders containing particles that do not adsorb the dye into internal pores. Bulky molecular dyestuffs such as methylene blue (organic anion with formula weight of 320) are chemisorbed on the surface of clay particles owing to an exchange mechanism. The clay suspension is first acidified to free montmorillonite particles

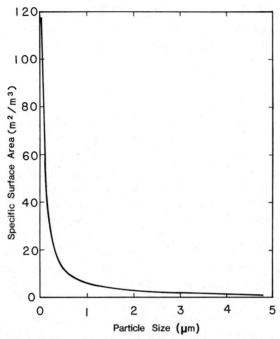

Fig. 8.10 Variation of specific surface area with mean particle size $\bar{a}_{V/A}$.

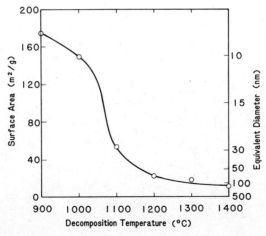

Fig. 8.11 Dependence of specific surface area of an alumina powder derived from $Al_2(SO_4)_3$ on the calcining temperature.

and replace surface-adsorbed ions with H^+. A standardized methylene blue solution is admixed incrementally in a dilute suspension of the clay. The end point of the titration is reached when blue tint initially migrates with the liquid medium, as when a drop of suspension is placed on a piece of filter paper. The methylene blue index (MBI), defined as the milliequivalent of methylene blue adsorbed per 100 g of dry sample, is commonly reported for clay materials (Table 5.5).

For pores larger than about 1 nm, the pore size distributions in particles may be estimated from the hysteresis in the physical adsorption and desorption behavior of a gas at a crygenic temperature (Fig. 8.12). The pore radius is estimated using the Kelvin equation (Eq. 2.7) and relative pressure values in the desorption loop, since the Kelvin equation describes liquid behavior. For a cylindrical pore shape, the radius of the pore R_p is $R_p = R_K + nX$, where R_K is the radius of the liquid in the pore, calculated from the Kelvin equation, containing n statistical layers of thickness X in the adsorbed film. Values of the Kelvin radius and adsorbed thickness are given in Table 8.3 for liquid N_2 at 77.4°K. The shape of the adsorption desorption hysteresis may provide qualitative information about the pore shape.

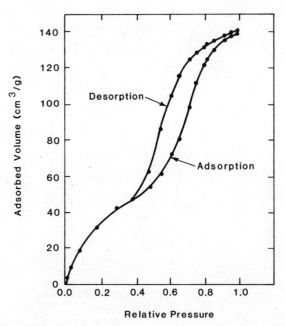

Fig. 8.12 Cryogenic adsorption and desorption of N_2 on a porous alumina powder at 77.4°K. (Courtesy of Quantachrome Corp, Greenvale, NY.)

Table 8.3 Values of Kelvin Radius and Absorbed Thickness
for Nitrogen Desorption

P/Ps	Kelvin Radius (nm)	Film Thickness (nm)
1.00		
0.98	47.3	5.0
0.97	31.3	4.2
0.96	23.4	3.6
0.95	18.6	3.1
0.90	9.1	2.0
0.85	5.9	1.6
0.80	4.3	1.2
0.70	2.7	0.9.
0.50	1.4	0.6
0.30	0.8	0.5

Source: F. M. Nelsen and F. T. Eggertsen, *Anal. Chem.* **30**,
1387 (1958).

SUMMARY

The density of a ceramic material depends on the proportion and density of
component phases. Internal voids in the material reduce the mass, and the
apparent density is less than the ultimate density of the solid phase. The
particle size of the sample and accuracy desired are factors in selecting a
technique for determining the density. Microscopy, penetration analysis,
and desorption behavior may be used to obtain the size and amount of open
porosity in particles. Specific surface area analysis complements the particle
size analysis and may provide information about the presence of a minor
amount of extremely fine particles not detected in the particle size analyses.

SUGGESTED READING

1. S. Lowell and Joan E. Shields, *Powder Surface Area and Porosity*,
   Chapman and Hall, New York, 1984.
2. C. Orr Jr., Physical Characterization Techniques for Particles, Chapter 6
   in *Ceramic Processing Before Firing*, George Y. Onoda Jr. and Larry L.
   Hench (eds), Wiley-Interscience, New York, 1978.
3. Arthur W. Adamson, *Physical Chemistry of Surfaces*, Wiley-Interscience,
   New York, 1976.
4. E.S. Palek, Specific Surface Area Measurements on Ceramic Powders, *J.
   Powder Tech.* **18**, 45–48 (1977).
5. S. Karp, S. Lowell, and A. Mustacciuolo, Continuous Flow Measurement of Desorption Isotherms, *Anal. Chem.* **44**(12), 2395–2397 (1972).
6. Clyde Orr Jr., Application of Mercury Penetration to Materials Analyses, *J. Powder Tech.* **3**, 117–123 (1969–70).

7. H. Gobrecht and A. Richter, Density Measurement and Disorder in Lead Selenide, *J. Phys. Chem. Solids* **26**, 1889–1893 (1965).

8. S. Brunauer, P. H. Emmett, and E. Teller, *J. Am. Chem. Soc.* **60**, 309 (1938).

PROBLEMS

8.1 Calculate the bulk, apparent, and ultimate particle densities from the data in Fig. 8.1.

8.2 The data obtained in a liquid pycnometry test are as follows: pycnometer mass, 32.383 g; pycnometer and powder, 41.923 g; pycnometer and water, 84.229 g; and pycnometer, powder, and water, 91.170 g. The ambient temperature was 19°C. Calculate the apparent density. What is the closed porosity if the ultimate density is $3.98 \, Mg/m^3$?

8.3 The unit cell parameters of the hexagonal ferrite $BaFe_{12}O_{19}$ are $c = 23.194$ Å and $a = 5.893$ Å. The unit cell contains two formula units. What is the x-ray density?

8.4 In problem 8.3, if 5% of the Ba are replaced with Pb ions, can the change in density be detected using an Archimedes' balance?

8.5 The following data was determined for a single crystal of periclase MgO using the Archimedes' balance technique: mass of wire 0.00241 g; mass suspended in air (25.8°C), 5.38810 g; mass suspended in water (at 25.8°C), 3.89109 g; diameter of wire, 0.00254; wetting angle, 70°. Calculate the ultimate density and compare it to the x-ray density. For MgO, the lattice constant is 4.2117 ± 0.0002 Å, and the unit cell contains four formula units.

8.6 Calculate the mean density of a powder composed of 90 wt % alumina ($D_a = 3.98 \, Mg/m^3$) and 10 wt % zirconia ($6.03 \, Mg/m^3$).

8.7 Calculate the interstitial porosity for the calcined alumina in Fig. 8.7. Estimate the intraaggregate porosity for the two special chemically prepared aluminas. The ultimate density for these three α-aluminas is $3.98 \, Mg/m^3$. What are the pore sizes for the powder aggregates? Sketch the fractional pore size distributions of each.

8.8 The specific surface areas of conventional and chemically prepared $BaTiO_3$ powders are 1.2 and $10.3 \, m^2/g$, respectively. Estimate and compare the mean particle size of each powder. Assume $D_a = 6.01 \, Mg/m^3$.

8.9 A $BaFe_{12}O_{19}$ powder is composed of hexagonal platelike particles with an aspect ratio of 4/1. Estimate the mean particle size $\bar{a}_{V/A}$ for a powder with $S_M = 19.3 \, m^2/g$. ($D_a = 5.31 \, Mg/m^3$.)

8.10 Nitrogen gas is adsorbed on an alumina powder at 77.4°K. The area occupied by an adsorbate molecule is $16.3 \, Å^2$. At STP, the volume of a mole of gas is $2.24 \times 10^4 \, ml$, and the volume of gas required for monolayer adsorption on 1 g of powder is 2.52 ml. Calculate the specific surface area of the powder.

8.11 Estimate the specific surface area of kaolinite and montmorillonite particles assuming the particles are square platelets. Assume that the dimensions for the kaolinite are $1 \times 1 \times 0.1$ μm and for montmorillonite $100 \times 100 \times 1$ nm, and the particle density is 2.60 Mg/m^3. How many grams of each material have an area equal to that of a soccer field $(50 \times 100$ m$)$?

8.12 In problem 8.10, calculate the water content if each material absorbs a film of water 2 molecules thick (about 0.5 nm).

8.13 When the absorptive capacity of other clay minerals is relatively insignificant compared to that of the bentonite present, the bentonite content can be estimated from the methylene blue index based on 75 mEq/100 g dry bentonite. Estimate the bentonite content in the plastic clay in Table 5.5. Why is this test not accurate for a clay containing halloysite?

8.14 Estimate the range of pore size using the results for the desorption of N$_2$ as shown in Fig. 8.12. Assume that the molar volume of adsorbed N$_2$ is 34.69 cm^3/mol.

PART IV

PROCESSING ADDITIVES

In processing ceramic materials, several different processing additives must be incorporated in the batch to produce the flow behavior and properties requisite for forming. These additives may be categorized conveniently as follows:

1. liquid/solvent medium
2. surfactant (wetting agent)
3. deflocculant
4. coagulant
5. flocculant/binder
6. plasticizer
7. lubricant
8. bactericide/fungicide

Although most of these substances are added in a relatively small amount and some may be eliminated in a later stage of processing and do not appear in the final product, from a processing perspective they are essential materials. The selection and control of these additives are often the key to successful processing or the development of an improved process.

To understand the nature and structure of organic additives, it is necessary to have an elementary understanding of organic chemistry. The selection of these additives is now based on their chemical characteristics and functions as well as empirical experience. We will consider the structure and functions of these important additives in Chapters 9–12.

9

Liquids and
Wetting Agents

Liquids are used in ceramic processing to wet the ceramic particles and provide a viscous medium between them and to dissolve salts, compounds, and polymeric substances in the system. The admixed liquid changes the state of dispersion of the particles and alters the mechanical consistency.

A surfactant is a substance added to reduce the surface tension of the liquid or the interfacial tension between the surface of the particle and the liquid to improve wetting and dispersion. In this chapter we will consider the composition and properties of water, nonaqueous liquids, and surfactants used in ceramic processing.

9.1 WATER

The principal liquid used in ceramic processing is water, and until recently its availability, low cost, and consistency have been taken for granted. Process water must be of satisfactory quality, and water used in industrial processing is now monitored routinely. At some manufacturing sites, the available water must be treated to make it usable or consistent, and process water is sometimes treated and recycled in the plant. Water is now viewed by some plant engineers as a raw material that must be refined to improve processing and productivity.

Pure water contains polar H_2O molecules and the ions H_3O^+ and OH^-; it has a pH 7 and a specific conductivity of 0.055 μmho/cm at 20°C. Water exposed to air combines with CO_2 to form a weak acid, i.e.

$$CO_{2(gas)} + H_2O \rightleftharpoons H_2CO_3 \tag{9.1}$$

and

$$H_2CO_3 + H_2O \rightleftharpoons H_3O^+ + HCO_3^- \tag{9.2}$$

123

The additional H_3O^+ decreases the pH, and the greater concentration of ions increases the specific conductivity of the water.

Well water and municipal water may contain suspended organic and inorganic substances and dissolved mineral salts, as indicated in Table 9.1. The hardness of water is generally due to the dissolved calcium, magnesium, and ferrous salts. Calcium carbonate is relatively insoluble, but calcium bicarbonate $Ca(HCO_3)_2$ is significantly soluble. Heating reduces the hardness according to the reaction

$$Ca^{2+} + 2HCO_3^- \rightarrow CaCO_{3(solid)} + H_2O + CO_{2(gas)} \tag{9.3}$$

However, heating does not reduce hardness due to dissolved sulfates and chlorides.

The total dissolved solids (TDS) is estimated from the specific conductivity C using the equation

$$TDS\ (ppm) = \frac{C(\mu mho/cm) - 0.055}{2.5} \tag{9.4}$$

The constant 2.5 is a weighted average of the specific conductivities of common ionized solids in a water sample.

In-plant treatment of city or well water using a commercial filtration-deionizing system can reduce the dissolved salts to trace concentrations (Table 9.1). Chlorine and suspended matter may be removed in a carbon filter. Cation and anion impurities are removed using rechargeable exchange resins.

Table 9.1 Composition of Water (ppm)

Component	Well	City	Treated
Ca	300	78	Trace
Mg	172	42	Trace
Na	8	360	Trace
Bicarbonate	350	350	0
Carbonate	0	0	0
Sulfate	100	125	Trace
Chloride	30	5	Trace
Nitrate	0	0	0
pH	7.8	7.9	4.5–5.5
Sp conductivity ($\mu mho/cm$)	670	650	10–40
TDS (ppm)	268	260	4–16

Source: R. Thomas *"Water—Its Processing Consequences"* in *Processing Consequences of Raw Material Variables*, J. Varner et al. (eds), Alfred University Press, Alfred, NY, 1985, p. 53.

Table 9.2 Representative Properties of Liquids (20°C)

Liquid	Formula	Dielectric Constant	Surface Tension (mN/m)	Viscosity (mPa's)	Boiling Pt. (°C)	Flash* Pt. (°C)
Water	H_2O	80	73	1.0	100	None
Methyl alcohol	CH_3OH	33	23	0.6	65	18
Ethyl alcohol	C_2H_5OH	24	23	1.2	79	8
n-Propyl alcohol	C_3H_7OH	20	24	2.3		
Isopropyl alcohol	C_3H_7OH	18	22	2.4	49	21
n-Butyl alcohol	C_4H_9OH	18	25	2.9	100	38
n-Octyl alcohol	$C_8H_{17}OH$	10	28	10.6	171	
Ethylene glycol	$C_2H_6O_2$	37	48	20	>197	>116
Glycerol	$C_3H_8O_2$	43	48	20	290	
Trichloroethylene	C_2HCl_3	3			87	
Methylethyleneketone	C_4H_8O	18	25	0.4	80	2

Source: Handbook of Chemistry and Physics, R. C. Weast (ed), CRC Press, Boca Raton, FL, 1958.
*Note: more recent values provided by suppliers are somewhat lower.

Table 9.3 Variation of Viscosity (mPa s) of Water and Alcohols with Temperature

Liquid	Temperature (°C)				
	0	20	30	50	70
H_2O	1.8	1.0	0.8	0.55	0.4
CH_3OH	0.8	0.6	0.5	0.4	
C_2H_5OH	1.8	1.2	1.0	0.7	0.5
C_3H_7OH	3.9	2.3	1.7	1.1	0.8
C_4H_9OH	5.2	3.0	2.3	1.4	0.9

Reagent-grade water should have a specific conductance below 5.0 μmho/cm, and treated water of this purity can be produced by filtration and ion exchange. Higher-quality reagent-grade water with a conductivity of about 1 μmho/cm is produced by distillation and deionization.

Water is a polar liquid and has a high dielectric constant and surface tension relative to other liquids, as indicated in Table 9.2. It is a good solvent for polar and ionic compounds. Water can form a hydrogen bond with substances containing an –OH or –COOH group, and substances such as polymerized alcohols and carbohydrates dissolve in water if the molecule is not too large. The viscosity of water decreases significantly on heating from 20 to above 50°C, as shown in Table 9.3.

In the production of ceramic tile, it has been reported that recycling of all waste water can reduce the external water requirement by about one-half. Recycled water is sometimes conveniently used for the wet grinding of raw materials, to add moisture to the dry body, and for washing equipment and the work areas.

9.2 ORGANIC LIQUIDS

Nonaqueous liquids such as trichloroethylene, alcohols, ketones, and refined petroleum oil or liquid wax are used as the liquid and solvent medium in suspensions of materials that react with water or when dispersion or drying is a particular problem such as in casting electronic substrates and for printing resistive or conductive films on surfaces of formed ceramics. Organic liquids may be flammable and toxic, and special precautions and monitoring are required where they are used and disposed. The flash point is the lowest temperature in air at which ignitable or explosive vapors are given off and must be high enough to provide a safe range of working temperatures. Because of their higher cost and the environmental concerns, recycling is often cost-effective.

Nonpolar molecules tend to be soluble in nonpolar solvents. The dissolving power of the liquid used should be sufficient to dissolve processing additives but should not prevent the adsorption of these additives on the particles. Organic liquids used in ceramic processing have a lower surface tension and dielectric constant than water, as indicated in Table 9.2. The lower surface tension γ_{LV} aids in the wetting of solids if γ_{SL} is not too high. A solution of solvents may often provide the best compromise of dielectric constant and surface tension for dispersion and dissolving additives and a satisfactory viscosity, boiling point, and flash point for ease in handling and drying. Nonaqueous solutions that have been used in preparing a slurry for casting include ethyl alcohol and trichloroethylene or methylethyl ketone. Mixtures of alcohols may offer a satisfactory compromise of chemical and physical properties and be more acceptable for health and safety reasons. These solutions have a low boiling point and dissolve any water impurity that might cause hydration of the solids. Terpineol and butyl carbitol are liquids of low volatility used in paste printing compositions.

Petroleum oils are composed of hydrocarbon molecules that may vary in molecular weight. Longer molecules have a higher viscosity and boiling range. Waxes are relatively high molecular weight hydrocarbons that are solid at room temperature but that soften to form a low-viscosity liquid (see Chapter 11).

The viscosity η_s of ideal solvent solutions may be calculated using the equation

$$\ln \eta_s = \sum f_i^w \ln \eta_i \qquad (9.5)$$

where f_i^w is the weight fraction and η_i the viscosity of each liquid component. For solutions of interacting solvents such as the alcohols in hydrocarbon solvents, the solution viscosity is lower than that calculated using Eq. 9.5, and an effective viscosity of the minor component must be used.

9.3 POLAR LIQUIDS NEAR OXIDE SURFACES

A polar liquid may be both physically and chemically adsorbed on the surface of dispersed oxide particles. For oxides such as SiO_2, Al_2O_3, and TiO_2 in water, evidence of chemical surface hydration is provided by infrared adsorption studies, heats of immersion, and the thermal behavior of the adsorption-desorption kinetics. Reactions of the type

$$O_{(surface)} + H_2O_{(1)} \rightarrow O-H_2O_{(surface)} \qquad (9.6)$$
$$\text{physical adsorption}$$

and

$$O_{(surface)} + H_2O_{(1)} \rightarrow 2OH_{(surface)} \qquad (9.7)$$
$$\text{chemical adsorption}$$

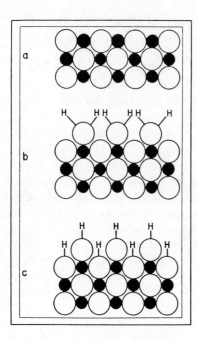

Fig. 9.1 Cross section of an oxide showing (a) dry surface and a surface with (b) physically adsorbed H₂O, and (c) chemically adsorbed water (chemically hydrated surface). (Reproduced with permission from W. H. Morrison, *J. Coatings Tech.* **57**(721), 56 (1985).)

have been suggested (see Fig. 9.1). The polar hydroxyl (–OH) groups may cause the surface to attract and physically adsorb one to several additional layers of polar water molecules. This physically bonded water is immobilized and is of a different structure from bulk water. Alcohols and liquids containing the carboxyl (–COOH) group are also chemisorbed on oxide surfaces. The absorption behavior depends significantly on the thermal history and prior hydration of the surface.

9.4   SURFACTANTS

Methyl (–CH₃) and ethyl (–C₂H₅) molecular groups are nonpolar. But hydroxyl, carboxyl, sulfonate (–SO₃⁻), sulfate (–OSO₃⁻), ammonium (–NH₄⁺), amino (–NH₂), and polyoxyethylene (–CH₂OCH₂) groups are polar in nature. A polar group attracts polar liquid molecules and is called a lyophilic group. A nonpolar group such as a hydrocarbon chain (–C$_x$H$_y$) is a lyophobic group. In an aqueous solution, these groups are often referred to as hydrophilic and hydrophobic groups, respectively.

Nonionic surfactants are molecules having polar and nonpolar ends, and they do not ionize when dissolved in the liquid. Anionic surfactants have a relatively large lyophobic group that is commonly a long-chain hydrocarbon and a negatively charged lyophilic group that is the surface-active portion of the molecule. Anionic surfactants are widely used in industry. Family types

Table 9.4 Examples of Surfactants

Type	Generic Name	Composition
Nonionic	Ethoxylated nonylphenol	$C_9H_{19}(C_4H_4)O(CH_2CH_2O)_{10}H$
	Ethoxylated tridecyl alcohol	$C_{13}H_{27}O(CH_2CH_2O)_{12}H$
Anionic	Sodium stearate	$C_{17}H_{35}COO^-Na^+$
	Sodium disopropylnaphtalene sulfonate	$(C_3H_7)_2C_{10}H_5SO_3^-Na^+$
Cationic	Dodecyltrimethylammonium chloride	$[C_{12}H_{25}N(CH_3)_3]^+Cl^-$

include the alkali sulfonates, sulfates, lignosulfonates, carboxylates, and phosphates. Alkali and ammonium anionic surfactants are more soluble in water than in organic liquids. Cationic surfactants have a positively charged lyophilic group. They are often toxic. Examples of ionic and nonionic surfactants are listed in Table 9.4.

When added to a polar liquid, molecules that are polar at one end and nonpolar at the other will tend to concentrate at the surface with the lyophilic end adsorbed in the polar liquid (Fig. 9.2). An apparent concentration of only 0.01–0.2% may reduce the surface tension γ_{LV} of the liquid very significantly (Table 2.1) and greatly improve the wetting of suspended solids. Accordingly, surfactants are often called wetting agents.

Surfactant molecules may also improve the compatibility of the solid with the liquid medium when they are adsorbed at the interface. When added along with an oxide powder into a nonpolar liquid, the surfactant will be adsorbed with the lyophobic end in the liquid. When the adsorption preference is not strong, such as on the addition of stearic acid $CH_3(CH_2)_{16}COOH$ to a suspension of clay in water, the lyophilic end will project into the water, and the lyophobic end will be attached to the clay surface.* Cationic surfactants are strongly adsorbed on the negative surfaces of clay minerals. Anionic surfactants may be adsorbed on positive particles. The adsorption of ionic surfactants is discussed further in Chapter 10.

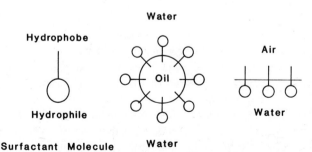

Fig. 9.2 Structure of a surfactant molecule and its oriented adsorption at water-oil and water-air interface.

* W. E. Worral, *Clays and Ceramic Raw Materials*, Halsted Press, New York, 1975.

After being mixed mechanically, water and oil separate spontaneously, because the interfacial tension is quite high. The addition of a surfactant, which lowers the interfacial tension, may produce a stable dispersion of fine droplets of one liquid in the other. In this role, the surfactant is called an emulsifier. The Bancroft rule says that the continuous phase is the one in which the emulsifier is more soluble. Stability is also enhanced by the emulsifier forming a physical barrier to the coalescence of droplets during stirring or mixing.

One method for rating the relative strengths of the hydrophilic and hydrophobic groups of a surfactant molecule is the HLB method. Using an empirical scale ranging from 0 to 20, a high value is assigned to a strongly hydrophilic group and a low value to a strongly hydrophobic group. The HLB number is an index of the balance between these groups. HLB numbers are reported for commercial surfactants, and an arithmetic mean is used for a mixture of surfactants. The stability of a dispersion should be higher when the HLB number of the surfactant approximates the weighted mean for the dispersed phase and liquid. For dispersion in water, the HLB of the surfactant should exceed $8^{\dagger}$; for water in oil, HLB is less than 6.

Surfactants are effective in very small concentrations, and when added in excess of that required for adsorption, clusters of surfactant molecules called micelles form in the solution. Above the critical concentration at which clusters form, called the critical micelle concentration, the surface tension is essentially constant. Other properties such as the electrical conductivity and refractive index exhibit a different dependence on surfactant concentration above the critical micelle concentration.

Polymer molecules with repeating ionizable groups, called polyelectrolytes, which promote dispersion without necessarily reducing the surface tension or interfacial tension, are sometimes included in the category of surfactants. These additives are discussed under the topic of deflocculants in Chapter 10. In this text, polyelectrolytes used for dispersion will be called deflocculants.

SUMMARY

Liquids used in ceramic processing provide a viscous medium for dissolving chemical additives and dispersing particles. A surfactant is an additive that is preferentially adsorbed at a surface or interface and promotes wetting and dispersion by reducing the surface or interfacial tension.

Water is the major liquid used in ceramic processing, and it is often refined to improve its purity or consistency. It is polar and has a high surface tension. Nonaqueous liquid systems are less polar in nature and are good solvents for relatively nonpolar substances; their solutions provide greater ranges in properties. The liquid system used must dissolve additives yet permit their adsorption at interfaces to assist in dispersion. Precautions must be exercised in the handling and disposing of nonaqueous liquids.

---

† Milton J. Rosen, *Surfactants and Interfacial Phenomena*, Wiley-Interscience, 1978.

SUGGESTED READING

1. Temple C. Patton, *Paint Flow and Pigment Dispersion*, Wiley-Interscience, New York, 1979.
2. Milton J. Rosen, *Surfactants and Interfacial Phenomena*, Wiley-Interscience, New York, 1978.
3. J. C. Williams, Doctor Blade Process, in *Treatise on Materials Science and Technology*, Vol. 9: *Ceramic Fabrication Processes*, Franklin F. Y. Wang (ed), Academic Press, New York, 1976, pp. 173–198.
4. Ronald J. Thomas, Water—Its Processing Consequences, in *Processing Consequences of Raw Materials Variables*, James R. Varner et al. (eds), Alfred University Press, Alfred, NY, 1985, pp. 47–54.
5. ASTM D1193-77, Part 31, Water, in *1980 Annual Book of ASTM Standards*, American Society for Testing Materials, Philadelphia, 1980.
6. *Detergents and Emulsifiers*, McCutcheon's Division, Allured Publishing Corp., Ridgewood, NJ, 1974.
7. M. V. Parish, R. R. Garcia, and H. K. Bowen, Dispersions of Oxide Powders in Organic Liquids, *J. Mater. Sci.* **20**, 996–1008 (1985).

PROBLEMS

9.1 Construct a graph showing the variation in surface tension of solutions of water and ethyl alcohol at 20 and at 50°C. Explain the behavior observed.

9.2 A sample of water used in a laboratory experiment has a specific conductivity of 15 μmho/cm. What is the TDS of this water compared to reagent-grade water with a specific conductivity of 2 μmho/cm?

9.3 Draw the structure of isopropyl alcohol and the nonionic surfactants in Table 9.4.

9.4 Estimate the viscosity of a solution composed of 70 wt% trichlorethylene and 30 wt% ethyl alcohol. The effective viscosity of ethanol is 1.1 mPa s.

9.5 Explain the difference in pH values for the untreated and treated water in Table 9.1.

9.6 Sketch the structure of a micelle of nonionic surfactant molecules in water.

9.7 The HLB numbers of three different surfactants are 3.7, 6.7, and 13.5. Which would you select for dispersion of an oxide in an aqueous medium, and which for water in oil?

9.8 Write the reactions for the chemisorption of water on the surface of silica SiO_2.

10

Deflocculants and Coagulants

Additives called deflocculants and coagulants are very important in ceramic processing. Dispersion produced by an additive adsorbed on the particles which increases the repulsive forces by electrical charging and/or by sterically hindering the close approach of particles is called deflocculation, and the additive is called a deflocculant. A coagulant is a simple electrolyte that promotes particle agglomeration by reducing the particle repulsion forces or steric hindrance. Deflocculation and coagulation processes must be carefully controlled to produce a satisfactory process system.

Particle agglomeration may also be produced by the bridging action of adsorbed polymer molecules and coagulated colloidal particles. This type of agglomeration is called flocculation and is discussed in Chapter 11.

10.1 PARTICLE CHARGING IN LIQUID SUSPENSIONS

Ceramic powders are of high surface area and relatively low solubility, and surface chemistry tends to control their charging behavior. The surface of a particle may become charged by (1) the desorption of ions at the surface of the material, (2) a chemical reaction between the surface and the liquid medium changing the composition of the surface, and (3) the preferential adsorption of specific additive or impurity ions from the chemical solution adjacent to the particle.

Desorption and Dissolving

A typical example of charging by desorption is the liberation of alkali from the surface of a clay mineral such as kaolinite. When the kaolinite mineral is formed, lattice substitution of the type

Clay Stack

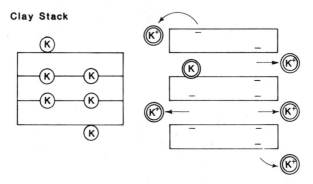

Fig. 10.1 Clay particles become charged on dispersion when adsorbed alkali ions are liberated.

$$Al^{3+}_{(lattice)} + K^+_{(surface)} = Si^{4+}_{(lattice)} \qquad (10.1)$$

and

$$Mg^{2+}_{(lattice)} + K^+_{(surface)} = Al^{3+}_{(lattice)} \qquad (10.2)$$

occurs to some extent. Surface-adsorbed alkali and large alkaline earth ions required for charge neutrality are weakly bonded to the crystallite and are exchangeable ions. The cation exchange capacity (CEC) is the maximum amount of exchangeable cations per 100 g of clay material. In clay minerals, the exchangeable cations are adsorbed at interfaces between crystallites in the stacked clay aggregates. The dispersion of the clay aggregates in water will liberate alkali into the aqueous medium, leaving the surface of the particle negatively charged, as shown in Fig. 10.1.

Chemical Reaction with an Aqueous Medium

For oxides with a hydrated surface, the surface chemistry in water is dominated by the chemical reactions

$$MOH^+_{2(surface)} \overset{K_1}{\rightleftharpoons} MOH_{(surface)} + H^+_{(solution)} \qquad (10.3)$$

and

$$MOH_{(surface)} \overset{K_2}{\rightleftharpoons} MO^-_{(surface)} + H^+_{(solution)} \qquad (10.4)$$

where M represents a metal ion at the surface, such as Ba^{2+}, Al^{3+}, Si^{4+}, Ti^{4+}, Zr^{4+}, etc. Each M is also bonded to bulk oxygen ions to satisfy its bonding needs. The point of zero charge (PZC) of the surface may be defined in terms of the pK's of reactions (10.3) and (10.4), i.e.

$$PZC = \frac{pK_1 + pK_2}{2} \qquad (10.5)$$

and indicates the average acid-base character of the surface. When bonded to a tetravalent cation, the surface hydroxyls are more acidic, and the PZC is lower. Approximate values of the PZC of several ceramic materials are listed in Table 10.1.

The charge on a hydrated surface of a pure oxide particle dispersed in water is determined by its reaction with H_3O^+ or OH^- ions, as is shown for alumina in Fig. 10.2. The addition of H_3O^+ ions will reduce the pH and cause the uncharged surface to become protonated and positively charged. The addition of OH^- ions will remove hydrogen from the surface and

Table 10.1 Nominal Isoelectric Points of Oxides

Material	Nominal Composition	IEP
Quartz	SiO_2	2
Soda lime silica glass	$1.00Na_2O \cdot 0.58CaO \cdot 3.70SiO_2$	2–3
Potassium feldspar	$K_2O \cdot Al_2O_3 \cdot 6SiO_2$	3–5
Zirconia	ZrO_2	4–6
Apatite	$10CaO \cdot 6PO_2 \cdot 2H_2O$	4–6
Tin oxide	SnO_2	4–5
Titania	TiO_2	4–6
Kaolin (edges)	$Al_2O_3 \cdot SiO_2 \cdot 2H_2O$	5–7
Mullite	$3Al_2O_3 \cdot 2SiO_2$	6–8
Chromium oxide	Cr_2O_3	6–7
Hematite	Fe_2O_3	8–9
Zinc oxide	ZnO	9
Alumina (Bayer process)	Al_2O_3	8–9
Calcium carbonate	$CaCO_3$	9–10
Magnesia	MgO	12

Source: Temple C. Patton, *Paint Flow and Pigment Dispersion*, Wiley-Interscience, New York, 1979, and Errol G. Kelly and David J. Spottiswood, *Introduction to Mineral Processing*, Wiley-Interscience, New York, 1982.

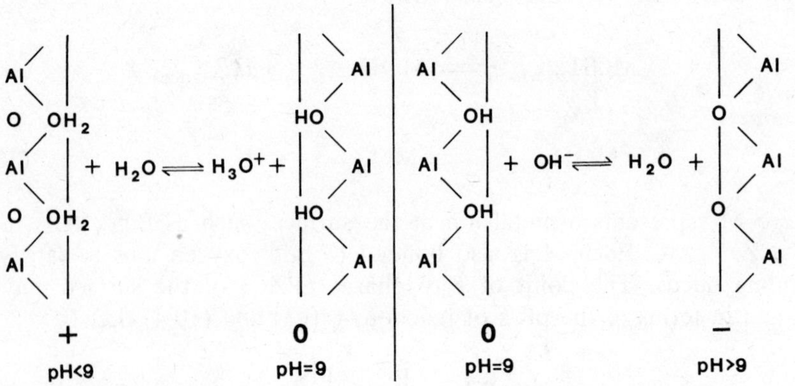

Fig. 10.2 Reaction of the hydrated surface of alumina with H_3O^+ or OH^-.

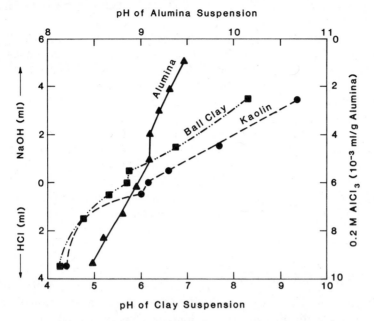

Fig. 10.3 Titration curves for alumina, kaolin, and ball clay.

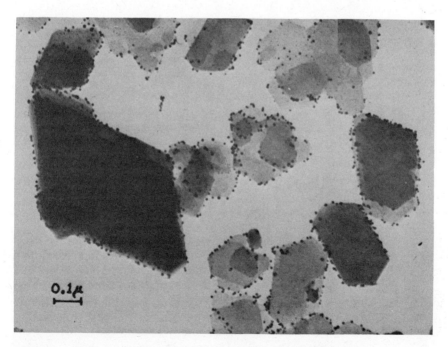

Fig. 10.4 A negatively charged gold colloid is adsorbed on the edges of kaolin in an acidified suspension. (From H. van Olphen, *An Introduction to Clay Colloid Chemistry*, Wiley-Interscience, New York, 1977.)

produce a negative surface charge when the pH is greater than the PZC of the surface. Exchange at the surface is reversible but time-dependent, and the relative concentration of OH^- or H_3O^+ ions is potenial determining. Generally, the rate of dissolving of a particle is a minimum at the PZC.

In layer minerals such as kaolinite, the exchange process occurs predominantly at the edges of the crystals. The faces are not surfaces with broken bonds (as for the surfaces of quartz or gibbsite), and exchange on the faces is limited. The faces of the kaolinite crystal dispersed in water are usually considered to have a negative charge. On titrating of a clay with an acid or base, ion exchange with little change in pH occurs near the PZC of the edges of the particles (Fig. 10.3). In the acidified suspension, negative gold sol particles are adsorbed preferentially on the edges of kaolinite crystals, as is shown in Fig. 10.4.

Adsorption of Specific Ions

Ions in solution that are not a component of the pure liquid medium may also interact with the surface. Simple ions are adsorbed on oppositely charged surfaces. Complete adsorption of the simple ions M^+ or A^- can neutralize the surface charge as indicated by the reactions

$$MO^-_{(surface)} + M^+_{(solution)} \rightarrow MOM_{(surface)} \tag{10.6}$$

and

$$MOH^+_{2(surface)} + A^-_{(solution)} \rightarrow MOH_2A_{(surface)} \tag{10.7}$$

Complex ions such as sodium pyrophosphate (Fig. 10.5) may adsorb on neutral or charged surfaces. The adsorption of a multivalent ion may reverse the surface charge;

$$MO^-_{(surface)} + M^{n+}_{(solution)} \rightarrow MOM^{(n-1)+}_{(surface)} \tag{10.8A}$$

$$MOH^+_{2(surface)} + A^{n-}_{(solution)} \rightarrow MOH_2A^{(n-1)-}_{(surface)} \tag{10.8B}$$

Studies of the adsorption of simple hydrated ions on hydrated oxide surfaces indicate that the specific ion adsorption from the aqueous solution occurs rather abruptly at a particular pH.

Ions such as Al^{3+} and Fe^{3+} may hydrolyze in aqueous solutions, and the adsorption is more complicated. The hydrolysis of aluminum salts is represented by the following equations:

$$Al(H_2O)_6^{3+} \rightleftharpoons Al(OH)(H_2O)_5^{2+} + H_3O^+ \tag{10.9A}$$

$$Al(OH)(H_2O)_5^{2+} \rightleftharpoons Al(OH)_2(H_2O)_4^+ + H_3O^+ \tag{10.9B}$$

$$Al(OH)_2(H_2O)_4^+ \rightleftharpoons Al(OH)_3(H_2O)_3 + H_3O^+ \qquad (10.9C)$$

$$Al(OH)_3(H_2O)_3 \rightleftharpoons Al(OH_3) + 3H_2O \qquad (10.9D)$$

The ions $Al(OH)(H_2O)_5^{2+}$ and $Al(OH)_2(H_2O)_4^+$ occur at $pH < 7$. In the process of hydrolysis, hydrated aluminum ions lose protons, and the final neutral complex loses water to form the hydroxide precipitate. The hydroxide $Al(OH)_3$ is amphoteric, and at $pH > 8.5$ the aluminate ion $Al(OH)_4^-$ is formed.

Using Auger spectroscopy for surface analysis, Horn and Onoda[*] observed that aluminum is rapidly adsorbed from an aqueous $AlCl_3$ solution onto a silica glass surface in the pH range from about 5 to 9; particles were observed to be positively charged for pH below 9 and negatively charged for pH greater than 9. Their results indicated that a rather stable aluminum hydroxide coating was formed between pH 5 and 9. Similarly, adsorption onto alumina occurs at a pH of about 9, as indicated in Fig. 10.3. James and Healy[†] also showed that the adsorption of hydrolyzable ions can cause a reversal in polarity at (1) the PZC of the substrate surface, (2) the pH of the nucleated metal hydroxide, and (3) the PZC of the adsorbed metal hydroxide coating.

An ionic polymer has regular ionizable side groups, as shown in Fig. 10.5; ionization in solution produces charged side groups and oppositely charged counterions. The adsorption of a polymer electrolyte may reverse the polarity and increase the charge on the particle, as indicated by Eqs. 10.8 and 10.9. Molecules are adsorbed at multiple points on the surface of the particle (Fig. 10.6). Polymer electrolytes of low molecular weight may be powerful deflocculents in systems containing polar liquids. Relatively short polymer molecules are used where milling or high-shear mixing may degrade the polymer molecules.

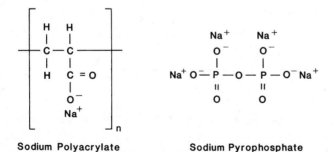

Sodium Polyacrylate   Sodium Pyrophosphate

Fig. 10.5 Structure of sodium polyacrylate and sodium pyrophosphate deflocculants.

[*] J. M. Horn and G. Y. Onoda Jr., *J. Am. Ceram. Soc.* **61**(11), 523–527 (1978).
[†] R. O. James and T. W. Healy, *J. Colloid Interface Sci.* **40**, 52–64 (1972).

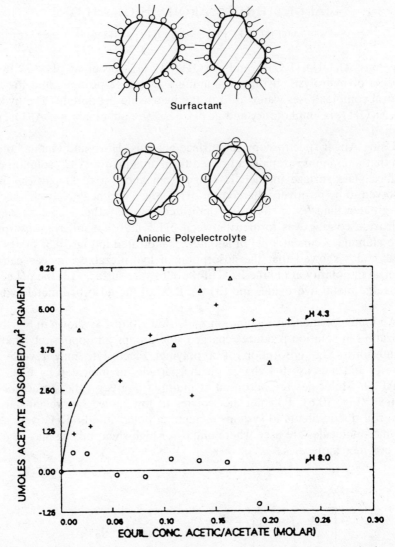

Fig. 10.6 Adsorbed polymer electrolyte and surfactant molecules and adsorption isotherm for acetate ions on titania. (From W.H. Morrison, *J. Coatings Technol.* **57**(721), 57 (1985).)

Compositions of common polyelectrolytes used for deflocculation are listed in Table 10.2. Polyelectrolytes containing the sulfonate ($-SO_3^-$) group have an extreme affinity for water, and alkali polysulfonates are powerful surfactants and deflocculents.

A dilute solution of sodium silicate (water glass) is an inorganic polyelectrolyte that is widely used for deflocculating slurries of clays and other industrial minerals. The sodium silicate acts as a buffer, emulsifies oil, and

Table 10.2 Common Deflocculants Used in Water

Inorganic	Organic
Sodium carbonate	Sodium polyacrylate
Sodium silicate	Ammonium polyacrylate
Sodium borate	Sodium citrate
Tetrasodium pyrophosphate	Sodium succinate
	Sodium tartrate
	Sodium polysulfonate
	Ammonium citrate

precipitates highly charged cations from the solution. Grades are available with a SiO_2/Na_2O weight ratio of about 1.6–3.3; the chain length increases as the weight ratio increases. Sodium phosphates may also be used as deflocculants, but their disposal in waste water is now restricted for environmental reasons.

10.2 DEVELOPMENT OF AN ELECTRICAL DOUBLE LAYER

Ions and polar molecules in the solution surrounding a particle will respond to the charged surface. Coulombic forces will repel like-charged ions but attract polar liquid molecules and oppositely charged ions into a region near the surface, increasing their concentration relative to that in the bulk. Although overall charge neutrality must obtain, a difference in electrical potential between the surface and the bulk solution can be effected, as shown in Fig. 10.7. The potential gradient will not be sharp, because thermal vibration of molecules of the liquid medium causes the diffusion of some counterions. Our model is a charged particle with a relatively immobile adsorbed layer of counterions and a concentration gradient of counterions and polar liquid molecules in a diffuse layer which attains some steady-state configuration. This model is now commonly called the diffuse electrical double-layer model.

   Gouy[*] and Chapman[†] independently derived a model for the potential gradient in the diffuse double layer based on a uniformly charged surface, an adjacent solution with uniform dielectric constant ε_r, and point charges. Assuming the distribution of charge is described by the Boltzman equation, the concentration of counterions N_i in the diffuse layer relative to the concentration adsorbed on the surface N_{i_0} is

$$N_i N_{i_0} = \exp\left(-U_i/k_B T\right) \qquad (10.10)$$

[*] G. Gouy, *Ann. Phys.* **1**(9), 129 (1917).
[†] D. L. Chapman, *Phil. Mag.* **25**(6), 475 (1913).

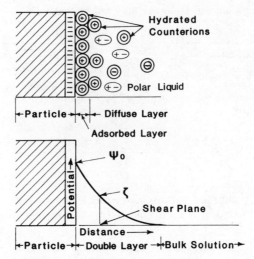

Fig. 10.7 Electrical double-layer model for particle charging in a polar liquid and Gouy-Chapman profile of the electrical potential.

The potental energy of the ion U_i is a function of the valence of the ion Z_i and the electrical potential ψ at that position;

$$U_i = Z_i(e\psi) \qquad (10.11)$$

For the case of a surface potential $\psi_0 \leq 100\,\mathrm{mV}$,

$$\psi = \psi_0 \exp - (x/\kappa^{-1}) \qquad (10.12)$$

where κ^{-1} is the distance from the charged surface to the plane where $\psi = \psi_0/2.718$ and is called the thickness of the double layer. The double-layer thickness is calculated from the equation

$$\kappa^{-1} = \left(\frac{\varepsilon_r \varepsilon_0 k_B T}{F^2 \Sigma N_i Z_i^2} \right)^{1/2} \qquad (10.13)$$

where N_i is the concentration of each ion type in the solution phase and F is the Faraday constant. The variation of the potential with position predicted by the Gouy-Chapman model is shown in Fig. 10.8. For a 0.01 M aqueous solution of 1:1 electrolyte at 20°C, the effective double-layer thickness κ^{-1} is calculated to be about 3 nm. Increasing the concentration or valence of the counterions compresses the double layer and increases the potential gradient. The use of a liquid of a lower dielectric constant or temperature also compresses the double layer.

We will see in the next section that the stability of a suspension

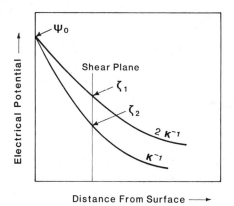

Distance From Surface ⟶

Fig. 10.8 The zeta potential is larger when the double-layer thickness is larger.

deflocculated with a simple electrolyte depends importantly on $d\psi/dx$. Although the Gouy-Chapman model does not consider the specific structure of adsorbed ions of finite size, it is very useful for qualitatively understanding and predicting deflocculation behavior.

10.3 ELECTROKINETIC PROPERTIES

Charged particles in a suspension will respond to an imposed potential difference, and the particle velocity is called the electrophoretic velocity. In a similar manner, the ionic solution adjacent to the wall of a capillary with an electrical double layer will be motivated to flow if a potential difference is imposed; this is called electroosmotic flow. During flow, a hydrodynamic plane of slippage must occur somewhere in the double layer. The location of the slippage plane X_s is a matter of some disagreement but is certainly beyond the first adsorbed layer of simple or polymer counterions (called the Stern layer) and perhaps additional layers of adsorbed polar liquid molecules and ions. The potental at the slippage plane is called the zeta potential ξ and can be calculated from an electrokinetic property. For the electrophoretic transport of nonconducting particles,

$$\xi = \frac{f_H \eta \nu_e}{\varepsilon_r \varepsilon_0 E} \tag{10.14}$$

where η is the viscosity of the electrolyte and ν_e is the electrophoretic velocity for an imposed electrical field E. The ratio ν_e/E is the electrophoretic mobility ν_e. The Henry constant f_H is equal to 1 when the product of the particle diameter a and κ is greater than 100 and 3/2 when $a\kappa$ is less than 1.

In systems with simple counterions, the zeta potential is an indication of the gradient in electrical potential when the surface potential remains constant. The pH at which the zeta potential is zero is termed the isoelectric point IEP. For charging of a hydrated surface due to the reaction with OH^- or H_3O^+, raising or lowering the pH from the isoelectric point will initially increase the absolute value of the zeta potential (Fig 10.9). However, a further addition of OH^- or H_3O^- will cause a reduction in the zeta potential because of the compression of the double layer. When a polyelectrolyte is adsorbed on the surface, the zeta potential increases rapidly, initially, and then more slowly, as indicated in Fig. 10.10. The distance from the surface to the slippage plane may be expected to vary somewhat with the molecular size and the conformity of the adsorbed molecules. Molecules are expected to protrude farther from the surface when the relative solvent power of the liquid medium is higher.

The electrophoretic mobility of particles in a dilute suspension may be determined by measuring the migration velocity of individual particles in a known potential gradient. This technique, called microelectrophoresis, is fatiguing unless automated and is not suitable when settling occurs during

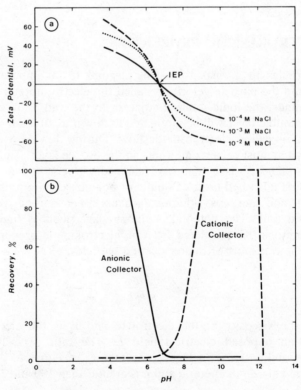

Fig. 10.9 Variation of zeta potential of geothite $\alpha Fe_2O_3 \cdot H_2O$ with pH. (From E.G. Kelly and David J. Spottiswood, *Introduction to Mineral Processing*, Wiley-Interscience, New York, 1982.)

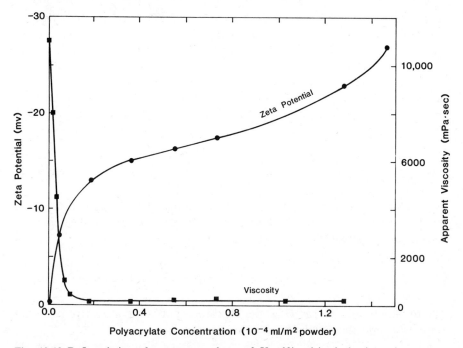

Fig. 10.10 Deflocculation of an aqueous slurry of 50 vol% calcined alumina using an ammonium polyacrylate deflocculant.

the test. An average mobility of the particles in a suspension is conveniently determined using a bulk technique, in which the electrophoretic migration causes an increase in solids content in a simple cell that is large relative to the particle size. The average zeta potential is calculated from the average mobility using Eq. 10.14.

10.4 DEFLOCCULATION AND THE STABILITY OF SUSPENSIONS

Powders suspended in a liquid spontaneously agglomerate unless they are suitably deflocculated by creating mutually repelling charged double layers or by physically preventing the close approach of particles due to the steric hindrance of adsorbed molecules (Fig. 10.6). Electrical charging may stabilize slurries in polar liquids.

The interaction of two particles with identical charged double layers was examined by Derjaguin and Landau* and Verwey and Overbeek,† and their

---

* B. Derjaguin and L. Landau, *Acta. Physicochim.* **14**, 633 (1941).
† E. Verwey and J. Th. G. Overbeek, *Theory of the Stability of Lyophobic Colloids*, Elsevier, Amsterdam, 1948.

combined theories are now referred to as the DLVO theory. The motivating force for coagulation is the ever-present Van der Waals attractive force which is a function of the dielectric constant of the medium and the mass and separation of the particles (see Eq. 2.17). Repulsion is provided by the interaction of two electrical double layers. The form of the repulsion depends on the size and shape of the particles, the distance h between their surfaces, the double-layer thickness κ^{-1} and ε_r of the liquid medium. For two spherical particles of diameter a the potential energy of repulsion is*

$$U_R = \frac{\varepsilon_r a^2 \psi_0^2}{4(h+a)} \exp - \left(\frac{h}{\kappa^{-1}} \right) \qquad (10.15)$$

when $\kappa a \ll 1$—i.e., small particles with a relatively large double layer. A greater surface potential and double-layer thickness increases the repulsive force. For the situation where $\kappa a \gg 1$*

$$U_R = \frac{\varepsilon_r a \psi_0^2}{4} \ln \left(1 + \exp - \left(\frac{h}{\kappa^{-1}} \right) \right) \qquad (10.16)$$

The total potential energy U_T is the algebraic sum of the attractive potential energy U_A and the repulsive potential energy U_R

$$U_T = U_A + U_R \qquad (10.17)$$

The dependence of U_A, U_R, and U_T on the separation of spherical particles is illustrated in Fig. 10.11. For each system there is some critical

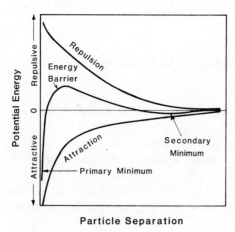

Particle Separation

Fig. 10.11 Potential energy of interaction between two particles with electrical double layers. The secondary minimum is not expected when $\kappa a \ll 1$.

---

* M. J. Rosen, Surfactants and Interfacial Phenomena, Wiley-Interscience, New York, 1978.

zeta potential and range of double-layer thickness for which the repulsive potential energy exceeds the attractive potential energy, producing an energy barrier to flocculation. A secondary minimum in the potential energy U_T occurs for large flat particles when the separation is of the order of the particle size, and the primary minimum occurs when the separation is approching molecular dimensions. The DLVO interaction diagram is very useful for understanding much of the behavior of suspensions.

The kinetic energy of colloidal particles due to Brownian motion is of the order of $10k_BT$, and at 20°C a repulsion barrier corresponding to a zeta potential of about 25 mV is requisite to minimize coagulation by means of electrical charging. A larger repulsion barrier or other mechanism for stability is needed to retard agglomeration during pouring and mixing processes which produce a greater kinetic energy.

An adsorbed nonionic surfactant with its associated water molecules or an adsorbed polyelectrolyte may stabilize suspensions at an apparent zeta potential less than 25 mV because the adsorbed layer can provide a steric hindrance to the close approach of the particles. If the repulsive potential energy due to electrical charging is U_{Re} and steric stabilization is U_{Rs}, then

$$U_T = U_A + (U_{Re} + U_{Rs}) \tag{10.18}$$

For colloidal particles, U_A is relatively small, and with an adsorbed polyelectrolyte, the distance from the hydrated surface to the slippage plane may be greater than for deflocculation using a simple electrolyte. Deflocculants such as sodium polyacrylate can produce a very stable suspension at an apparent zeta potential of 15 mV (Fig. 10.10); a contribution of U_{Rs} to the stabilization is expected. The dissociation of the polyelectrolyte and its adsorption vary with pH.

The Gouy-Chapman theory indicates that the thickness of the double layer is proportional to ε_r of the liquid and a polar liquid with a higher dielectric constant should promote dispersion. This effect of ε_r has been observed in preparing suspensions of MgO and CaO in alcohols where water must be avoided.

In nonpolar liquids of low dielectric constant, the specific adsorption of oriented polar molecules may provide deflocculation and stability. Tartaric, benzoic, oleic, stearic, and trichloroacetic acids, glyceryl trioleate, and some natural oils produce deflocculation of oxide powders in several nonpolar liquids. Oleic acid ($C_{18}H_{34}O_2$) is a liquid at 25°C and has a dielectric constant of only about 2.5, but the molecular groups at the two ends of the molecule are very different, as is shown in Fig. 10.12. The –COOH group is strongly adsorbed on the surface of an oxide. Steric hindrance provided by the kinked molecule may produce a U_{Rs} sufficient for the deflocculation of fine particles. In general, the steric repulsion is proportional to the thickness of the adsorbed layer and the chemical nature and concentration of the adsorbed molecules.

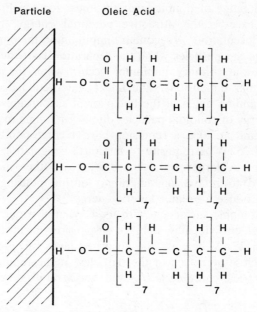

Fig. 10.12 Model for the adsorption of oleic acid on an oxide particle. (The molecule is kinked at the C=C bond.)

10.5 COAGULATION AND FLOCCULATION

A suspension may coagulate because of one or more of at least three mechanisms as shown in Fig. 10.13. Slow coagulation occurs when the $\zeta < 25$ mV and steric hindrance is negligible. Particles of quartz in an aqueous suspension have a negative potential when the pH is higher than the IEP. Slow coagulation occurs for a $2 < pH < 3$ because U_{Re} is low. In the range $3 < pH < 12$ the zeta potential and U_{Re} are high and a relatively stable suspension is formed.[*] However, at an extreme pH, the concentration of ions is high, and the system may coagulate because the double layer is compressed.

Electrolytic coagulation occurs when the concentration of counterions is sufficient to reduce the thickness of the double layer and reduce the zeta potential and U_{Re}. In the Gouy-Chapman theory, the concentration N of counterions of valence Z in the double layer is in the proportion of about N^Z. Counterions of higher valence have a strong effect in causing coagulation; this observation is known as the Schulze-Hardy rule. The order of coagulation power for monovalent cations is $Li^+ > Na^+ > K^+ > Rb^+ > NH_4^+$, and for divalent cations $Mg^{2+} > Ca^{2+} > Sr^{2+} > Ba^{2+}$. This sequence is often referred to as the Hofmeister series and can be rationalized in terms of

[*] B. Yarar and J. A. Kitchener, *Trans. Inst. Min. Metall.* **79**(3), 23–33 (1970).

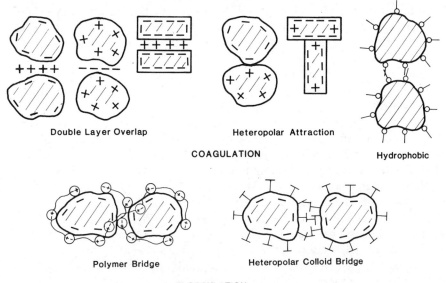

Double Layer Overlap   Heteropolar Attraction

COAGULATION

Hydrophobic

Polymer Bridge   Heteropolar Colloid Bridge

FLOCCULATION

Fig. 10.13 Models of coagulation and flocculation.

the charge/radius of the counterion, as indicated in Table 10.3. For negative counterions, the observed coagulation value is $SO_4^{2+} > Cl^- > NO_3^-$.

Negative oxide particles in an aqueous system are commonly coagulated using an additive such as $CaCl_2$, $CaCO_3$, $MgCl_2$, or $MgSO_4$. The adsorption of the more highly charged cation from the dissolved salt reduces the zeta potential. Some coagulants may also change the pH of the system in the direction of the IEP. $AlCl_3$ is a powerful coagulant and acidifier that can be

Table 10.3 Hydrated Ionic Radii

Ion	Radius (Å)	Hydration (mol H_2O)	Hydrated Radius (Å)
Li^+	0.78	14	7.3
Na^+	0.98	10	5.6
K^+	1.33	6	3.8
Rb^+	1.49	0.5	3.6
NH_4^+	1.43	3	—
Mg^{2+}	0.78	22	10.8
Ca^{2+}	1.06	20	9.6
Ba^{2+}	1.43	19	8.8
Al^{3+}	0.57	57	—

Source: W. E. Worrall, *Clays and Ceramic Raw Materials, Halsted Press*, New York, 1975.

used to reduce the pH to less than 4. When coagulating a clay slurry using a salt, an agglomerate with a denser particle packing may occur when the salt concentration exceeds some particular concentration. Moderately powerful coagulants are more commonly used, and the pH must be controlled, because the solubility of the fine particles is dependent on pH.

Basic organic compounds such as an amine that ionizes in solution may also be used as a coagulant in an aqueous system.

Adsorption of ethylamine $C_2H_5-NH_3^+$ may cause coagulation by causing the surface to become hydrophobic and less hydrated. Coagulation is motivated by contact between hydrophobic surfaces, which reduces the area of the hydrophobic surface-water interface.

Heteropolar coagulation occurs when the particles in suspension have surfaces of opposite charge. Suspensions of clay particles with negative faces and positive edges are coagulated as shown in Fig. 10.13. Admixing a deflocculated suspension of negative particles in a deflocculated suspension of positive particle may produce a coagulated system.

In general, particles in a dilute coagulated suspension settle rapidly in bulk, and the supernatant solution above is clear; the sediment is of relatively low packing density and is easily redispersed. Particles in a well-deflocculated suspension settle relatively slowly, and the sediment is relatively dense and less easily dispersed. Sediment height is used as a qualitative index of coagulation (Fig. 10.14).

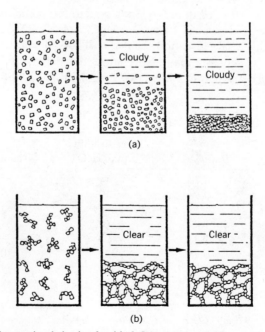

Fig. 10.14 Sedimentation behavior for (a) deflocculated and (b) coagulated suspensions.

SUMMARY

Deflocculants and coagulants are essential additives that modify the inter-particle forces, agglomerate structure, and consistency of the processing system. Agglomerating forces of the Van der Waals type may be offset by repulsive forces produced by electrical charging and steric hindrance. Particle charging and steric hindrance may produce the deflocculation of particles in polar liquids, whereas steric hindrance produces deflocculation in nonpolar liquids. Deflocculants producing particle charging may be either simple or polymer electrolytes; the type and concentration must be controlled to control the interparticle repulsion indicated by the zeta potential. The pH relative to the IEP may affect the adsoprtion of a polyelectrolyte. Coagulation is produced by crowding counterions in the double layer, sometimes called "overdeflocculation," or by adding counterions of a higher valence; both compress the double layer and reduce the repulsion between particles. Surfactants that produce a lyophobic surface on the particles may also produce coagulation. The bonding of oppositely charged surfaces that may occur in mixture of minerals or clay systems is called heteropolar coagulation. The role of deflocculants and coagulants in real systems is sometimes difficult to analyze because of aging caused by the effects of soluble impurities and time-dependent dissolving and adsorption processes.

SUGGESTED READING

1. Robert D. Vold and Marjorie J. Vold, *Colloid and Interface Chemistry*, Addison-Wesley, Reading, MA, 1983.

2. G. D. Parfitt (ed), *Dispersion of Powders in Liquids*, 3d Ed., Wiley-Interscience, New York, 1981.

3. Milton J. Rosen, *Surfactants and Interfacial Phenomena*, Wiley-Interscience, New York, 1978.

4. Michael D. Sacks and Chandrashekhar S. Khadilkar, Milling and Suspension Behavior of Al_2O_3 in Methanol and Methyl Isobutyl Ketone, *J. Am. Ceram. Soc.* **66**(7), 488–494 (1983).

5. Arthur W. Adamson, *Physical Chemistry of Surfaces*, Wiley-Interscience, New York, 1976.

6. Paul Sennet and J. P. Olivier, Colloidal Dispersions, Electrokinetic Effects, and the Concept of Zeta Potential, in *Chemistry and Physics of Interfaces*, American Chemistry Society, Washington, DC, 1965.

7. H. van Olphen, *An Introduction to Clay Colloid Chemistry*, Wiley-Interscience, New York, 1963.

8. Paul D. Calvert et al., Fish Oil and Triglycerides as Dispersants for Alumina, *Am. Ceram. Soc. Bull.* **65**(4), 669–672 (1986).

9. William H. Morrison Jr., Stabilization of Aqueous Oxide Pigment Dispersions, *J. Coatings Technol.* **57**(721), 55–65 (1985).

10. J. M. Horn Jr. and G. Y. Onoda Jr., Surface Charge of Vitreous Silica and Silicate Glasses in Aqueous Electrolyte Solutions, *J. Am. Ceram. Soc.* **61**(11), 523–527 (1978).

11. R. H. Yooa, T. Salmon, and G. Donnay, Predicting Points of Zero Charge of Oxides and Hydroxides, *J. Colloid Interface Sci.* **70**(3), 483 (1979).

12. R. F. Long, Electrophoresis as Method of Investigating Electrical Double Layer, *Ind. Engr. Chem.* **57**(8), 58–71 (1969).

PROBLEMS

10.1 Illustrate the structure of silica with a hydrated surface (silanol groups). How does the surface change on increasing or decreasing the pH from the PZC?

10.2 Sketch the variation of electrical potential with pH for an aqueous suspension of silica, alumina, and mullite. Indicate the PZC for each material.

10.3 Write an equation for the reaction between an uncharged surface of hydrated alumina and sodium pyrophosphate, producing a particle with a negative surface charge. What is the counterion?

10.4 Write the reaction for hydrated alumina surface with a positive charge and ammonium polyacrylate, producing a deflocculated slurry at a pH 9.0.

10.5 Calculate κ^{-1} for an aqueous solution of a $1:1$ electrolyte of 0.01, 0.1, and 1.0 M concentration at 20°C.

10.6 What is the compression of κ^{-1} in problem 10.5 if a $2:2$ electrolyte is used?

10.7 Should heating increase or decrease the tendency of a "charge-stabilized" slurry to coagulate? Consider both repulsion forces and Brownian motion.

10.8 Compare κ^{-1} for water at 20°C and ethyl alcohol at 40°C. Assume that the electrolyte concentration is equal in each.

10.9 In macroelectrophoresis, the amount of solid material that moves into the collection chamber is determined by weighing the collection chamber filled with suspension before and after applying an electric field E. The electrophoretic mobility ν_e is related to the change in mass Δw in a time interval t by the equation

$$\nu_e = \frac{\Delta w}{t} \frac{1}{EAD} \frac{1}{(1 - f_p^v) f_p^v}$$

where A is the area for transference, D is the density of solids minus the density of liquid in suspension, and $1/(1 - f_p^v)$ is a correction for counterflow of the liquid. Calculate v_e from the following data: $E = 10.8 \, \text{V/cm}$, $A = 0.33 \, \text{cm}^2$, $D = 2.0 \, g/cm^3$, $f_p^v = 0.30$, $\Delta w/t = 1.0 \, g/h$.

10.10 Calculate the zeta potential for particles in an aqueous suspension at 25°C if $v_e = 0.23 \times 10^{-3} \, \text{cm}^2/(\text{V} \cdot \text{s})$. Assume that $f_H = 1$ in Eq. 10.14.

10.11 Compare the Hofmeister series with the ratio of valence to radius of the ions.

10.12 Explain why sodium polyacrylate and sodium silicate are often effective deflocculants when cations of high valence are present in the solution.

10.13 Estimate the number of repeating units in a molecule of sodium polyacrylate if the molecular weight is 5000 g/mol.

10.14 Estimate the length of a molecule of oleic acid and compare this to the values of κ^{-1} determined in problem 10.5. Indicate the form of U_{RS} on Fig. 10.11.

10.15 Draw the curves for the concentration of positive ions and negative ions in the diffuse double layer as a function of distance from the surface of a negatively charged particle. Stern considered that the first layer of counterions was chemically adsorbed in manner similar to Langmuir adsorption, and the first layer is often called the Stern layer.

10.16 Explain why $AlCl_3$ reduces the pH of an aqueous solution much more rapidly than $CaCl_2$. Write the chemical reactions.

10.17 Draw curves indicating the potential energy of repulsion and the net interaction when the zeta potential is very low. How is the repulsion curve modified by steric stabilization?

10.18 Deflocculated suspensions of alumina and silica powders are prepared at pH 6. When these aqueous suspensions are mixed, a coagulated suspension is formed. Explain.

11

Flocculants, Binders, and Bonds

Polymer molecules and coogulated coloidal particles that are adsorbed and bridge between ceramic particles may provide interparticle flocculation and are commonly called flocculants or binders. However, these additives may provide several different practical functions in ceramic processing. An adsorbed flocculant may improve the wetting of the particles (wetting agent), significantly increase the apparent viscosity (thickener), retard the settling rate of the particles (suspension aid), and alter the dependence of the apparent viscosity on the flow rate or temperature (rheological aid). In more condensed systems such as granules for pressing, extrusion bodies, cast bodies, and unfired ceramic coatings, the flocculant may improve the plasticity (body plasticizer), reduce the liquid migration rate (liquid retention agent), alter the liquid requirement (consistency aid), and improve the green strength (binder). Often a workable system is produced on using only one flocculating additive, but in some cases two or more different types are used. In ceramic processing these additives are referred to as binders more commonly than as flocculants, and the term binder will be used here as a general term for these additives.

Other types of binders are waxes, resins, gels, low-temperature reaction bonds, and hydraulic cements. A film of a wax or resin or an additive that polymerizes into a gel may bind the particles together. Additives that react with the ceramic particles and polymerize or crystallize are called reactive bonds. Hydraulic cements are inorganic bonds that react with water and form very fine needlelike, hydrated calcium silicate or calcium aluminate minerals which coat the particles and form a bonded network. In this chapter we will examine the compositions and structure of both inorganic and organic binder materials.

11.1 BINDER COMPOSITIONS

Colloidal clay minerals were the binders used in processing classical ceramics, and these binders are used today in processing a wide variety of traditional ceramics and more advanced ceramic systems containing a minor amount of silica. Examples of clay bonds are fine kaolin, ball clay, and bentonite (Table 11.1). Clay binders and bentonite are refined and characterized to obtain a more consistent behavior. Microcrystalline cellulose is an organic colloidal particle binder. It is manufactured from high-purity cellulose pulp and is widely used where submicron pores are desired such as in producing ceramic catalyst carriers.

Molecular binders range widely in composition and may be either natural or synthetic substances (Table 11.1). The refined natural materials are more

Table 11.1 Binder Materials

Colloidal Particle Type			
Organic		Inorganic	
Microcrystalline Cellulose		Kaolin Ball clay Bentonite	

Molecular Type			
Organic	Examples	Inorganic	Example
Natural gums	Xanthan gum, gum arabic	Soluble silicates	Sodium silicate
Polysaccharides	Refined starch, dextrine	Organic silicates	Ethyl silicate
Lignin extracts	Paper waste liquor	Soluble phosphates	Alkali phosphates
Refined alginate	Na, NH_4 alginate	Soluble aluminates	Sodium aluminate
Cellulose ethers	Methyl cellulose, hydroxyethyl cellulose, sodium carboxymethyl cellulose		
Polymerized alcohols	Polyvinyl alcohol		
Polymerized butyral	Polyvinyl butyral		
Acrylic resins	Polymethyl methacrylate		
Glycols	Polyethylene glycol		
Waxes	Paraffin, wax emulsions, microcrystalline wax		

expensive than clay binders but less expensive than the refined and synthetic organic polymers. Organic molecular binders may be introduced as aqueous or nonaqueous solutions, liquid emulsions, or liquid melts. Common binders such as polyvinyl alcohol and the cellulose types are purchased as a powder and must be dissolved in water before admixing. Inorganic binders are commonly introduced into a system as an aqueous solution or colloidal sol. The more popular organic molecular binders may be purchased in grades differing in the range of molecular sizes, and molecular binders provide a wide flexibility in modifying rheological behavior. When decomposed under oxidizing conditions, synthetic organic binders introduce relatively little inorganic impurity.

Molecular organic binders are often used in concert with refined clay binders when alumina and silica are present in the ceramic composition. Ball clay contains variable, natural organic matter of both a colloidal and molecular nature which affects its flocculating properties.

11.2 CLAY BINDERS

Coagulated clay colloids may be adsorbed and bridge the larger ceramic particles, effectively flocculating them (Fig. 10.13). The edge-face agglomeration of colloidal clay particles with positive edges and negative faces is a special case of hetero-coagulation that occurs at a pH less than the PZC of the edges of the particles. Kaolin is not deflocculated at a low pH using simple acids, because the faces remain negative. The coagulated clay agglomerates in the suspension are usually called clay flocs.

If a sufficient concentration of highly charged cations such as Ca^{2+} are present in the suspension prior to the coagulation, a measurable degree of parallelism in the packing of small groups of clay particles is observed, termed salt-type coagulation. Agglomerates formed in deflocculated suspensions due to double-layer overlap have a high degree of particle orientation, as is shown in Fig. 11.1.

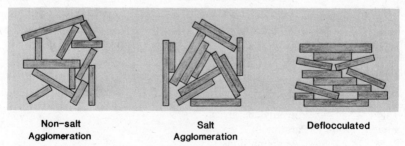

| Non–salt | Salt | |
| Agglomeration | Agglomeration | Deflocculated |

Fig. 11.1 Particle structures in clay agglomerates formed in coagulated suspensions and on sedimentation of a deflocculated suspension.

Finer clays containing more anisometric particles and a low alkali content and clays containing more of the mineral montmorillonite are more powerful flocculants and binders, as indicated by the higher MBI index and the plastic strength (Table 5.5). Montmorillonite is a colloidal mineral of very high specific surface area and is a scavenger for cations. Clays containing montmorillonite tend to coagulate naturally, and their behavior is less sensitive to fluctuations in concentrations of electrolytes. Bentonite is a claylike material containing a relatively high percentage of the mineral montmorillonite and is a relatively powerful flocculant and binder.

11.3 MOLECULAR BINDERS

Molecular binders are low- to high-molecular-weight polymer molecules that adsorb on the surfaces of particles and bridge them together, as shown in Fig. 10.12. The functionality of the polymer molecule may be nonionic, anionic, or cationic (Fig. 11.2). Most of the polymer binders used in ceramic processing are nonionic or mildly anionic. We will first consider the adhesion type of binders which are adsorbed on particles. The film type such as the waxes are considered later.

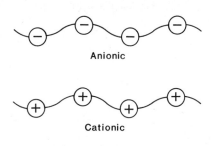

Anionic

Cationic

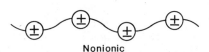

Nonionic

Fig. 11.2 Nonionic binders have polar functional groups, and ionic binders have charged functional groups after dispersion in the liquid and ionization.

Vinyl Type

Consider the molecular structure of a very common binder, polyvinyl alcohol (PVA), which is illustrated in Fig. 11.3. The basic repeating unit in the structure indicated by brackets is called the mer. The number of mers in the molecule, indicated by n, is the degree of polymerization. Accordingly, the molecular weight of a polymer increases as the degree of polymerization increases. The carbon-carbon linkage is referred to as the vinyl backbone, and the H and OH are referred to as the side groups. The polar OH side

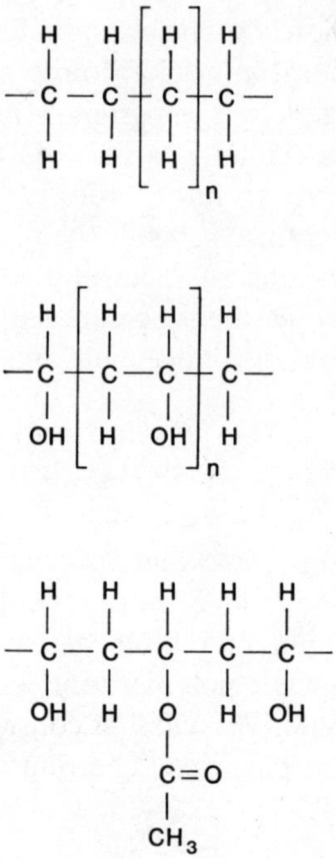

Fig. 11.3 Molecular structures of hydrocarbon chain and fully and partially hydrolyzed polyvinyl alcohol.

Table 11.2 Side Groups in Vinyl Binders

Binder	Group	Type
Soluble in water		
Polyvinyl acetate	—COOCH$_3$	Nonionic
Polyvinyl alcohol	—OH	Nonionic
Polyacrylamide	—CONH$_2$	Nonionic
Polyvinyl pyrolidone	—NC$_2$H$_4$C$_2$OH$_2$	Nonionic
Carboxylic polymer	—COOH	Anionic
Soluble in non-polar liquids		
Polyvinyl butyral	O—C—O HC$_3$H$_7$	Nonionic
Polymethylmethacrylate	—CH$_3$ —COOCH$_3$	Nonionic

group is hydrophilic, which promotes initial wetting and dissolving in a polar liquid. Hydrogen bonding of the OH side group to the surface of a particle provides adhesion, and the dipolar attraction of the OH side groups produces intermolecular bonding. Polyvinyl alcohol is manufactured by the hydrolysis of polyvinyl acetate in the presence of a catalyst. Fully hydrolyzed PVA contains less than 4% residual acetate and is soluble only in hot water. Partially hydrolyzed material (Fig. 11.3) contains more than 20% of the bulkier acetate groups, which precludes dense packing and aids in dissolving. Partially hydrolyzed PVA is soluble in cold water. Other vinyl-type binders containing different side groups are indicated in Table 11.2.

Cellulose Type

The cellulose ethers are a second important family of binders and have the polymeric backbone of cellulose. Cellulose is a natural carbohydrate that contains the basic repeating structure of anhydroglucose units (Fig. 11.4). Hydroxyethyl cellulose is produced by reacting cellulose with ethylene oxide. Methyl cellulose is manufactured by treating cellulose with caustic solution and then methyl chloride to produce the methyl ether of cellulose. These treatments cause substitution for some of the OH groups. Other forms of cellulose ethers are produced by substituting other molecular groups in the anhydroglucose unit, as is indicated in Table 11.3. Each anhydroglucose ring has three hydroxyls where substitution may occur, and the degree of substitution DS indicates the average number of substituent groups in the ring. A DS of about 1.6–2.0 provides maximum water solubility and is typical of commercial refined cellulose binders. The cellulose binder molecule is less flexible than the vinyl type.

When two different types of groups are substituted, the average molar substitution MS of the second group is also controlled. Differences in the

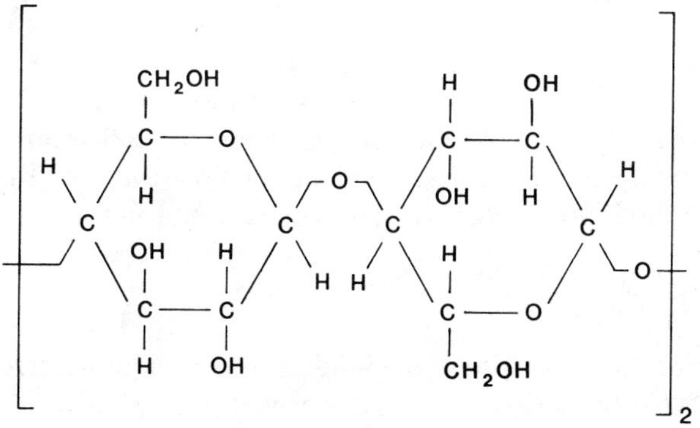

Cellulose

Fig. 11.4 Molecular structure of two cellulose units.

Table 11.3 Side Groups in Cellulose Derivatives[a]

Cellulose Type	Group	Type
Hydroxyethyl	$-CH_2OCH_2CH_2OH$	Nonionic
	$-CH_2OCH_2CH_2OCH_2CH_2OH$	
Methyl	$-OCH_3$	Nonionic
	$-CH_2OCH_3$	
Sodium carboxymethyl	$-CH_2OCH_2COONa$	Anionic
	$-CH_2OH$	
Hydroxypropyl methyl	$-OCH_2CHOHCH_3$	Nonionic
	$-CH_2OCH_3$	
	$-OCH_3$	
Sodium alginate	$-COONa$	Anionic
Starch, dextrine	$-CH_2OH$	Nonionic
Soluble in nonpolar liquids		
Ethyl cellulose	$-CH_2OCH_2CH_3$	Nonionic

[a] Some unsubstituted CH_2OH remains.

degree of polymerization of the cellulose backbone and the type, DS, and MS of the substituent side groups produce different behavior when dissolved in solution or adsorbed on ceramic particles. Some binders may contain ionizable side groups, and the ionized group can change the adsorption behavior and flocculation. A charged molecule is less twisted and effectively longer.

Starch is a natural binder that is cold-water-insoluble; it is relatively active biologically and subject to breakdown when heated and agitated. Refined starches are more easily dissolved and controlled. Solium carboxymethyl cellulose is an anionic binder sometimes used to increase viscosity and control filtration properties of more traditional ceramics. Natural gums are relatively unrefined substances derived from plants and are very complex branched polymers. Lignosulfonates are complex polymeric by-products obtained in producing paper from wood pulp using the sulfite process. They are prepared as sodium and calcium salts with a molecular weight range of about 1000 to 20,000, are used as inexpensive dispersants, and can serve as binders on drying. Many natural gums are ionic in nature. In general, the behavior of ionic binders is more sensitive to electrolytes.

Waxes and Glycols

Common waxes used as film-type binders are paraffin derived from petroleum, candelella and carnuba waxes derived from plants, and beeswax of insect origin. Paraffins are mixtures of straight-chain saturated hydrocarbons which tend to crystallize as plates or needles. Microcrystalline waxes are branched saturated hydrocarbons also derived from petroleum; they are somewhat tougher than paraffins. The plant waxes are more complex mixtures of straight-chain hydrocarbons, esters, acids, and alcohols that are relatively hard and have a relatively high melting point of 85–90°C. Beeswax is a complex mixture of esters and saturated and unsaturated hydrocarbons and has a much lower melting point than the plant waxes. The mechanical properties of waxes are directly related to secondary bonding between the molecules. Intermolecular bonding is weaker between the straight and branched hydrocarbons, and they are softer and more plastic. Well below the hardening point, waxes are brittle. Heating disrupts the bonding, and at a higher temperature plastic flow occurs under a relatively small stress. Internal stress frozen into the wax during cooling and relieved by molecular rearrangment causing distortion is minimized by using a microcrystalline wax.

Polyethylene glycol is a binder that has a third type of backbone; it is polymerized ethylene oxide with the formula $HOCH_2(CH_2OCH_2)_n\text{-}CH_2OH$. Polyethylene glycols of a low molecular weight are relatively heat-stable viscous liquids. High-molecular-weight forms are waxy solids used as binders and lubricants.

11.4 DISSOLVING AND ADMIXING BINDERS

The dissolving of a binder depends on overcoming nonpolar and polar bonding between molecules. Some binders are surface treated to improve dissolving. Binders with polar side groups attract water molecules, and may become considerably hydrated. Binders with nonpolar side groups are soluble in nonpolar solvents. Liquid solutions may dissolve binders that have only weakly polar side groups. Binders of a nonpolar nature dissolved in organic solvents and used in ceramic processing include polyvinyl butyral, ethyl cellulose, and acrylic polymers such as polymethyl methacrylate; their functional groups are shown in Tables 11.2 and 11.3. Waxes may be melted and admixed into hot powder or first dissolved into a nonpolar solvent before admixing. Aqueous, stabilized wax emulsions with a wax content of about 40% are available for admixing at room temperature. Polyethylene glycols used as binders are quite soluble in water and a variety of organic solvents including alcohols and trichloroethylene.

11.5 MOLECULAR WEIGHT AND BINDER GRADE

Commercial binders of a particular grade contain molecules with a particular range and distribution of molecular weight. The number average molecular weight ($\bar{M}_N$) of a polymer binder is determined experimentally from the osmotic pressure, melting point depression, and boiling point elevation of extremely dilute solutions. The mass average molecular weight ($\bar{M}_W$) is

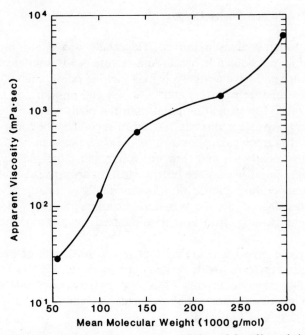

Fig. 11.5 Variation of the viscosity of an aqueous solution of 2 wt % methylcellulose with number average molecular weight (20°C, $\dot\gamma = 10\,s^{-1}$). (Data courtesy of Dow Chemical Co., Midland, MI.)

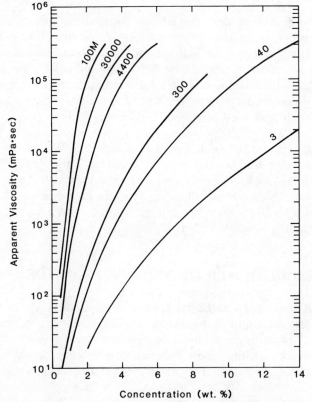

Fig. 11.6 Dependence of the viscosity of aqueous solutions of different viscosity grades of hydroxyethyl cellulose on concentration (25°C, $\dot\gamma = 14\,s^{-1}$) (Cellosize, Union Carbide Corp., New York).

determined from the slope of viscosity/concentration of dilute solutions of less than 1% concentration, using a very sensitive viscometer. The molecular weight distribution in a binder may also be determined by gel permeation chromatography. For the more popular binders, a series of grades are available. However, molecular-weight information is seldom presented in specifications of commercial binders. Because the apparent viscosity of a binder solution increases with the mean molecular weight (Fig. 11.5), the apparent viscosity of a solution of some standard concentration is commonly used as a relative index of the mean molecular weight. As indicated in Fig. 11.6, the apparent viscosity of a 2 wt % solution at 20°C at a shear rate representative of pouring conditions is commonly used as a nominal designation of the viscosity grade of cellulose binders.

11.6 GELATION

For some types of binders, a thermal or chemical change in the system or a loss of solvent increasing the concentration of the binder may cause cross-linking of the polymer molecules into a three-dimensional configuration with interspersed liquid, which is called a gel;

$$x \text{ polymer}_{(solution)} \rightarrow [\text{polymer}]_{n(gel)} \qquad (11.1)$$
$$+ (x - n)\text{polymer}_{(solution)}$$

where n is an index of gelation.

A thermally reversible gel may be formed in an aqueous solution of methylcellulose or hydroxypropyl methyl cellulose* (Fig. 11.7). Initially the viscosity of the binder solution decreases on heating. But above some particular temperature, additional heating causes the molecules to become more hydrophobic. At a temperature where the molecules lose water of hydration and three-dimensional bonding occurs, the viscosity increases rapidly. Shearing of the gel can occur only if the shear stress exceeds the yield strength of the gel. The gel strength and elasticity increase with the concentration and molecular weight of the binder and attain a maximum value at a particular temperature. Electrolytes and glycerol, which have an affinity for water, reduce the gelation temperature; ethanol and propylene glycol raise the gelation temperature. This thermal gelation is reversible and disappears on cooling.

Gelation may occur with no loss of liquid when the liquid is occluded in capillary pores or in micelles. Syneresis is the spontaneous shrinking of a gel to an equilibrium volume with the expulsion of liquid. Imbibition of liquid or delayed condensation that produces liquid in the gel may cause the gel to expand.

* N. Sarkar and G. K. Greminger Jr., *Am. Ceram. Soc. Bull.* **62**(11), 1280–1283, 1288 (1983).

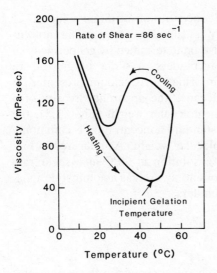

Fig. 11.7 Change in viscosity and gelation of an aqueous solution of 2 wt % methylcellulose on heating at 0.25°C/min. (Courtesy of Dow Chemical Co., Midland, MI.)

Gelation may also be produced by chemical means. Using a simple acid or base, ionic binders may be coagulated at their PZC (Fig. 11.8). Highly charged cations may coagulate (cross-link) high-molecular-weight anionic polymers into a gel. Partially hydrolyzed polyvinyl alcohol may be gelled by soluble carbonates, sulfates, and borates. Sodium alginate is used as a binder and dental impression material. The aqueous alginate sol may be caused to gel in the presence of alkaline earth ions such as Ca^{2+};

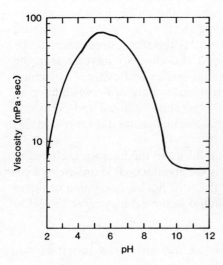

Fig. 11.8 Dependence of gelation on the pH of a 0.1 wt % solution of a cationic binder.

$$\text{Na}_n \text{alginate}_{(\text{sol})} + \frac{n}{2} \text{Ca}^{2+}_{(\text{solution})} \rightarrow n \text{Na}^+_{(\text{solution})}$$

$$+ \text{Ca}_{\frac{n}{2}} \text{alginate}_{(\text{gel})} \tag{11.2}$$

A controlled slow set is achieved by introducing the calcium as a low-solubility salt such as calcium sulfate and a retarder such as PO_4^{3-} that is more mobile than the bulky alginate and forms an insoluble tricalcium phosphate. Chemical gelation is temperature-sensitive, and the temperature of the liquid should be controlled to control the setting time. The reversal of chemical gelation in an undisturbed system occurs very slowly, and chemical gelation is often practically irreversible.

Soluble and colloidal silicates are popular gel-type binders used in construction and refractory products. The dehydration and polymerization of esters of silicon, titanium, and zirconium as described in Chapter 4 may produce a binding phase. Aqueous solutions of sodium silicate with a molecular formula $Na_2O \cdot 2.5SiO_2$ to $Na_2O \cdot 4SiO_2$ are basic (pH 11–13) and contain polymerized silicate anions. They are important bonds because a gel structure forms on loss of a small amount of water and because of their high wetting powder. The proportion and size of the larger ring structures increase as the ratio of SiO_2/Na_2O increases. A gel structure is also produced in a concentrated silicate solution on adding polyvalent cations and by the addition of an acid or the hydrolysis of an organic ester such as ethyl acetate, alcohol, or glycerol diacetate, which produces an acid. Antimigration agents such as alkali carbonates, phosphates, perchlorates, or fluorides, which cause local polymerization, are added when air-drying alkali silicate binders. Dry alkali silicate binder films must be gelled to prevent rehydration and dissolving in the presence of moisture.

The chemically assisted coagulation of colloidal clay minerals and colloidal cellulose particles is also a form of chemical gelation. The gelling time is inversely proportional to the concentration of colloidal particles and coagulant and is dependent on the particle size and shape and the type of coagulant.

11.7 GENERAL EFFECT OF BINDERS

Flocculation

Ionic and nonionic binders may flocculate particles in suspension under particular conditions.* When the pH is greater than 4, negative quartz particles in an aqueous suspension are flocculated by a variety of high-molecular-weight cationic binders. With an increasing concentration of

* B. Yarar and J. A. Kitchener, *Trans. Inst. Min. Metall.*, **79**(3), 23–33 (1970).

absorbed binder, the zeta potential of the agglomerates becomes positive, and the binder may function as a protective colloid. Suspensions of negative quartz particles may also be flocculated by anionic acrylamide polymers if assisted by activating cations such as Ca^{2+} that reduce the zeta potential. Flocculation occurs at a lower pH when there is a higher concentration of Ca^{2+} ions in the suspension.

For a constant concentration of Ca^{2+} ions, the pH for the onset of flocculation is lower when the anionic character of the polymer is higher. Highly charged cations can coagulate many anionic polyelectrolytes and apparently reduce the repulsion between negatively charged particles and negative polymer chains. In suspensions of mixtures of minerals, specific adsorption and flocculation of only one mineral type may sometimes occur.

Most oxide particles are flocculated by a variety of nonionic polymers owing to hydrogen bonding or an attraction between the hydrated surface and polar groups on the polymer chain. Flocs formed by polymer or colloid bridging may be considerably stronger than coagulated particle agglomerates.

Thickening, Suspending, and Rheological Control

The substitution of a viscous binder solution for a simple liquid increases the mean viscosity and effectively thickens the system. In a more concentrated suspension, flocculation may cause the suspension to have the consistency of a paste. The settling rate is reduced in the higher-viscosity liquid and may become zero when the viscosity or yield point is sufficiently high. The presence of the binder solution or gel between the particles can greatly alter the flow properties of the system. The important topic of rheology control is discussed in several sections that follow.

Liquid Requirement, Elasticity, and Liquid Retention

The steric hindrance provided by colloidal or molecular bridges between particles decreases the particle packing density and increases the liquid required to saturate the body. The steric hindrance of polymer molecules generally increases with the molecular weight. The cellulose ether binders are less flexible than the vinyls, butyrals, and acrylics. Chemical effects such as increased solvation and greater ionization of an ionic polymer tend to increase the effective length of the molecule. Flocculated colloids or polymer molecules between particles may increase the elastic deformation at low pressures and can increase the elastic springback on decompression. The matrix of colloidal particles or molecules greatly reduces liquid migration rates. Binder migration due to capillary effects or applied pressure is lower for binders of higher molecular weight and gelled binder solutions.

Lubricity and Film Properties

Binder adsorbed on a surface may decrease the surface roughness and coefficient of friction. The lubricity of the surface is dependent on the hydrophilicity of the film, adsorbed layer displacement, surface tension, and molecular orientation.* The mechanical properties of binder films are very dependent on adsorbed plasticizing molecules, and this topic is discussed in Chapter 12.

11.8 POLYMER RESINS

Thermoplastic resins such as polyethylene and polystyrene, acrylic resins, and silicones are used as a bonding matrix in injection molding systems and in dental applications. Synthetic resins are polymers of high molecular weight. In ceramic bodies they are usually added to form a bonding matrix, which requires that they must be moldable at an early stage and later capable of being hardened to form a structural matrix.

Resins such as polyethylene and polystyrene are formed by addition polymerization which occurs by the addition of mers to an infant polymer molecule, increasing its size. Reaction of the monomer with the chain involves the breaking of an unsaturated $C{=}C$ bond and requires free radicals. Free radicals are intermediate molecules that have a highly reactive, unpaired electron and are usually formed by the fracture of an initiator molecule produced by heating, light, or a chemical reaction. The polymerization of ethylene occurs by reaction of the monomer CH_2CH_2 with the free radical $A\dot{}$, i.e.,

$$
\begin{array}{ccc}
\text{H} \quad \text{H} & \quad & \text{H} \quad \text{H} \\
\text{A}\dot{} + \text{C}{=}\text{C} \rightarrow & \text{A}{-}\text{C}{-}\text{C}\dot{} \\
\text{H} \quad \text{H} & & \text{H} \quad \text{H}
\end{array} \tag{11.3}
$$

followed by the activated monomer reacting with additional monomer, i.e.,

$$
\begin{array}{cccc}
\text{H} \ \ \text{H} \ \ \text{H} \ \ \text{H} & \quad & \text{H} \ \ \text{H} \ \ \text{H} \ \ \text{H} \\
\text{A}{-}\text{C}{-}\text{C}\dot{} + \text{C}{=}\text{C}\dot{} \rightarrow & \text{A}{-}\text{C}{-}\text{C}{-}\text{C}{-}\text{C}\dot{} \\
\text{H} \ \ \text{H} \ \ \text{H}{=}\text{H} & & \text{H} \ \ \text{H} \ \ \text{H} \ \ \text{H}
\end{array} \tag{11.4}
$$

until the chain propagation is terminated. Termination can occur by the reaction of two growing chains, which destroys free radicals. A high concentration of free radicals will tend to limit chain propagation. Chain transfer is a termination process that occurs when an activated polymer transfers the free radical to another specie for growth of a new polymer

* N. Sarkar and G. K. Greminger, *Am. Ceram. Soc. Bull.* **62**(11), 1280–1283, 1288 (1983).

molecule. A retarder is a substance that can react with a free radical to form a product incapable of polymerizing.

Vinyl polymers are formed from vinyl monomers having the structure CH_2CHR where the R group determines the specific compound (Table 11.2). Polystyrene is polyvinyl benzene and the R is a benzene ring. A copolymer is produced when two functional R groups are substituted along the chain.

A polymethyl methacrylate (acrylic) bond may be formed by reacting the monomer methyl methacrylate $CH_2{=}C(CH_3){-}COOCH_3$ in the presence of an initiator such as benzoyl peroxide $C_6H_5COO{-}OOC_6H_5$ which forms free radicals when heated or exposed to ultraviolet light. Heating may be produced by the addition of a tertiary amine such as dimethyl-p-toluidine; spontaneous polymerization is inhibited by adding a small amount of an antioxidant such as hydroquinone. The acrylic resin softens on heating and can be molded at 125°C; it is also soluble in a variety of organic solvents.

In stepwise polymerization, both the monomer and growing chains are active. Stepwise polymerization is referred to as condensation polymerization when a nonpolymerizable molecule such as water is a by-product of the reaction. Silicone polymers have the siloxane backbone $-S_i-O-S_i-$ and are formed by stepwise polymerization. Epoxy resins have an epoxide

$$\overset{\displaystyle O}{\overset{\displaystyle \triangle}{-C-C-}}$$

linkage and polymerize by both addition and condensation polymerization.

The strength and softening temperature of polymer material increases with the average molecular weight. Ordering of the –R groups along the chain increases the crystallinity and brittleness. Linear polymers soften on heating and are termed thermoplastic. Chain-branching connecting linear chains and cross-linking between chains having unsaturated carbon bonds strengthen and stiffen the polymer structure. These materials must be molded to shape before or during polymerization and are referred to as thermosetting resins. Elastomers such as rubbers and silicones are coherent elastic solids composed of interwined chains with limited cross-linking. Under a load, the chains straighten, lenghten, and then slip; they return to the unloaded state when the load is removed.

11.9 REACTION BONDS

An important family of bonds used in the refractories and construction industries are formed by the reaction of alumina with orthophosphoric acid H_3PO_4 or an acidic aqueous solution of mono-aluminum phosphate $Al(H_2PO_4)_3$ or mono-magnesium phosphate $Mg(H_2PO_4)_2$. The reaction of alumina or hydrated alumina with orthophosphoric acid is exothermic and produces mono-aluminum phosphate, $Al(H_2PO_4)_3$, or an amorphous bond, i.e.,

$$Al_2O_3 + 6H_3PO_4 \leftarrow 2Al(H_2PO_4)_3 + 3H_2O \qquad (11.5)$$

$$Al(OH)_3 + H_3PO_4 \rightarrow \text{amorphous bond} \qquad (11.6)$$

Heating of the bond causes dehydration and the formation of the metaphosphate $Al(PO_3)_3$, but the availability and further reaction with alumina produce a phosphate with a higher Al_2O_3/P_2O_5 ratio:

$$Al(H_2PO_4)_3 + Al_2O_3 \rightarrow 3AlPO_4 + 3H_2O \qquad (11.7)$$

The mono-aluminum phosphate bond is an acid solution in equilibrium with $AlPO_4$ and $AlPO_4 \cdot H_3PO_4$:

$$Al(H_2PO_4)_3 \rightleftarrows AlPO_4 \cdot H_3PO_4 + H_3PO_4 \qquad (11.8)$$

$$Al(H_2PO_4)_3 \rightleftarrows AlPO_4 + 2H_3PO_4 \qquad (11.9)$$

Reaction with a base that consumes free H_3PO_4 or heating drives the reaction to the right, and aluminum phosphate phases precipitate in a cross-linked polymeric structure. For mono-magnesium phosphate $Mg(H_2PO_4)_2$ bonds, the reactions are

$$Mg(H_2PO_4)_2 \rightleftarrows MgHPO_4 + H_3PO_4 \qquad (11.10)$$

$$2MgHPO_4 \rightleftarrows Mg_3(PO_4)_2 + H_3PO_4 \qquad (11.11)$$

Reaction with a base also produces precipitates having a polymeric structure.

Phosphate bonds exhibit aspects of both gel-type and reaction bonds. The viscosity of the phosphoric acid or aluminate solution decreases on heating, but very little migration of the bond on particle surfaces occurs on drying, because of gelling. The reaction bond may be formed at a temperature as low as 250–300°C, but heating above 500°C is required for complete dehydration and the formation of a stable aluminum phosphate phase.

The reaction of phosphorous acid and zinc oxide has been used to form a dental cement at room temperature. Phosphoric bonds have been produced with a wide variety of metal oxides. The reaction of a phosphoric acid solution with silicate compounds also produces a gel structure which forms silicyl phosphates such as the metaphosphate $SiO(PO_3)_2$ on heating above 250°C. Patented cements for dental ceramics have been prepared using zinc oxide and an acid such as polyacrylic acid and polycarboxylic acid. The polyacrylic acid $CH_2CHCOOH$ is soluble in water but reacts with zinc oxide to form a nearly insoluble, hard bond. Setting occurs by the cross-linking of the –COOH side groups by zinc ions:[*]

$$
\begin{array}{c}
\overset{\displaystyle H}{\underset{\displaystyle COOH}{-C-}} \quad \overset{\displaystyle H}{\underset{\displaystyle H}{C}}-\overset{\displaystyle H}{\underset{\displaystyle COOH}{C}} \quad \overset{\displaystyle H}{\underset{\displaystyle H}{-C-}} \\[2ex]
\diagdown \qquad \diagup \\
Zn \\
\diagup \qquad \diagdown \\[2ex]
\overset{\displaystyle COOH}{\underset{\displaystyle H}{-C-}} \quad \overset{\displaystyle H}{\underset{\displaystyle H}{C}}-\overset{\displaystyle COOH}{\underset{\displaystyle H}{C}} \quad \overset{\displaystyle H}{\underset{\displaystyle H}{-C-}}
\end{array}
\tag{11.12}
$$

Un-cross-linked –COOH groups may bond to an oxide surface such as the surface of a tooth by chemisorption or chelation. The properties of the bond depend on the degree of polymerization of the acid, the particle size of the zinc oxide, and the presence of other ions such as magnesium or calcium in the zinc oxide or added as a second solid phase.

11.10 HYDRAULIC CEMENTS

Hydraulic bonds commonly called hydraulic cements are prereacted compounds of calcium oxide and silica, alumina, and iron oxide that have been ground to a fine powder. When mixed with water, the compounds harden into a strong material in air or under water, with relatively little change in volume. Cements containing natural hydrating materials such as volcanic ash or shale are called pozzolana cements.

The principal inorganic phases in Portland cement are listed in Table 11.4. Water is mixed with the cement powder to produce a paste or slurry, and the cement sets when dissolving and reaction with water occur at the surfaces, and a gellike hydrated precipitate phase is formed. The hydration reaction is exothermic, and the water bound chemically as a hydroxide produces a hard product.

Portland cements are multiphase systems, and the reactions leading to their setting and hardening are very complex. Silicate and aluminate phases higher in calcium tend to dissolve and react more rapidly. Tricalcium aluminate $3CaO \cdot Al_2O_3$ reacts rapidly in water with or without the presence of calcium hydroxide, causing a fast set:

[*]E. H. Greener et al., *Materials Science in Dentistry*, Williams and Wilkins, Baltimore, 1972.

Table 11.4 Phases in Type II Portland Cement and Calcium Aluminate Cement

Phase	Oxide Formula	Amount (wt %)
Type II Portland cement		
Tricalcium silicate	$3CaO \cdot SiO_2$	46
Dicalcium silicate	$2CaO \cdot SiO_2$	28
Tricalcium aluminate	$3CaO \cdot Al_2O_3$	11
Tetracalcium aluminoferrite	$4CaO \cdot Al_2O_3 \cdot Fe_2O_3$	8
Gypsum	$CaSO_4 \cdot 2H_2O$	3
Magnesia	MgO	3
Calcium oxide	CaO	0.5
Sodium, potassium oxides	Na_2O, K_2O	0.5
Calcium aluminate cement		
Tricalcium aluminate	$3CaO \cdot Al_2O_3$	0–minor
Monocalcium aluminate	$CaO \cdot Al_2O_3$	major
	$12CaO \cdot 7Al_2O_3$	minor
Calcium dialuminate	$CaO \cdot 2Al_2O_3$	major

Source: H. W. Hayden et al., *The Structure and Properties of Materials*, Wiley-Interscience, New York, 1965.

$$3CaO \cdot Al_2O_3 + 6H_2O \rightarrow 3CaO \cdot Al_2O_3 \cdot 6H_2O \qquad (11.13)$$

and

$$3CaO \cdot Al_2O_3 + Ca(OH)_2 + 12H_2O \rightarrow 4CaO \cdot Al_2O_3 \cdot 13H_2O$$
$$(11.14)$$

Gypsum anhydrite is added to reduce the rate of these reactions; a fine, needlelike crystalline film of the complex mineral ettringite is deposited on the aluminate particles (Fig. 11.9) which retards further hydration and causes the cement to be more plastic:*

$$3CaO \cdot Al_2O_3 + 3(CaSO_4 \cdot 2H_2O) + 26H_2O \qquad (11.15)$$
$$\rightarrow 3CaO \cdot Al_2O_3 \cdot 3CaSO_4 \cdot 32H_2O$$

Hydration of $3CaO \cdot SiO_2$ and $2CaO \cdot SiO_3$ by several of many possible reactions causes hardening, e.g.

$$3CaO \cdot SiO_2 + xH_2O \rightarrow 2CaO \cdot SiO_2 \cdot xH_2O + Ca(OH)_2 \quad (11.16)$$

* W. Baumgart, A. C. Dunham, and G. C. Amstutz (eds), *Process Mineralogy of Ceramic Materials*, Elsevier, New York, 1984.

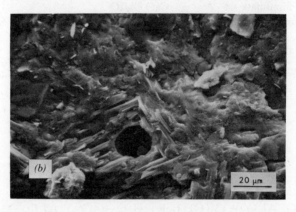

Fig. 11.9 (a) Portland cement paste 1 day after mixing with water; fine amorphous colloidal calcium silicate hydrate covers cement particles, and long thin needles of ettringite have formed in water-filled pores. (b) After 64 days, pores from air bubbles remain, and the matrix is a mixture of calcium silicate hydrates of indistinct morphology, large straited calcium hydroxide crystals, and hydrated cement particles (far left and top center). (Photos courtesy of J.F. Young, University of Illinois, Urbana-Champaign.)

$$2CaO \cdot SiO_2 + xH_2O \rightarrow 2CaO \cdot SiO_2 \cdot xH_2O \qquad (11.17)$$

Studies of the hardening of cement indicate that the hydration of $3CaO \cdot SiO_2$ continues beyond 30 days and that the hydration time for $2CaO \cdot SiO_2$ exceeds 1 year. The setting time can be reduced by grinding the cement finer or increasing the temperature of the water. Free calcium and magnesium oxides in the cement powder must be minimal because of the large volume expansion that accompanies their hydration:*

* W. Baumgart, A. C. Dunham, and G.C. Amstutz (eds), *Process Mineralogy of Ceramic Materials*, Elsevier, New York, 1984.

Calcium aluminate cements are used in castable refractory products. These cements set more rapidly than Portland cement and may be ready for service in 24 h. The hydration process is exothermic and sensitive to the temperature of hydration.*

$$CaO \cdot Al_2O_3 + 10H_2O \xrightarrow{T<22°C} CaO \cdot Al_2O_3 \cdot 10H_2O \qquad (11.18)$$

$$2(CaO \cdot Al_2O_3) + 11H_2O \xrightarrow{22-35°C} 2CaO \cdot Al_2O_3 \cdot 8H_2O + Al_2O_3 \cdot 3H_2O$$
$$(11.19)$$

$$3(CaO \cdot Al_2O_3) + 12H_2O \xrightarrow{T>35°C} 3CaO \cdot Al_2O_3 \cdot 6H_2O + 2(Al_2O_3 \cdot 3H_2O)$$
$$(11.20)$$

Cement mixed and cured between 22 and 35°C is stronger and more explosion-resistant (more permeable) apparently because of the particular hydrated phase formed and the more crystallized alumina gel. Calcium aluminate cements dehydrate on heating and lose most of their strength by about 800°C. They are used in many refractory applications, because a dimensionally stable refractory material of low permeability can be formed in place and be ready for use in about 24 h.

Other hydraulic cements include gypsum anhydrite $CaSO_4$ and plaster of Paris $CaSO_4 \cdot 0.5H_2O$ which hydrate rapidly and are used in building materials and for gypsum molds. Fine CaO mortars harden by the reaction of the hydroxide $Ca(OH)_2$ with carbon dioxide in air or dissolved in water, forming calcium carbonate. Magnesium oxychloride cement $3MgO \cdot MgCl_2 \cdot 11H_2O$, known as sorrel cement, formed in a solution of magnesium chloride $MgCl_2$ containing magnesia MgO, is used as a temporary bond in some refractories.

SUMMARY

Colloidal clay minerals and organic or inorganic polymers are commonly used as binders in ceramic processing. Bond clays contain a relatively high concentration of coagulated colloidal particles which coat and bridge between larger ceramic particles. Organic polymer binders of the vinyl and cellulose type are widely used in ceramic processing. The solubility and properties of solutions of these polymers may be varied by changing the type and number of substituent side groups of the molecule and the molecular length. Gelation increases the cohesive strength and viscosity of a polymer system and is common in the binding action of inorganic polymers and some organic polymer systems. Polymerized glycols and waxes are also used as binders in ceramic processing. Polymer resins, reaction bonds, and hydraulic

* W. Baumgart, A.C. Dunham, and G.C. Amstutz (eds), *Process Mineralogy of Ceramic Materials*, Elsevier, New York, 1984.

cements are used in special systems. The binder or bond phase may alter many properties of the system. Decomposition of the binder during firing commonly produces gas and pores which must be eliminated during sintering to obtain a dense ceramic. For both technical and economic reasons, the binder must be conveniently dispersed and admixed into the system and sufficiently time-stable during subsequent processing.

SUGGESTED READING

1. Wolfgang Baumgart, A.C. Dunham, and G. Christian Amstutz (eds), *Process Mineralogy of Ceramic Materials*, Elsevier, New York, 1984.
2. Ralph J. Fessenden and Joan S. Fessenden, *Organic Chemistry*, Willard Grant Press, Boston, 1982.
3. G. Y. Onoda Jr., The Rheology of Organic Binder Solutions, in *Ceramic Processing Before Firing*, George Y. Onoda Jr. and Larry Hench (eds), Wiley-Interscience, New York, 1978, pp. 235–251.
4. W. E. Worral, *Clays and Ceramic Raw Materials*, Halsted Press Division, Wiley-Interscience, New York, 1975.
5. E. H. Greener et al., *Materials Science In Dentistry*, Williams and Wilkins, Baltimore, 1972.
6. J. F. Wygant, Cementitious Bonding in Ceramic Fabrication, in *Ceramic Fabrication Processes*, W.D. Kingery (ed), MIT Press, Cambridge, MA, 1963.
7. Nitis Sarkar and George K. Greminger Jr., Methylcellulose Polymers as Multifunctional Processing Aids in Ceramics, *Am. Ceram. Soc. Bull.* **62**(11), 1280–1283, 1288 (1983).
8. F. J. Gonzalez and J. W. Hallorn, Reaction of Orthophosphoric Acid with Several Forms of Aluminum Oxide, *Am. Ceram. Soc. Bull.* **59**(7), 727–731 (1980).
9. K. Robert Lange, Properties of Soluble Silicates, *Ind. Eng. Chem.* **61**(4), 29–44 (1969).
10. A. G. Pincus and T. E. Shipley, The Role of Organic Binders in Ceramic Processing, *Ceram. Ind. Magazine* **92**(4), 106–109, 146 (1969).

PROBLEMS

11.1 Draw the structure of a mer of cellulose and methyl cellulose with a degree of substitution of 2.0.

11.2 Draw the structure of a mer of sodium carboxymethyl cellulose with a DS of 1.0 and an ionized side group. (Note: Substitution occurs for H in CH_2OH group.)

11.3 Draw the structure of polyvinyl alcohol that has 20% residual acetate.

11.4 Draw the structure of hydroxypropyl methyl cellulose with a DS of 2.0 and a molar substitution of hydroxypropyl of 50%.

11.5 Compare the average degree of polymerization and average chain length for the five viscosity grades of methylcellulose in Fig. 11.5. Assume that DS = 2 and that the length of one anhydroglucose ring is 0.58 nm. (Note: Chains in solution are not linear.)

11.6 Draw the structure of polymethylmethacrylate. Note that the two groups indicated in Table 11.2 are bonded to the same carbon atom.

11.7 From the molecular structure, would you expect an aqueous solution of partially hydrolyzed PVA to have a lower surface tension than pure water? Explain.

11.8 From the molecular structure explain why ethyl cellulose is soluble in a nonpolar liquid but hydroxyethyl cellulose is soluble in water.

11.9 Explain why the aqueous solubility of methylcellulose increases with the DS to about 2 and then decreases when DS > 2.

11.10 A ceramic body is prepared using a 5 wt % solution of an organic binder in amounts of 4 and 8 wt % of solution in the mix. What is the concentration of organics in each mix after drying? What is the percent residue in the body after firing if the ash content of the binder is 0.75%, as for polyvinyl alcohol?

11.11 Compare the plastic strength and MBI of the clays in Table 5.5, and explain your results.

11.12 Write the reaction between methyl methacrylate monomer and benzoyl peroxide to form polymethyl methacrylate.

11.13 Write the reaction between Na_2SiO_3 and water forming silicic acid H_2SiO_3.

11.14 Mono-aluminum phosphate is used to form a protective coating on steel used in aqueous environments. Write the reaction assuming that the oxide Fe_2O_3 on the steel reacts more rapidly than the metal phase.

11.15 Explain how the particle size, alumina content, and uniformity of mixedness can influence the amount of $Al(PO_3)_3$ and $AlPO_4$ formed on reacting alumina with H_3PO_4.

12

Plasticizers, Foaming and Antifoaming Agents, Lubricants, and Preservatives

The processing engineer often uses small amounts of other types of additives, in addition to deflocculants, coagulants, and binders, to develop a satisfactory processing system. A plasticizer is added to modify the viscoelastic properties of a condensed binder-phase film on the particles. The tendency for bubbles to persist in a slurry may be reduced by adding a small amount of an antifoaming agent; a foaming agent may increase the stability of gas bubbles. A lubricant is a surfactant that is strongly adsorbed and especially effective in reducing the coefficient of friction of the surfaces of ceramic particles and metal dies. A preservative is used when enzymatic degradation of a binder must be controlled.

12.1 PLASTICIZERS

The moldability of a binder is very dependent on temperature. At 20°C polyvinyl alcohol is elastic and brittle, and two separate films bond very poorly when pressed together; movement of the molecules is very limited, and the brittle binder is said to be in a glassy state. At about 90°C, the thermal energy is sufficient to enable segments of the molecules to flow and realign when compressed, and bonding between films occurs. This is called the rubbery state. The temperature between these two states at which the deformation changes from elastic behavior (fracture at a strain greater than about 5%) to time-dependent viscoelstic deformation (no fracture at a strain exceeding 100%) is called the glass transition temperature (T_g) and is shown in Fig. 12.1. Polymer films exhibit a change in resistance to mechanical deformation, thermal expansion, and specific heat at the glass transition temperature, and mechanical deformation, dilatometry, and calorimetry can

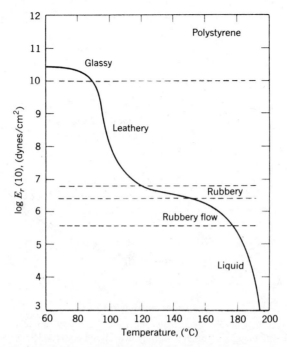

Fig. 12.1 Variation of relaxed elastic modulus of polystyrene with temperature and regions of differing viscoelastic behavior. (From H. W. Hayden et al., *The Structure and Properties of Materials*, Vol. III, *Mechanical Behavior*, Wiley-Interscience, New York, 1965.)

be used to determine the T_g of a binder material.* On heating to a temperature above the rubbery state, molding produces viscous flow behavior.

Ceramic systems containing a binder are commonly molded above the glass transition temperature of the binder. Small molecules distributed among the larger polymer molecules cause the polymers to pack less densely and reduce the Van der Waals forces binding the polymer molecules together. The presence of the small "plasticizing" molecules softens and increases the flexibility of the binder but also reduces its strength. The plasticizer effectively reduces the T_g of the polymer.

Common water-soluble binders are hygroscopic, as is indicated for polyvinyl alcohol in Fig. 12.2. Adsorbed water acts as a plasticizer, and polyvinyl alcohol plasticized by the adsorbed water has a lower Young's modulus of elasticity and strength and a greater elongation at rupture, as is indicated in Table 12.1. When stored at 50% relative humidity, the partially hydrolyzed polyvinyl alcohol, which contains about 12% acetate groups and is less

*J. P. Polinger and G. L. Messing, in *Advances in Materials Characterization II*, R. L. Snyder et al. (eds.), Plenum, New York, 1985.

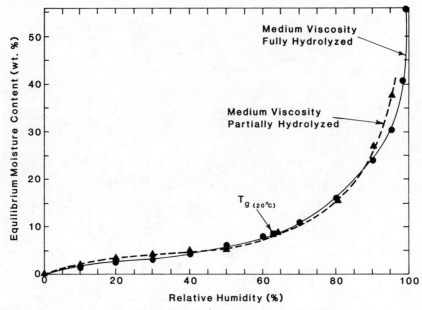

Fig. 12.2 Hygroscopic behavior of partially and fully hydrolyzed polyvinyl alcohols and the relative humidity producing a T_g of 20°C in fully hydrolyzed material. (Courtesy Du Pont Inc., Wilmington, DE.)

Table 12.1 Effect of Storage Humidity on Properties (20°C) of Polyvinyl Alcohol[a]

	Relative Humidity (%)		
Property	40	60	80
Tensile strength (MPa)	90	55	40
Elongation at rupture (%)	10	180	100
Young's modulus (MPa)	720	345	230

[a]Fully hydrolyzed, medium-low-viscosity grade. (Data courtesy of du Pont Inc., Wilmington, DE.)

densely packed, has a lower strength and elastic modulus, as indicated in Table 12.2. At each moisture content, T_g is 3–10°C lower for the partially hydrolyzed material. The glass transition temperature is higher for binders with stiff side groups that resist the movement of binder chains and for binders with more polar side groups that bond more strongly together. The glass transition temperature is also generally higher for a binder of higher molecular weight and when the amount of cross-linking between molecules is greater.

A liquid with low vapor pressure at the molding temperature is usually selected as a plasticizer and may be used in combination with water or a

Table 12.2 Typical Properties of Binder Films at 50% RH and 25°C

Property	Polyvinyl Alcohol[a] (Fully Hydrolyzed)	Polyvinyl Alcohol[b] (Partially Hydrolyzed)	Methyl Cellulose[c]
Density (Mg/m^3)		1.2–1.3	1.39
Tensile strength (MPa)	60–85	50–60	60–80
Elongation (%)	150–190	140–190	10–15
Glass transition temperature (°C)	31	23	
Young's modulus (MPa)	430–500	270–370	
Thermal expansion coefficient (cm/cm/°C)		1×10^{-4}	

[a]Du Pont Inc., Wilmington, DE.
[b]Monsanto Inc., St. Louis, MO.
[c]Dow Chemical Co., Midland, MI.

Table 12.3 Common Plasticizers

Plasticizer	Melting Point (°C)	Boiling Point (°C)	MW (g/mol)
Water	0	100	18
Ethylene glycol	−13	197	62
Diethylene glycol	−8	245	106
Triethylene glycol	−7	288	150
Tetraethylene glycol	−5	327	194
Polyethylene glycol	−10	>330	300
Glycerol	18	290	92
Dibutyl phthalate		340	278
Dimethyl phthalate	1	284	194

nonaqueous liquid. Most plasticizers also increase the hygroscopicity of the binder system, causing a greater concentration of adsorbed moisture at a particular storage humidity. An organic plasticizer may decrease the sensitivity of the plasticiszing action to relative humidity.

Liquids used to plasticize common binders are listed in Table 12.3. Ethylene glycol is very effective in lowering the glass transition temperature and is also relatively inexpensive. The plasticizing action of ethylene glycol decreases as its molecular weight increases. A plasticizer with a relatively high boiling point, such as polymerized ethylene glycol or glycerol, is important when the system is exposed to elevated temperatures, such as in spray-dying. The effect of glycerol and adsorbed water for plasticizing polyvinyl alcohol is shown in Fig. 12.3.

Plasticized viscoelastic binder films have a lower tensile strength but a greater impact strength, as is indicated for hydroxyethyl cellulose in Table 12.4. The stability of the properties of the binder film with time depends largely on the retention of the plasticizer. Viscoelastic binder films show

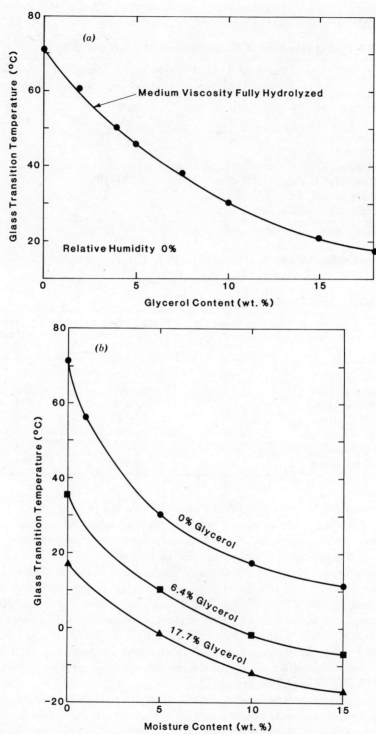

Fig. 12.3 Dependence of glass transition temperature of polyvinyl alcohol on glycerol content for (a) no adsorbed moisture and (b) adsorbed moisture. (Courtesy Du Pont Inc., Wilmington, DE.)

Table 12.4 Impact Strength of Hydroxyethyl Cellulose (HEC) Films (20°C)[a]

Composition	10% RH	50% RH
	Impact Resistance ($10^3 J/m^2$)	
HEC (low-MW)	15	27
80% HEC, 20% diethylene glycol	46	99

[a]Courtesy of Union Carbide Corp., New York.

elastic springback when the stress-producing deformation is removed. However, when the molding temperature is considerably above the glass transition temperature, the binder may exhibit viscous behavior with essentially no elastic springback.

Plasticizers for thermoplastic polymers such as polyethylene and polystyrene include oils and waxes that are molten at the molding temperature. A solvent type of plasticizer is retained more and resists exudation during deformation. Additives such as stearic and oleic acid are plasticizers for waxes.

Many of the general effects of plasticizers apply to colloidal particle binders as well as molecular binders. Glycerine and ethylene glycol may be added to plasticize clay bodies and to extend the plastic behavior below the freezing point of water in the system.

12.2 FOAMING AND ANTIFOAMING AGENTS

A foam is air or some other gas enclosed with a thin film of liquid. Gas introduced into a pure liquid does not cause a foam, because the liquid drains from the lamellae when bubbles make contact, and the bubbles rupture. A foaming agent should reduce the surface tension of the foaming solution, increase the film elasticity, and also prevent localized thinning. An aqueous foaming additive may contain a surfactant that makes the particles hydrophobic and reduces the surface tension, and a stabilizing component. Tall oil or sodium alkyl sulfate and polypropylene glycol ether are effective aqueous foaming agents. Foamed systems are used in fabricating lightweight concrete and in the beneficiation of some minerals.

When it is desirable to eliminate bubbles from a suspension, an additive called an antifoam, or defoaming aid, is used. Commercial antifoams are used in a wide variety of industries.

An effective antifoam is a surfactant of low surface tension. It must displace any foam stabilizer and have a positive spreading coefficient S_{AL} on the liquid or existing film, i.e.

$$S_{AL} = \gamma_L - (\gamma_A + \gamma_{AL}) \tag{12.1}$$

where γ_L, γ_A, and γ_{AL} are the surface tensions of the liquid, antifoam, and antifoam-liquid interface, respectively.

Commercial aqueous defoaming surfactants include fluorocarbons, dimethylsilicones, higher-molecular-weight alcohols and glycols, and calcium and aluminum stearate. Tributyl phosphate, which reduces surface viscosity, may aid the antifoam. Many polyelectrolyte deflocculents have no significant effect on the surface energy of water, but some binders and lubricants have a moderate effect. A change in temperature, which decreases the solubility of the antifoam surfactant, may increase its effectiveness.

12.3 LUBRICANTS

An interfacial phase that reduces the resistance to sliding is an effective lubricant. Fluid lubrication is provided by a thick film of a low-viscosity liquid such as water or oil, but these may migrate rapidly from an interface under compressive stress. Boundary lubrication is provided by an adsorbed film of high lubricity which improves the surface smoothness and minimizes adhesion between surfaces.

A plasticized binder film may act as a boundary lubricant between ceramic particles, and a gelled binder solution or suspension of colloidal particles may act as a fluid lubricant. Effective boundary lubricants must have a high adhesion strength but a low shear strength. Common boundary lubricants are listed in Table 12.5. Surfactants such as steric acid $CH_3(CH_2)_{16}COOH$ or its salts are particularly effective lubricants, because the carboxyl end of the molecule may be strongly bonded to an oxide surface, and the shear resistance between the first oriented adsorbed layer and successive layers is low. Molecular boundary lubricants are of a higher molecular weight than plasticizers, and they are more effective below their melting point.

Solid lubricants are fine particles with a laminar structure and smooth surfaces. Solid lubricants are effective on rough surfaces and are particularly effective at high pressure. Solid lubricants are sometimes mixed with molecular boundary lubricants that must be used at high temperatures.

Table 12.5 Common Lubricants

Paraffin wax	Steric acid
Aluminum stearate	Zinc stearate
Butyl stearate	Oleic acid
Lithium stearate	Polyglycols
Magnesium stearate	Talc (platey form)
Sodium stearate	Graphite, boron nitride

12.4 PRESERVATIVES

Binders not derived from polysaccharides are biologically inert. Natural organic binders and some cellulose derivatives are subject to biological degradation due to the enzymes produced by bacteria and fungi present in an industrial environment. This degradation usually causes a decrease in the viscosity of the binder solution.

Biological activity is minimized by using sterile materials and a closed environment or by adding a chemical preservative. Many preservatives are toxic, and their use and disposal must be carefully controlled.

SUMMARY

Binders used in ceramic processing must be plasticized to produce a moldable composition. A water-soluble binder is plasticized by adsorbed moisture, and the relative humidity must be controlled to control the plasticizing effect. An organic liquid with a lower vapor pressure than water is commonly used as the primary plasticizer. The plasticized binder is of lower strength but is more deformable and resistant to failure on impact. The strength of the binder may increase when the plasticizer is lost during drying. Special surfactants may be added to promote foaming or to reduce the stability of bubbles and eliminate a surface foam. A lubricant is a special surfactant added to reduce the friction between ceramic particles and the surface of a die used in forming. A molecular boundary lubricant must be capable of being strongly adsorbed but of such a molecular size and structure that a smoother, low-friction surface is produced. A chemical preservative is used to reduce enzymatic activity in binder systems.

SUGGESTED READING

1. Errol G. Kelly and David J. Spottiswood, *Introduction to Mineral Processing*, Wiley-Interscience, New York, 1982.
2. Milton J. Rosen, *Surfactant and Interfacial Phenomena*, John Wiley, New York, 1978.
3. L. E. Nielsen, *Mechanical Propertis of Polymers and Composites*, Marcel Dekker, New York, 1974.
4. P. Meares, *Polymers: Structure and Bulk Properties*, Van Nostrand, New York, 1965.
5. Ted Morse, *Handbook of Organic Additives for Use in Ceramic Body Formulation*, Montana Energy and MHD Research and Development Institute, Butte, MT, 1979.

PROBLEMS

12.1 Draw a graph approximating the stress-strain behavior of polyvinyl alcohol at 40, 60, and 80% RH. Use data in Table 12.1.

12.2 For a film of polyvinyl alcohol of a low molecular weight, the elongation to rupture increases on increasing the relative humidity from 40 to 60% and then decreases when the relative humidity increases further. For a polyvinyl alcohol of medium of higher molecular weight, the elongation at rupture increases monotonically. Explain.

12.3 Describe the change in strength and elasticity when polyvinyl alcohol binder films with and without a polyethylene glycol plasticizer are dried, eliminating adsorbed water.

12.4 Does the order of addition of the different processing additives influence their adsorption and effectiveness? Consider the sequential addition (1) deflocculant, binder/lubricant; (2) binder/lubricant, deflocculant.

12.5 What sterate should be selected to minimize chemical contamination in a manganese zinc ferrite, alumina, magnesia, or a whiteware body?

12.6 Can a lubricant added into a binder act as a plasticizer? Explain.

12.7 A large bubble is in contact with a small bubble in a foam. Which bubble will grow with time?

PART V

PARTICLE MECHANICS AND RHEOLOGY

The mechanical behavior of a system for processing is very dependent on the particle characteristics and the amount, distribution, and properties of the admixed liquid and processing additives. The selection and performance of machinery for processing and the microstructural changes effected are also very dependent on the structure and flow behavior of the processed system. Parameters used to describe the packing of particle systems are discussed in Chapter 13. We will examine the flow and deformation behavior of cohesive powders and plastic ceramic materials using principles of soil mechanics in Chapter 14. Principles of rheology and the flow behavior of slurries and pastes are discussed in Chapter 15.

13

Particle Packing Characteristics

Processing systems with particular distributions of particle sizes and shapes are produced by selecting and blending raw materials with different initial characteristics and by subsequent crushing, grinding, dispersion, classification, and granulation operations. The geometrical characteristics of the particle system have a significant impact on the particle arrangements and the packing density, size and shape of pore interstices, resistance to permeation by a fluid, bulk flow and deformation behavior, drying behavior, and microstructure development during firing.

Powders for high-density, fine-grained electronic and technical ceramics are usually submicron or finer than a few microns and approximately log-normal in size distribution to avoid exaggerated grain growth in sintering. Systems for whitewares and some technical products sintered in the presence of a liquid phase contain particles ranging from a maximum size of 10 to 100 μm down to a submicron size and have a distribution that packs relatively densely. Conventional dense refractory shapes and concrete contain particles ranging from about 1 cm down to a submicron size and are formulated using discrete sizes to achieve a high packing density. In refractory insulation where a high volume of fine pores is required, flocculated powders mixed with refractory fibers are commonly used.

Principles of particle characterization were discussed in Section III. In this chapter we will examine models for the packing of particles and experimental results for both coarse and fine systems. The packing of particles varying in size and shape, as is common in most ceramic systems, can be very complex; many different arrangements are possible, and several parameters must be defined to consider the packing and porosity. We will see that common working models are based on spherical particles and serve best to approximate the average packing density and pore size and to suggest changes in these when the distribution of sizes is changed. Particle and binder parameters hindering particle motion and packing are also briefly discussed.

185

13.1 CHARACTERISTICS OF PACKINGS OF UNIFORM SPHERES

The packing of nonporous spheres of uniform size is studied in beginning courses on the structure of solids. Those results provide an approximation of the geometrical characteristics of packed, monosize ceramic particles and provide reference values and a basis for comparing experimental results for real-particle systems.

Uniform spheres may pack in five different ordered arrangements, as shown in Fig. 13.1. The fraction and percent of the bulk volume occupied by the spheres are referred to as the packing fraction (PF) and the packing density (PF(%)), respectively. As seen in Table 13.1, the packing density ranges from 52% for simple cubic packing to 74% for tetrahedral and pyramidal close packing and is independent of the size of the spheres. The packing density of nonordered arrangements of uniform spheres larger than 1 mm has been determined experimentally to be about 60% without vibrations and 64% with vibrations.[*]

The interstitial pore fraction (ϕ), or interstitial porosity ($\phi(\%)$), indicates the portion of the bulk volume present as interstices among the nonporous spheres. The pore size is a function of both the size of the

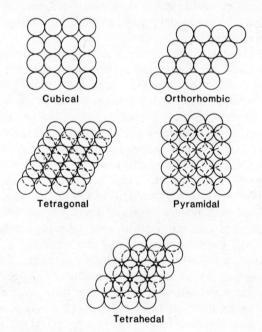

Fig. 13.1 Ordered packing arrangements of uniform spheres.

[*]G. D. Scott and D. M. Kilgour, *Br. J. Appl. Phys. Ser. 2* **2**, 863 (1969).

Table 13.1 Packing Density of Regular Configurations of Uniform Spheres

Configuration	Coordination Number	Packing Density (%)
Cubic	6	52.4
Orthorhombic	8	60.5
Tetragonal	10	69.8
Tetrahedral	12	74.0
Pyramidal	12	74.0

spheres and the packing arrangement. In a tetrahedral packing, the cross-sectional area A of the interstices ranges from $0.04a^2$ to $0.21a^2$, and for simple cubic packing $A = 0.21a^2$, as indicated in Table 13.2. For dense packings, the cross-sectional area of the interstices is a fraction of the cross-sectional area of a sphere, and the area decreases as the diameter of the spheres decreases.

The center line of a continuous pore through the packing may be tortuous, and the length of the center line relative to the edge length of the unit cell is the tortuosity (T_0). For a cubic packing, $T_0 = 1$, and in a tetrahedral packing, $T_0 = 1.3$; displacement of the particles into a less dense packing would reduce the mean tortuosity.

The number of particles per unit bulk volume (N_p) in a system of packed spheres of uniform size is

$$N_p = \frac{6(PF)}{\pi a^3} \tag{13.1}$$

The number of particle contacts per unit bulk volume (N_c) is the product of N_p and-half the coordination number (CN), i.e.

Table 13.2 Pore Size in Cubic and Tetrahedral Packings of Uniform Spheres

Parameter	Cubic	Tetrahedral
Entry pore area	$0.21a^2$	$0.04a^2$
Entry pore area$/\pi a^2/4$	0.26	0.05
Entry pore diameter$/a$	0.51	0.22
Entry sphere diameter$/a$	0.42	0.15
Void fraction	0.48	0.26
Vol. void/vol spheres	0.92	0.34
Radius ratio primary sphere/interstitial sphere	1.37	4.44

Sphere diameter $= a$.

$$N_c = \frac{3(PF)(CN)}{\pi a^3} \tag{13.2}$$

For regular packing arrangements, N_c is very dependent on the packing geometry.

For nonregular packing arrangements, the number of contacts is a function of the arrangement of local groups of spheres and the porosity. The coordination number (CN) is observed to range between 6 and 10; an empirical approximation of the average coordination number is $\overline{CN} = \pi/\phi$.[†] For random packing, the average number of contacts $\bar{N}_c$ is approximated by the equation

$$\bar{N}_c = \frac{3(1-\phi)}{\phi a^3} \tag{13.3}$$

McGeary studied the packing of coarse ($>37 \mu$m), monosize spherical particles experimentally using axial vibrations, and observed packing mostly in an orthorombic arrangement and a packing density of 62.5%. This arrangement appeared to be stabilized by slight lateral spreading and nesting of spheres in the vibration direction. Hindrance of the container wall was insignificant when the container diameter/particle diameter exceeded 10.

The packing density of nearly monosize spherical particles of silica and alumina of colloidal size packed by filter pressing a deflocculated slurry is about 60–65%. More densely packed domains in which the particles are ordered are separated by interdomain boundary regions containing a less ordered arrangement and larger pores.

13.2 PACKING IN INTERSTICES AMONG COARSER PARTICLES

Smaller particles introduced and distributed in the interstices of packed larger particles will reduce the porosity and pore size. Large particles added to finer particles displace fines and pores and reduce the porosity. This model of packing is called the Furnas model[*] and is shown in Fig. 13.2.

When the large particles in a packing are in contact, the theoretical maximum packing fraction PF_{max} for a mixture of coarse, medium, and fine particles is

$$PF_{max} = PF_c + (1 - PF_c)PF_m + (1 - PF_c)(1 - PF_m)PF_f \tag{13.4}$$

where PF_c, PF_m, and PF_f are the packing factors of the coarse, medium, and

[†]W. O. Smith et al., *Phys. Rev.* **34**, 1272 (1929).
[*]C. C. Furnas, *U.S. Bur. Mines Rep. Invest.* **2894** (1928).

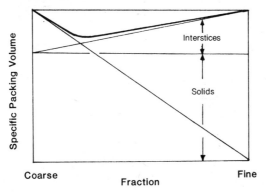

Fig. 13.2 Theoretical variation of packing volume for the packing of smaller particles among coarser particles. The straight lines correspond to ideal packing when the size ratio is infinite and the heavy line indicates typical behavior.

fine particles, respectively. The weight fraction f_i^w of each size at PF_{max} is

$$f_i^w = \frac{W_i}{W_{total}} \tag{13.5}$$

where W_i is the weight of each size and $W_{total} = \Sigma W_i$. For a three-component system the weight of coarse (W_c), medium (W_m), and fine (W_f) particles is

$$W_c = PF_c D_c \tag{13.6}$$

$$W_m = (1 - PF_c)PF_m D_m \tag{13.7}$$

$$W_f = (1 - PF_c)(1 - PF_m)PF_f D_f \tag{13.8}$$

For this model, the volume fraction of particles required decreases with particle size. In practice, the maximum packing fraction is achieved when the ratio between nearest sizes is greater than about 7 and the finer particles are dispersed uniformly. Finer particles must be small enough to enter into all regions of the interstices, as indicated in Fig. 13.3. Following this packing scheme, McGeary experimentally achieved a packing density of 95% for a quaternary system of vibrated steel spheres with the sizes indicated in Table 13.3.

The blending of discrete sizes to achieve a high packing density is common practice in formulating coarse-grained refractories and concrete that have very little dimensional change of firing. Refractory batches for "common heavy refractories" composed of particles of equal density contain approximately 60–65 wt % coarse particles about equally divided between 0.7–0.17 cm and 0.17–0.06 cm, and 35–40 wt % fines <74 μm of which

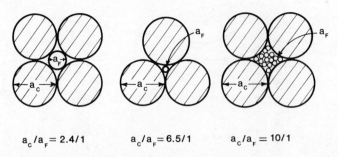

$$a_c/a_F = 2.4/1 \qquad a_c/a_F = 6.5/1 \qquad a_c/a_F = 10/1$$

Fig. 13.3 Packing of fine spheres in a planar interstice among coarse particles.

Table 13.3 Packing Density of Mixed Spheres of Different Size

Diameter (cm) (weight fraction of spheres)				Packing Density (%)	
1.28	0.155	0.028	0.004	Calculated	Experimental
1.000	—	—	—	60.5	58.0
0.726	0.274	—	—	84.8	80.0
0.647	0.244	0.109	—	95.2	89.8
0.607	0.230	0.102	0.061	97.5	95.1

Source: R.K. McGeary, *J. Am. Ceram. Soc.* **44**(10), 513–522 (1961).
Size ratio 320/39/7/1.

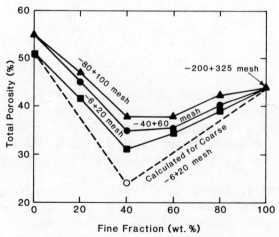

Fig. 13.4 Calculated and experimental total porosity for vibrated, two-component mixtures of tabular alumina fines (−200 + 325 mesh) and three different coarse fractions. (*Note*: particles are porous.)

about one-half are finer than 44 μm; the packing density achieved on vibrating dry particles is about 65–70% (Fig. 13.4) and on pressing this commonly increases 5–10%. The packing density increases as the size ratio increases. A smaller maximum particle size is used for some refractory products such as kiln furniture, where the flexural strength must be higher. When using particles of different density, the proportions by weight for maximum packing are different from the proportions stated above. The particle size distribution of castable refractories and concrete is similar to that for heavy refractories. Hydraulic cement composes a portion of the fines. The liquid requirement for working and the cured strength depend directly on the particle packing density.

In the packing model described above, coarse particles are assumed to be in contact; this type of microstructure is important in refractories where dimensional stability under a compressive load is requisite. Other compositions of coarse, medium, and fine sizes with the coarse particles dispersed can produce a relative maximum in packing density; these systems have received relatively little attention.

13.3 PACKING OF CONTINUOUS SIZE DISTRIBUTIONS

Most ceramics are produced from a material having a continuous distribution of particle sizes between some maximum and a finite minimum size. Calcined and milled calcined powders for fine-grained technical ceramics are often log-normal to a good approximation. Typical values of the geometric standard deviation σ_g are in the range 1.4 to 2.4. The calculated minimum pore fraction for the random packing of a log-normal distribution of spheres increases as the geometric standard deviation of sizes increases, but for an increase in σ_g from 1.4 to 2.4, the maximum packing density for spherical particles is calculated to increase from 65% to only 69%. The maximum packing fraction of four milled aluminas with a log-normal distribution of sizes, determined by mechanically pressing (deflocculated) filter cake, are listed in Table 13.4. The experimental maximum packing fraction is slightly higher for the powders of larger geometric standard deviation, as predicted. The packing density values are lower than is calculated for an LN distribu-

Table 13.4 Packing of Milled Calcined Aluminas Having a Log-Normal Size Distribution

Characteristic	Alumina Type			
$\bar{a}_{g_M}(\mu m)$	5.9	8.0	0.8	1.3
σ_g	1.7	1.8	2.2	2.5
$PF(\%)$	55	62	64	66

tion of spheres, because the larger particles in the coarse calcined aluminas are porous aggregates with a nonspherical shape.

One calcined alumina shown in Fig. 13.5 that is not log-normal in size distribution has a higher maximum packing fraction. The packed density depends significantly on the form of the size distribution as well as the range of particle size. In ceramic whitewares, the clay, quartz, and feldspar minerals, each having a continuous size distribution, are blended in proportions that produce a continuous distribution that packs more densely. The size distribution of the blend is approximated by the Andreasen* equation for dense packing

$$F_{\mathrm{M}}(a) = \left[\frac{a}{a_{\max}} \right]^n \tag{13.9}$$

and

$$f_{\mathrm{M}}(a) = \frac{n}{a} \left[\frac{a}{a_{\max}} \right]^n \tag{13.10}$$

where $a_{\max}$ is the maximum particle size and $1/n$ is the distribution modulus.

In this model, the amount of a particular size is a constant fraction of the total amount finer than that size; i.e., the proportion of fines increases as the

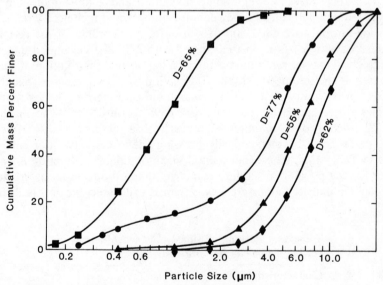

Fig. 13.5 Particle size distribution and maximum packing density of several calcined Bayer process aluminas.

*A. H. M. Andreasen and J. Anderson, *Z. Kolloid* **50**, 217–228 (1930).

particle size decreases, as indicated in Fig. 13.6. Andreasen recognized that for a particular a_{max}, the porosity of the packed particles would decrease as n decreased; his experiments indicated that a practical range of n was 0.33 to 0.50. Equation 13.9 with n equal to 0.33 and 0.50 and size data for an electrical porcelain body that has a maximum packing density of 75% are

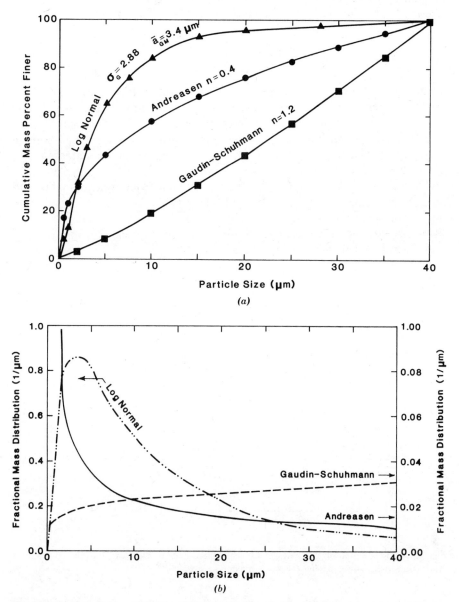

Fig. 13.6 (a) Cumulative and (b) fractional particle size distributions of log-normal, Andreasen, and Gaudin-Schuhmann materials having the same maximum size.

shown in Fig. 13.7. The experimental data for sizes larger than about 1 μm fall within the Andreasen bounds, but at smaller sizes the distribution becomes curved.

Andreasen assumed that the smallest particles would be infinitesimally small. Dinger and Funk* recognized that the finest particles in real materials are finite in size and rederived the Andreasen equation using a minimum size a_{min}. The Dinger-Funk equation for dense packing is

$$F_M(a) = \frac{a^n - a_{min}^n}{a_{max}^n - a_{min}^n} \tag{13.11}$$

In Fig. 13.7, the curvature of the size data is of the form predicted by Eq. 13.11. In general, for size distributions approximated by Eqs. 13.10 and 13.11, the packing density may increase as $1/n$ and the range of sizes increase, and the packing density may exceed 80%. The calcined alumina in Fig. 13.5 with a packing density of 77% is roughly approximated by Eq. 13.11, but with $n > 0.5$.

Fractional size distributions for the log-normal, Gaudin-Schuhmann, and Andreasen equations ($a_{max} = 40$ μm) are shown in Fig. 13.6. For the log-normal distribution, the fractional distribution has a relative maximum in contrast to the Andreasen distribution, where the relative proportion of mass increases with decreasing size.

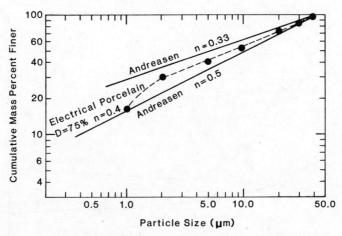

Fig. 13.7 Log-log plot of the particle size distribution of an electrical porcelain body and Andreasen distributions having the same maximum size.

*D. R. Dinger and J. E. Funk, *Particle Packing: Review of Packing Theories*, Fine Particle Society, 13th Annual Meeting, Sp. 1982, Chicago.

13.4 HINDERED PACKING

In real-particle systems, the movement of particles into the minimum porosity configuration during processing is hindered by both external and internal factors. Bridging of particles and agglomerates with rough surfaces between the walls of tooling is quite significant when the spacing between external surfaces is less than about 10 particle diameters. This is especially a problem when the surfaces of particles and aggregates are rough (Fig. 3.2) and when friction is high at the interface between the particles and the tooling. Mechanical vibration may momentarily reduce bridging forces and facilitate the rolling and sliding of dry particles into a denser configuration (Fig. 13.8). Lubrication of all surfaces may aid in achieving a higher packing density. However, mechanical forces causing fracture of some particles may be required to approach the minimal porosity configuration. The vibrated packing density of coarse angular particles of uniform size is 50–60%, in contrast to the 65% density for smooth spheres.

Coagulation, flocculation, and adhesion forces, which retard particle motion, may also hinder packing (Fig. 13.9). Random arrangements of anisometric particles typically have a higher porosity and a wider range of pore sizes than for an ordered arrangement, as is shown in Fig. 13.10. A higher porosity is generally observed when the aspect ratio of the particles is higher and when a higher content of fibers is added into a powder containing isometric particles (Fig. 1.2).

Adsorbed binder molecules and colloidal particles may hinder particle movement and reduce the packing density a few percent when the hindrance is not overcome by applied stress (Fig. 13.11). The steric hindrance in-

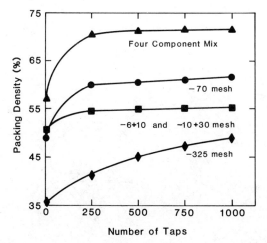

Fig. 13.8 Dependence of the packing density of different size fractions of alumina on the number of low-frequency vibration cycles.

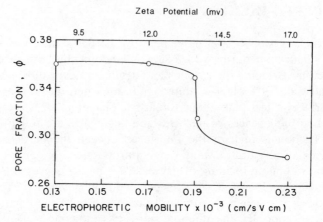

Fig. 13.9 Dependence of pore fraction of an alumina body formed by pressure-filtering alumina slurries on deflocculation using ammonium polyacrylate.

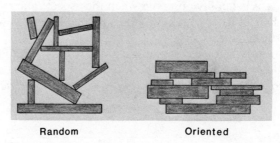

Random   Oriented

Fig. 13.10 Pore structures in random and ordered arrangements of anisometric particles.

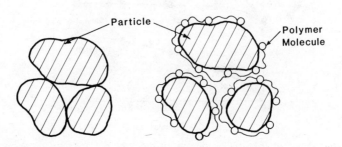

Fig. 13.11 Bulky binder molecules reduce the packing density at low effective stress.

196

creases as the molecular weight of the binder increases and when the molecules bridge particles. Fines mixed with a sticky binder may coat larger particles rather than entering interstices. The packing density in an undispersed agglomerate may be higher than in an agglomerate of dispersed particles with a film of binder and liquid; these packing inhomogeneities tend to be more pronounced when the particles are of colloidal size and when binder concentrations are higher.

SUMMARY

The distribution of particle sizes in a processing system has a significant effect on the particle packing and pore structure and the behavior during forming, drying, and firing. Models for the packing of monosize particles indicate that the packing density increases as the coordination number increases and that the mean pore size decreases as the particle size and porosity decrease. The packing density of about 62% for a monosize system can be increased above 75% by adding a specific proportion of a finer size that packs efficiently in the interstices among the coarse; this type of sizing is used to fabricate relatively dense, coarse-grained systems. Continuous particle size distributions of the log-normal type used for ceramics have a maximum packing density of about 65%; a continuous distribution of the Andreasen type containing a greater amount of finer particles may pack to a density exceeding 80%. The packing density of a continuous distribution of sizes varies with the form and breadth of the distribution.

Rough surfaces and adsorbed molecules which prevent particle motion reduce the packing density. In packings of platelike and fibrous particles, the packing density and pore structure are very dependent on the particle orientation and arrangements. Agglomeration and the addition of fibrous particles in a system of isometric particles tend to produce inhomogeneous porosity and a higher porosity.

SUGGESTED READING

1. N. G. Stanley-Wood, *Enlargement and Compaction of Particulate Solids*, Butterworth, London, 1983.
2. P. J. Sherrington and R. Oliver, *Granulation*, Heyden, Philadelphia, 1981.
3. A. N. Patankar and G. Mandel, The Packing of Solid Particles, *Trans. Indian Ceram. Soc.* **39**(4), 109–117 (1980).
4. G. P. Bierwagen and T. E. Saunders, Studies of the Effects of Particle Size Distribution on the Packing Efficiency of Particles, *Powder Tech.* **10**, 111–119 (1974).
5. H. Y. Sohn and C. Moreland, The Effect of Particle Size Distribution on Packing Density, *Can. J. Chem. Eng.* **46**, 162–167 (1968).

6. R. K. McGeary, Mechanical Packing of Spherical Particles, *J. Am. Ceram. Soc.* **44**(10), 513–522 (1961).

7. A. E. R. Westman and H. R. Hugill, The Packing of Particles, *J. Am. Ceram. Soc.* **13**(10), 767–779 (1930).

PROBLEMS

13.1 Calculate the void volume and entry pore diameter in packed monosize spheres of 5 μm diameter and compare your answers for a similar packing of 0.5 μm spheres. Assume orthorhombic packing.

13.2 Estimate the concentration of contacts for particles of 100 μm in orthorhombic packing and compare this to the value for particles of 1 μm in the same configuration.

13.3 Explain why one line in Fig. 13.2 extrapolatess to zero volume whereas the other line does not.

13.4 The model for Eq. 13.4 assumes that fines enter interstices between coarse and medium particles in a three-component mix. Can an equivalent maximum packing density be achieved by replacing some of the medium size with fines? Explain.

13.5 The bulk density after mechanical vibration (tap density) is 50% for particles of $-6 + 10$ mesh and $-10 + 30$ mesh and 62% for particles of -70 mesh. calculate the theoretical PF_{max} and the weight fraction of each size, assuming all particles are of equal density.

13.6 Estimate the theoretical packing density for the particles in problem 13.5 when the fines are increased by 15% above that required to achieve PF_{max}.

13.7 Calculate the weight fractions of particles in problem 13.5 when the coarse and medium particles have a density of 3.65 Mg/m^3 and the density of the fines is 3.80 Mg/m^3.

13.8 Calculate the theoretical PF_{max} and weight percent composition for a body consisting of coarse α silicon carbide particles and α alumina fines. Assume PF for orthorhombic packing for components.

13.9 Spray-dried granules may be nearly spherical. Estimate the porosity of monosize spray-dried alumina agglomerates of α-alumina when the vibrated bulk density is 32%.

13.10 Estimate the average coordination number $\overline{CN}$ for randomly packed particles when $\phi = 0.4$.

13.11 For random packing and $\phi = 0.4$, what is the increase in the number of particles (N_p) when 10% of a material with a size of 0.1 μm is added to a powder with a size of 1 μm?

13.12 Graph and compare the Andreasen and Dinger-Funk distributions for $a_{max} = 200$ μm, $a_{min} = 0.5$ μm, and $n = 0.4$.

13.13 Compare the matrix phase in the following two-component systems when the proportions correspond to PF_{max}: ideal Furnas model; dry fines are mixed with dry coarse and vibrated; dry fines mixed with dry binder are mixed with wetted coarse; and ideal Andreasen model.

13.14 Explain why the maximum tap density is achieved more rapidly for the -70 mesh material than for the -325 mesh material in Fig. 13.8.

13.15 Explain the cause of the different pore fractions in Fig. 13.9.

13.16 The difference between the calculated and experimental PF_{max}, ΔPF_{max}, was observed to vary with the ratio of mean size of coarse and fine angular refractory particles according to the equation $\Delta PF(\%) = 64 a_f / a_c$. Estimate the experimental PF_{max} in problem 13.5.

14

Consistency, Particle Mechanics, and Deformation Behavior

The consistency of a ceramic system is commonly described using a term such as bulky powder, an agglomerated powder, granular material, cohesive plastic body, thick sticky paste, thick slurry, pourable suspension, etc. These descriptions indicate something about the nominal particle sizes, the apparent liquid content, and the general mechanical behavior. On a microscale, the consistency of a particular material is a function of the amount, distribution, and properties of the liquid phase; the size range and packing of particles; and the interparticle forces of attraction and repulsion. A small change in one or more of these parameters (due to the effect of processing additives or mechanical agitation) may sometimes change the consistency rather significantly.

Agglomerates may form spontaneously in a ceramic powder owing to interparticle attraction caused by Van der Waals, electrostatic, or magnetic forces. The capillary effect of a wetting liquid in the region of particle contacts and adsorbed flocculating additives produce cohesive strength. Ceramic bodies containing a pore phase incompletely saturated with a liquid may compress when deformed. Volume dilation may precede the deformation of the close-packed bodies containing large particles. For these materials the resistance to deformation is dependent on the cohesive strength and the frictional resistance between particles which depends on the interparticle stress.

In this chapter we will consider the general effect of a low-viscosity liquid on apparent consistency. The very complex topic of the deformation and flow behavior of unsaturated systems is then examined using principles from the field of soil mechanics to provide theoretical insights.

14.1 CHANGE IN CONSISTENCY ON ADMIXING A LIQUID

When a wetting liquid of low viscosity is sprayed into a powder and the torque required for stirring is measured, the behavior is of the general form shown in Fig. 14.1. Particles with wet surfaces initially agglomerate into more densely packed agglomerates. Dry powder containing dispersed agglomerates is easily stirred. Further spraying and stirring produces additional agglomerates in the dry powder, and the apparent resistance to flow remains low. But eventually, at a point where very little dry powder remains, a small amount of liquid causes total agglomeration into a rather nonsticky, cohesive mass.

The formation of a cohesive mass causes a very sharp increase in the resistance to shearing. The cohesive mass may be rather plastic if the system is flocculated and contains a significant fraction of particles of colloidal size. A granular material may be poorly cohesive unless considerable binder has been added. This cohesive state may extend over a range of liquid content for which the fraction of pore volume filled with liquid, the degree of pore saturation (DPS), is about 0.9 to 1.0. A plastic state may extend over a 10 vol % range of liquid content when the additional liquid causes flocculated particles to rearrange into a less closely packed configuration.

Admixing additional liquid into a plastic body eventually produces a sticky body with only moderate resistance to shear, which is called a paste. The stiffness of the paste is very dependent on the particle packing, content of colloids, the liquid content, and the coagulation/flocculation forces and may vary with time. When the coagulated/flocculated structure becomes discontinuous, from adding a deflocculent or additional liquid, and the material cannot support its own weight, the system is called a suspension. A concentrated suspension that is relatively stable over time, (i.e., settling is hindered and the fluid properties are retained after standing for a considerable time) is called a slurry. The term slip is traditionally used for a slurry containing clay minerals.

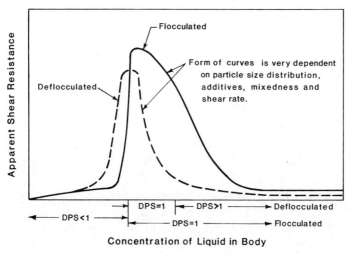

Fig. 14.1 Change in apparent shear resistance on admixing a wetting liquid into a powder.

Table 14.1 Nominal Characteristics[a] of Ceramic Systems

State	Gas	Liquid	Particle Structure
Agglomerate	C	D	C–D[b]
Plastic	D	C	D[b]
Paste	D/A	C	D[b]
Slurry	D/A	C	D[b]

[a]C, continuous; D, dispersed; A, absent.
[b]May be continuous in microscopic regions when poorly mixed; particles typically separated by an adsorbed film.

Nominal microscale characteristics of the bodies described above are listed in Table 14.1. Microscale characteristics may vary in a nonuniform material and in a paste or slurry that is not uniformly coagulated/flocculated.

Systems containing a high-viscosity liquid are less familiar. Terms such as rubbery, plastic, viscous, or asphaltic imply something about the consistency in a particular application but are lacking in a general context. The apparent behavior of materials containing a viscous liquid is relatively more dependent on the deformation rate and temperature.

14.2 CONCEPT OF EFFECTIVE STRESS

When an unsaturated particle system is compressed, the imposed pressure is supported by the mechanical stress at particle contacts and by the pore fluid pressure. The difference between the applied pressure P_a and the fluid pressure u is called the effective stress $\bar{\sigma}$, i.e.

$$\bar{\sigma} = P_a - u \tag{14.1}$$

The effective stress is sometimes called the interparticle stress but should not be confused with the local stress σ_c at contacts between particles. In general, the contact stress is a function of the effective stress, the pore pressure, and the net pressure due to repulsive and attractive electrical interactions $(R - A)$,

$$\sigma_c = [\bar{\sigma} - (R - A)] \frac{A_T}{A_c} + u \tag{14.2}$$

where A_c is the average area of interparticle contacts in the cross section of area A_T.

In an unsaturated system, the elastic strain, permanent strain, and yield strength of the body are dependent on the effective stress. The effective stress may vary with time or temperature when the pore pressure varies with time or temperature.

14.3 PARTICLE INTERACTIONS

Consider the uniaxial compression of a system of particles confined in a rigid die, as shown in Fig. 14.2. The forces at contacts between particles may be resolved into normal (N) and tangential (T) components. Particles under load are nonuniformly compressed, and particles that are fibers or platelets may bend. The translation and rotation of particles into a different configuration may decrease or sometimes increase the bulk volume and distort the

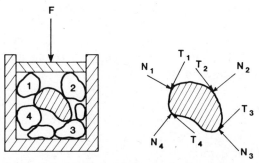

Fig. 14.2 Loading produces normal and shear forces at contacts between particles.

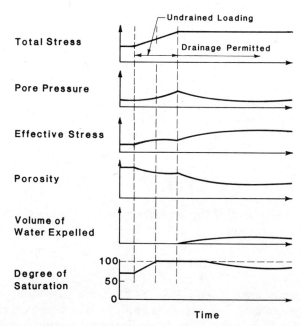

Fig. 14.3 Variation of stresses and porosity on loading a compressible, partially saturated particle system. (After T.W. Lambe and R.V. Whitman, *Soil Mechancis*, Wiley-Interscience, New York, 1969.)

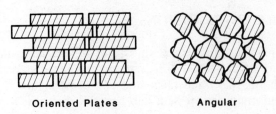

Oriented Plates   **Angular**

Fig. 14.4 The contact area between oriented platey particles is greater than between angular particles.

shape. A very intense stress at a particle contact may cause local plastic deformation and fracture.

Adsorbed surfactants and pore fluids also affect the compaction and flow of particle systems. Particle sliding is enhanced by an interparticle phase which reduces the coefficient of friction. The effective stress is lower when the pore phase is saturated with a relatively incompressible fluid and becomes pressurized, as is indicated in Fig. 14.3. The flow of viscous surfactants and pore fluids is time-dependent. Resistance to flow may depend on the loading rate when the effective stress ($\bar{\sigma}$) varies with time during the period of deformation.

The mean interparticle contact stress (σ_c) varies with the shape and orientation of the particles. For oriented platelets, the contact area ratio A_c/A_T is about 1, and σ_c is approximately equal to $\bar{\sigma}$. But as shown in Fig. 14.4, A_c/A_T may be much less than 1 in a packing of spherical or angular particles and σ_c may be much greater than $\bar{\sigma}$.

14.4 INTERPARTICLE FRICTION

Interparticle sliding is resisted by chemical adhesion and physical resistance produced by microscopic steps and asperites on particle surfaces (Fig. 3.2). The classical law of Amonton states that the resistive force (T_R) is proportional to the normal force (N) and independent of the contact area;

$$T_R = fN \tag{14.3}$$

where f is the coefficient of friction. Alternatively, the concept of an angle of friction ϕ is used where

$$\tan \phi = T_R/N = f \tag{14.4}$$

Sliding may occur when $T > T_R$ (Fig. 14.5). For a partcle in contact with several particles, sliding may not occur until $T > T_R$ at each contact or until translation or rolling reduces the shear force at a critical contact. Bulk powder flow involves the collective motion of several particles.

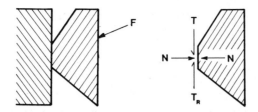

Fig. 14.15 Shear force T and friction resistance T_R at a contact.

The surface roughness of particles can be estimated from scanning electron micrographs of particles or for coarse particles by using a profilometer. In the profilometer technique, a stylus connected to a electromechanical transducer is dragged along the surface, and the center-line average CLA roughness is calculated from the mean amplitude of the fluctuations of the amplified electrical signal. The surfaces of most particles are rough on a microscopic scale, and the contact occurs at high points called asperites (Fig. 14.6). The contact stress on these asperites is much higher than the effective stress, even for small loads. Asperites of some rather brittle materials and ductile materials such as metals, alkali halides, and polymers may deform under load. The apparent contact area A_c is

$$A_c = \frac{N}{\sigma_y} \tag{14.5}$$

where σ_y is the yield strength of the deformed material for the imposed state of stress. The high-contact stress may cause chemical bonding between two very clean surfaces but less adhesion between unclean surfaces. If τ_c is the shear strength of the contact, then

$$T_R = \tau_c A_c = \frac{\tau_c N}{\sigma_y} \tag{14.6}$$

and

$$f = \tau_c / \sigma_y \tag{14.7}$$

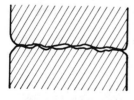

Granular Particles   **Powder Particles**

> 1 Interparticle   1 Interparticle
Contact   Contact

Fig. 14.6 Contacts between large and small particles.

For a contact between a hard and a soft material, friction is dominated by the mechanical properties of the softer material.

The static shear force at the onset of sliding is somewhat greater than the shear force required to maintain sliding. This difference may be due to a time dependence of the bonding in contact regions and the viscous flow of surface contaminants from the contact region. Rolling friction is much lower than kinetic sliding friction. During rolling, the bonding at contacts is broken in tension rather than shear. Rolling friction is generally quite independent of surface contamination. Sliding friction may sometimes be reduced by intense mechanical vibration.

Layer silicate minerals and pristine amorphous solids have relatively smooth surfaces with surface irregularities finer than about 100 nm. Crushed minerals have surfaces with steplike features coarser than a few microns, and aggregates have very rough surfaces with irregularities approaching the crystallite size. Equation 14.7 predicts that friction is independent of surface roughness. However, at a contact between brittle materials, the apparent friction angle is dependent on the surface roughness and the interspersion of asperites.

An adsorbed materal can reduce the coefficient of friction quite significantly if it is tenaciously bonded and resists extrusion under the high-contact stress. The friction coefficient of highly polished and chemically cleaned but wet surfaces of quartz and feldspar was shown* to be reduced from about 0.5 to 0.7 (26° to 35°) to about 0.1 to 0.2 (6° to 12°) for dry surfaces with a residual film of oil. But for surfaces with a CLA roughness of about 2 μm, the friction coefficient was about 0.5 and did not depend on surface cleaning.

The contact stress between oriented smooth sheet minerals is much lower than between angular particles. Surface-adsorbed contaminants and water are not completely extruded from the interface at an applied pressure below 500 MPa. The friction coefficient is observed to be about 0.3–0.4 (17–22°) for dry mica but only 0.15–0.2 (8.5–11.5°) for wet mica, because the water acts as a lubricant.

Results for clay systems suggest that the frictional behavior of oriented, smooth kaolinite platelets and clays containing a significant content of montmorillonite and illite is similar to that for oriented mica. The behavior of clays with rough surfaces (cleavage steps) or edge-face contact is probably more similar to that for angular particles.

14.5 LUBRICATION

Lubrication is produced by an interfacial fluid or an adsorbed boundary layer or low friction (Fig. 14.7). In fluid lubrication (sometimes termed

*T. William Lambe and Robert V. Whitman, *Soil Mechanics*, Wiley-Interscience, New York, 1969.

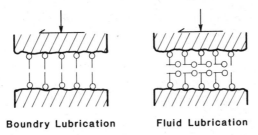

Boundry Lubrication   **Fluid Lubrication**

Fig. 14.7 Molecular orientation of a boundary lubricant and in fluid lubrication.

hydrodynamic lubrication), the interfacial lubricating film is thicker than irregularities on the surface of the particle and the friction is dependent on the viscosity of the fluid medium. Fluid lubrication is produced by liquids such as water, oil, aqueous emulsions, a colloidal suspension, or a binder solution. Parameters affecting fluid lubrication are indicated in Fig. 14.8. When displacement is slow and the load is high, fluid escapes, and a transition to boundary lubrication occurs. The coefficient of friction rises, because a low-viscosity fluid is not an effective boundary lubricant.

Oriented adsorbed polar compound of the type described in Chapter 12 are more efficient boundary lubricants. An adsorbed lubricant reduces the contact between surfaces, which reduces friction and wear. Boundary lubricants must be tenaciously adsorbed but of low shear resistance in the temperature range of interest. Evidence indicates that near the melting point of the lubricant, the long molecules may become disoriented, causing the friction to increase.

Finely divided solids with a laminar molecular structure can also act as

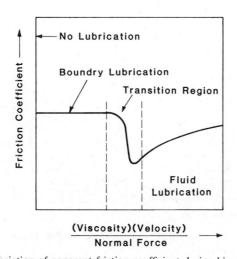

Fig. 14.8 Variation of apparent friction coefficient during kinetic loading.

lubricants between abrasive particles. Their platelike structure provides oriented smooth surfaces and low-friction coefficients. They can be admixed dry or in suspension form. Examples are talc, graphite, and graphitic boron nitride. Adsorbed water vapor or other gases may affect their lubricity. Solid lubricants resist breakdown under high-compressive stress and are good lubricants at high pressures and temperatures.

14.6 AVERAGE STRESSES IN A POWDER MASS

Ceramic systems are usually considered to be macroscopically homogeneous and isotropic for purposes of describing their deformation and flow. If the stress varies little over the dimension of the largest particle in the system, the average stress at some position is visualized as the sum of the forces between particles per unit area of the system at that position.

The analysis of plane stress and strain is discussed in undergraduate texts on the mechanics of materials. When the magnitudes and directions of normal stresses σ_x and σ_y on an element are known, as indicated in Fig. 14.9, the normal σ_θ and shear τ_θ stresses on a plane whose normal is at an angle θ are calculated from the equations

$$\sigma_\theta = \frac{\sigma_x + \sigma_y}{2} + \frac{\sigma_x - \sigma_y}{2} \cos 2\theta + \tau_{xy} \sin 2\theta \qquad (14.8)$$

and

$$\tau_\theta = \frac{-(\sigma_x - \sigma_y)}{2} \sin 2\theta + \tau_{xy} \cos 2\theta \qquad (14.9)$$

The average stress at a point is a function of the orientation of the plane used to define the stress. At any point, there are three orthogonal planes called the principal planes of stress on which the maximum and minimum normal stresses occur and for which the shearing stress is zero, as is shown

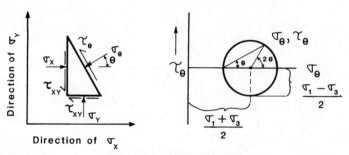

Fig. 14.9 Stress element and a general state of stress used to derive equations for the Mohr circle of stress.

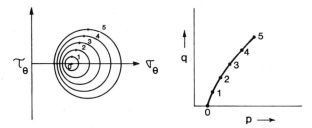

Fig. 14.10 Stress path diagrams.

in Fig. 14.9. The largest normal stress is termed the major principal stress (σ_1) and the smallest the minor principal stress (σ_3). The quantity $(\sigma_1 - \sigma_3)$ is called the deviator stress.

Squaring and combining Eqs. 14.8 and 14.9 yields

$$(\sigma_\theta - a)^2 + \tau_\theta^2 = b^2 \tag{14.10}$$

where

$$a = \frac{(\sigma_1 + \sigma_3)}{2} \text{ and } b = \frac{\sigma_1 - \sigma_3}{2}$$

Equation 14.10 is the familiar equation of a circle of radius b centered at $(+a, o)$ as shown in Fig. 14.9 and is known as the Mohr circle of stress. A plane with a normal oriented at an angle θ are indicated by a point on the Mohr circle.

The maximum shearing stress τ_{max} is equal to $(\sigma_1 - \sigma_3)/2$ and occurs on planes with $\theta = \pm 45°$ from the direction of the planes of principal normal stress. A normal stress equal to $(\sigma_1 + \sigma_3)/2$ acts on each plane of maximum normal stress.

When a material is loaded, the successive states of stress can be represented by a series of Mohr circles, as indicated in Fig. 14.10. A simpler alternative is to plot the maximum shear stress and mean principal stress for each loading as the stress points $q = (\sigma_1 - \sigma_3)/2$ and $p = (\sigma_1 + \sigma_3)/2$, respectively. The curve connecting the points on the q–p diagram is called the stress path produced during loading.

14.7 SHEAR RESISTANCE OF GRANULAR MATERIALS AND POWDERS

The shear strength of unsaturated particulate systems may be measured using the triaxial compaction and direct shear tests as shown in Fig. 14.11. In the triaxial compaction test, an elongated specimen enclosed in a flexible,

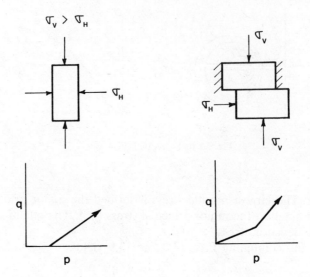

Fig. 14.11 Loading and stress path diagram for triaxial and direct shear loading tests.

impermeable membrane is first subjected to a confining fluid pressure; axial pressure is applied mechanically at constant fluid pressure to produce shear stresses and deformation. A split mold box is used in the direct shear test. Vertical pressure is applied using a piston. The mechanical displacement of the mobile portion of the shear box produces shear in the compressed material.

The shear resistance of granular materials and nearly dry powders is very dependent on the packing density and shape and surface characteristics of the particles. Pore pressures are generally insignificant except when the material is very compressible and the compaction rate is extremely fast. Differential shear between planes of particles requires either volume dilation or the fracture of particles when the particles are close-packed, because of the effective interlocking of nested particles. In loosely packed arrays, rolling and sliding may enable shear flow without an increase in volume. Studies of the flow of sand indicate that sands that pack more densely exhibit a greater volume dilation and a greater shear resistance during flow (Fig. 14.12).* Superimposed vibrations may momentarily reduce the contact stress between some particles and aid in particle sliding.

The strength of a granular material for a particular situation of loading may be defined at small strains by the maximum shear stress, as indicated by point P in Fig. 14.12 and after considerable straining by the ultimate shear stress τ_u. The principal stresses at failure σ_{1f} and σ_{3f} may be used to

*T. William Lambe and Robert V. Whitman, *Soil Mechanics*, Wiley-Interscience, New York, 1969.

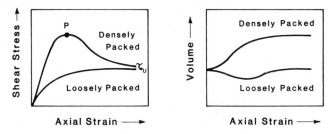

Fig. 14.12 General dependence of shear stress and specimen volume on axial strain during triaxial compaction.

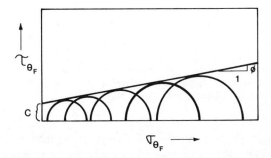

Fig. 14.13 Sequence of Mohr circles for the triaxial compaction of a cohesive body and the Mohr-Coulomb failure line.

construct the Mohr circle. The principal stresses are similarly determined for other states of confining stress, and a series of Mohr circles are plotted on a single graph as shown in Fig. 14.13. The line tangent to the Mohr circles is called the Mohr failure envelope. The significance of the failure envelope is that only states of stress lying outside the envelope will cause flow.

The Mohr envelope is often linear to a good approximation and is commonly expressed as

$$\tau_f = c + \sigma_f \tan \phi \tag{14.11}$$

where c is the cohesion intercept, ϕ is the angle of internal friction which indicates the shear resistance of the material, and the subscript f indicates flow conditions. Equation 14.11 is commonly called the Mohr-Coulomb failure law. For materials described by Eq. 14.11, the shear strength increases as the confining stress increases when $\tan \phi > 0$.

The parameters $\tan \phi$ and c may be considered to be properties of a particular material in a particular state. The friction angle for rounded, loosely packed sand is about 30° (Fig. 14.14). When the interlocking of particles increases owing to an increase in particle size or angularity, at constant porosity, or for a decrease in porosity for the same particle system,

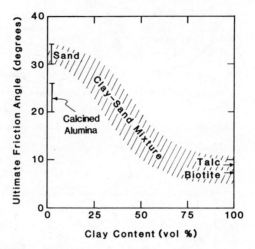

Fig. 14.14 Ultimate friction angle for wet clay $<2\ \mu$m, wet sand, sand-clay mixtures, alumina powder, and friction angle of talc and mica surfaces. (After T. W. Lambe and R. V. Whitman, *Soil Mechanics*, Wiley-Interscience, New York, 1969.)

the friction angle increases to 40–45°. Interparticle lubricants can lower the friction angle. The shear resistance of fine powders depends on the compressibility of the powder and the drainage of pore fluid as well as particle and packing parameters.

14.8 STRESS-STRAIN BEHAVIOR DURING COMPRESSION

The compression of an unsaturated plastic ceramic body or powder may cause permanent consolidation because of sliding and rearrangement of particles; the elastic compression of particles, binders, lubricants, and liquid; the compression or dissolving of gas in the liquid; particle orientation; and the migration of liquid and gas through the pore interstices. The compressibility may be determined under isostatic or uniaxial confined compression conditions, as we see in Fig. 14.15. In the isostatic compression test the specimen is enclosed in a thin, impermeable membrane and compressed by fluid. True isostatic compression causes a decrease in volume but no shear and distortion, as indicated on the q–p diagram.

In uniaxial compression, the specimen is compressed while confined in a smooth, rigid die; compression is the dominant source of strain, but die wall friction causes shear stresses to occur as indicated on the q–p diagram. The ratio of horizontal stress σ_h to vertical stress σ_v, $K_{h/v}$, i.e.

$$K_{h/v} = \frac{\sigma_h}{\sigma_v} \tag{14.12}$$

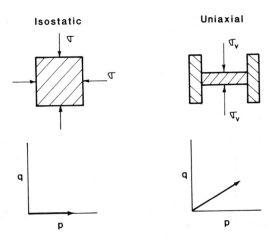

Fig. 14.15 Loading and stress path diagrams for isostatic and confined compression.

may be determined as a function of pressure by using a pressure transducer in the die wall. Permeable pistons may be substituted to determine the effect of drainage of the pore fluid on the mechanical behavior.

The rolling, sliding, and rearrangement of particles during isostatic or uniaxial compression of loosely packed powers may cause large volumetric compaction. Although the average shear stress is zero during isostatic compression, shear forces at individual particle contacts may be quite large.

Dense packings of sand exhibit less than 1% volumetric strain on uniaxial compression to 3MPa.* Initial strain occurs owing to the collapse of more loosely packed arrays and the elastic compression of particles. Above about 15 MPa, particle fracturing at contacts may enable sliding into a denser configuration. The ratio of horizontal stress to vertical stress $K_{h/v}$ is about 0.4 to 0.5. Elastic springback on decompressing from about 50 MPa is less than 2%. Similar results are observed for packings of other hard materials. The stress causing fracture will tend to increase as the particle strength increases.

The volumetric strain on uniaxially compressing filter cake of a flocculated electrical porcelain body (50% clay, 50% quartz and feldspar) between permeable pistons is shown in Fig. 14.16. Both permanent compaction and the reversible axial strain (springback) up to about 10% are apparent.* The primary compression of filter cake up to about 1 MPa is typical of the collapse of the coagulated clay coating hard particles which decreases the spaces between particles. Above 1 MPa, the behavior is typical of clay with a more oriented structure. Elastic springback is very significant below 1 MPa

*T. William Lambe and Robert V. Whitman, *Soil Mechanics*, Wiley-Interscience, New York, 1969.
*D. Price, MS thesis, Alfred University, Alfred, NY, 1981.

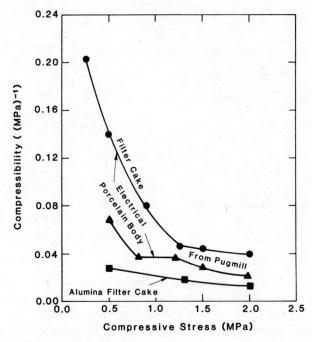

Fig. 14.16 Axial compressibility of flocculated porcelain and alumina bodies during confined compression.

but is only a fraction of the total volume change due to consolidation. Shear flow of the body in the vacuum pug mill mixer increases the orientation and packing density of clay in microscopic regions, which reduces the average compressibility and springback. The reversible compression of filter cake cast from a deflocculated suspension is lower and similar to the behavior of the pugged material. The clay matrix is also partially oriented in material cast from a deflocculated suspension. When liquid is unable to drain from the pores during compression, the volumetric strain will be similar up to the pressure causing pore saturation. Bodies close to saturation will show little total compression.

On initial compression, behavior similar to that exhibited in Fig. 14.12 may be expected for highly porous materials containing a significant amount of a molecular binder. Green ceramic tape containing 30 vol % binder may show an initial compressibility of 20%. However, the elastic springback is relatively low and less than about 2% for ordinary cast tape and organically plasticized extrusion bodies. A dense cast of micron-size alumina pressure cast from a deflocculated slurry exhibits relatively little compressibility, as is indicated in Fig. 14.16. In general, the stress ratio $K_{h/v}$ is higher for a saturated but highly porous body, and $K_{h/v}$ approaches 1 when the friction angle ϕ approaches 0°.

14.9 STRENGTH OF AGGLOMERATES

Agglomerates may form spontaneously in a dry, nonmagnetic powder owing to either Van der Waals forces or electrostatic attraction. Electrostatic forces occur because of surface-adsorbed ions or in transfer of electrons between particles in regions of contact. For particles that are nonconductors, charge mobility is limited to the surface of particles, and charges persist until contact with an electrically conducting material occurs. Electrostatic charges may also be diminished by an adsorbed film that enhances that charge mobility. Electrostatic and Van der Waals forces produce relatively fragile agglomerates. However, these fragile agglomerates can impede the flow and packing of dry powders and colloids.

Agglomeration is also caused by the capillary force of a wetting liquid at contacts between particles. We can estimate the effect of the capillary forces on the agglomerate strength using a model of packed equisize spheres containing a liquid distributed as is shown in Fig. 14.17. In the film state the tensile strength (S_t) is the product of the number of contacts (bonds) per unit area and the capillary force per contact b:

$$S_t = \frac{K(1 - \phi)}{\phi} \frac{b}{a^2} \qquad (14.13)$$

where K is a numerical constant.

For randomly packed uniform spheres separated by an adsorbed liquid Rumpf* has derived the equation

$$S_t = \frac{9(1 - \phi)b}{8\phi a^2} \qquad (14.14)$$

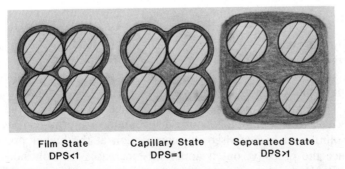

Film State	Capillary State	Separated State
DPS<1	DPS=1	DPS>1

Fig. 14.17 Liquid distribution in film, capillary, and separated particle states.

*H. Rumpf, in *Agglomeration*, W. A. Knepper (ed), Wiley-Interscience, New York, 1962.

The capillary force b is equal to the force produced by surface tension and the capillary pressure (Eq 2.6) multiplied by the area of the liquid in the neck. The force b increases as the liquid content is reduced. When the neck radius is very small compared to the radius of the circular liquid in the plane of contact, and the wetting angle is $0°$, $b = \pi \gamma_L a$. Equation 14.14 indicates that the tensile strength varies directly with the surface tension and packing density and inversely with the particle size, as is observed experimentally.

The formation of a hard, brittle mass from a soft mass on drying from the capillary state to the film state is also commonly observed. Capillary liquid bonding can produce a tensile strength ranging up to about 70 kPa for water in an oxide powder of about 1 μm particle size. When pores are saturated, internal capillary liquid bonding disappears. Agglomerates in the capillary state are usually of lower strength and more plastic.

Stronger agglomerates (and cohesive bodies) are produced by flocculation bonding, as was discussed in Chapter 11. Onoda* calculated the bond strength between two uniform spheres assuming that the binder was either uniformly concentrated at particle contacts (pendular model) or uniformly coating the particles (film model). Onoda's equation for the pendular model is

$$S_t = \frac{3}{8} \left(\frac{3\pi}{\phi} \right)^{1/2} (1 - \phi) S_o \left(\frac{V_b}{V_p} \right)^{1/2} \tag{14.15}$$

and for uniformly coated particles,

$$S_t = \frac{3\pi}{16} \left(\frac{1 - \phi}{\phi} \right) S_o \frac{V_b}{V_p} \tag{14.16}$$

The volume of binder and particles are V_b and V_p, respectively, and S_o is the strength of the binder in the plane of fracture.

Lukasiewicz* observed that a low-viscosity binder solution that wetted the particles initially formed a thin coating and then concentrated in the contact regions; the dependence of strength on V_b/V_p was more closely approximated by Eq. 14.15. An inhomogeneous distribution of the binder may reduce the apparent strength and alter the dependence on V_b/V_p. The uniform admixing of a binder solution with a powder is more difficult when the binder solution increases in viscosity and when the particle size is smaller. The nonlinear increase of tensile strength with V_b/V_p in Fig. 14.18 is probably due to the inhomogeneous distribution of the more concentrated binder solution of higher viscosity; compaction at the higher pressure did not reduce the porosity, but it apparently increases the adhesion strength (S_o). Although no dependence of agglomerate strength on particle size is

*G. Y. Onoda Jr., *Am. Ceram. Soc. Bull.* **58**(5–6), 236–239 (1978).
*S. Lukasiewicz, PhD thesis, Alfred University, Alfred, NY, 1983.

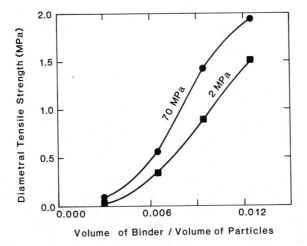

Fig. 14.18 Tensile strength of dried compacts of −6 mesh tabular alumina containing a polyvinyl alcohol binder (pressure indicated is compaction stress; packing fraction is 0.69.)

implied by Eqs. 14.15 and 14.16, a particle size effect is often observed in real systems. Materials with a tensile strength ranging up to 10 MPa may be produced using common binders. The tensile strength of clay bonded materials is very dependent on the liquid content (Fig. 14.19), because the strength is dependent on both the coagulating forces between particles and the capillary effect expressed by Eq. 14.13.

The apparent strength of individual agglomerates under a shear or compressive load may be higher for smaller agglomerates which are often of

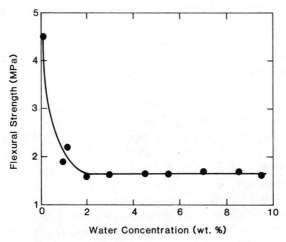

Fig. 14.19 Apparent tensile strength of an industrial electrical porcelain body as function of liquid content.

lower pososity. The strength of an agglomerate is described by Eq. 14.13. However, the force to produce rupture of an agglomerate varies directly with the product of its strength and the square of its diameter. The apparent yield strength of agglomerates containing a binder is very dependent on the yield strength of the binder phase.

14.10 SHEAR RESISTANCE OF NEARLY SATURATED POWDERS AND PLASTIC BODIES

The dependence of the shear resistance on straining an unsaturated plastic clay body is of the general form shown in Fig. 14.12. If the clay body has been consolidated at greater than 10 MPa, the maximum shear resistance is usually greater than the ultimate shear resistance. Highly consolidated bodies exhibit a greater initial resistance to the formation of an oriented clay structure facilitating shear, although the ultimate shear stresses for sustained deformation may be comparable. Orientation of platey clay particles occurs at an angle of about $(45° + \phi/2)$ relative to the plane of the major principal stress. For clay-sand mixtures, the friction angle calculated from the ultimate stress conditions decreases as the content of submicron clay increases (Fig. 14.14), and a friction angle of about 15° is achieved when the clay content exceeds about 40 vol %. The friction angle for clay may vary from 10° to as low as 4–5° for a clay with a relatively high content of montmorillonite. The ratio of lateral to axial stress $K_{h/v}$ is approximately $0.95 - \sin \phi$.

The shear resistance and deformation of fine powders nearly saturated with a liquid are very dependent on pressurized liquid in pores, which reduces the effective stress in the particle skeleton. Pore pressure effects are especially important in systems containing submicron particles and pores, because the liquid permeability of these systems is very low, and liquid migration occurs very slowly.

When molding ceramic materials, the term plasticity is commonly used. The term plasticity refers to a particular mode of mechanical behavior. A plastic ceramic material will exhibit permanent deformation without rupture when the applied compressive load produces a shear stress that exceeds the apparent shear strength, often called the yield strength of the material. For a plastic ceramic body, measurable elastic behavior is observed before yield occurs and when the applied load is removed (Fig. 14.20). In terms of the stress-strain behavior, the area under the stress-strain curve indicating energy adsorption is large, and the body is said to exhibit toughness.

The compressibility of unconsolidated fine powders and plastic ceramic bodies containing a significant fraction of colloidal particles or a molecular binder is significantly higher than the compressibility of liquid water. For a nearly saturated body, an increase in the applied stress first increases the effective stress and the packing density of the particles. As the degree of pore saturation increases, liquid may become pressurized in microscopic

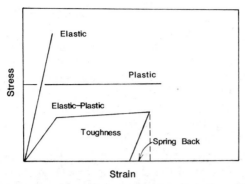

Fig. 14.20 Stress-strain behavior for ideal bodies. Toughness is indicated by the area under the stress-strain curve, and springback is indicated by the change of strain on unloading.

regions. Eventually, the pore structure becomes saturated if the system is sufficiently compressible. Air in the pores can dissolve in the pore liquid. Once saturated, an increment in the applied stress will increase the pore pressure but not the ultimate shear strength, as is shown in Fig. 14.21. When the pore pressure is not zero, the Mohr-Coulomb equation is

$$\tau_f = \bar{\sigma}_f \tan \bar{\phi} + \bar{c} \tag{14.17}$$

The material parameters $\bar{c}$ and $\bar{\phi}$ are effective parameters calculated using the effective stress $\bar{\sigma}_f$. If the pore pressure is not known, c and ϕ determined using the total stress are total stress parameters.*

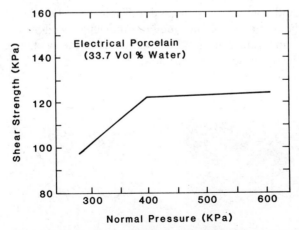

Fig. 14.21 Dependence of shear stress causing flow in the direct shear test on applied pressure.

*T. William Lambe and Robert V. Whitman, *Soil Mechanics*, Wiley—Interscience, New York, 1969.

In plastic clay bodies, the submicron coagulated colloidal clay coats the larger particles and forms a matrix, as shown in Fig. 14.22. The body consolidates during deformation and is not dilatant. Water separates the particles and acts as a lubricant. Shear flow occurs between larger particles in zones containing oriented clay and finer, nonplastic particles. A high concentration of shear zones is required for plastic flow. Liquid migration from a shear zone will increase the effective stress and the shear resistance in that zone, and alter the distribution of stress. A concentrated stress may cause the nucleation of a crack.

In nonclay bodies plasticized with a molecular binder, the macromolecules play a role similar to that of the clay (Fig. 14.22). Higher-molecular-weight molecules may range up to 0.5 μm in length, and hydrophillic groups attract water molecules. The higher-molecular-weight molecules are also more effective in reducing the liquid permeability. Higher-molecular-weight molecules may provide relatively more initial compressibility needed to increase the pore pressure. A critical volume of coagulated colloidal particles or molecular binder is required for plastic flow.

The cohesive strength (c) of a plastic body depends on capillary liquid bonding and flocculation bonding. Saturation eliminates internal capillary effects; however, on decompression of the plastic body, elastic expansion (springback) produces internal porosity, and the capillary effect alone does not provide the behavior characteristic of a plastic body. Plastic clay bodies are flocculated. Systems of submicron alumina and water develop the characteristic cohesiveness and plasticity of a plastic body when flocculated with an organic binder solution. General compositions of organically plasticized and clay plasticized bodies are presented in Table 14.2. The internal friction and ultimate shear strength of a plastic clay body decrease with increasing liquid content, as indicated in Table 14.3. For the alumina body, increasing the content of the medium-molecular-weight methylcellulose increased the apparent cohesion but did not significantly change the internal friction for this range of concentration.

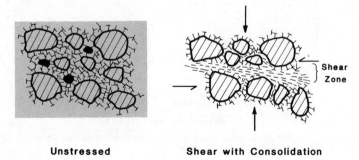

Unstressed   **Shear with Consolidation**

Fig. 14.22 Model for distribution of filler, finer plasticizing particles/molecules of high aspect ratio, pores, and liquid in a flocculated plastic body.

Table 14.2 Representative Compositions of Plastic Bodies

Clay Plasticized	Component (vol %)	Organic Plasticized
50–75	Filler	40–60
20–30[a]	Plasticizer	2–24
0–5	Pores	0–5
balance	Liquid	balance
100%		100%

[a]Coarser clay is considered to be a filler.

Table 14.3 Coulomb Parameters Determined for Flocculated Industrial Extrusion Bodies

Body	Binder (vol %)	Water (vol %)	ϕ (deg)	τ_y (kPa)
Electrical	31 (clay)	39	4	39
porcelain[a]	33 (clay)	34	12	122
Alumina[b]	6.0	40	5	129
	4.0	40	5	92
	2.0	40	5	76

[a]Basis includes liquid; total clay content is indicated.
[b]Ultimate values (R. Locker, PhD thesis, Alfred University, Alfred, NY, 1985).

The general dependence of the ultimate shear strength (τ_y) of the saturated body on the volume fraction of liquid f_L^v for a clay plasticized body is*

$$\tau_y = A \exp - \alpha(f_L^v - f_{cr}^v) \qquad (14.18)$$

where f_{cr}^v is the critical concentration of liquid at which particle flow is blocked and A and α are constants that depend on parameters such as the packing characteristics, the content and aspect ratio of colloidal particles, the content and molecular weight of a binder, and chemical flocculation. The dependence of the shear strength of industrial clay bodies on the parameter ($f_L^v - f_{cr}^v$) is approximated well by Eq. 14.18, as is seen in Fig. 14.23.

Relatively little has been published on the mechanical properties of technical ceramics plasticized with organic molecular binders. The behavior in Fig. 14.23 for the alumina plasticized with a medium-molecular-weight methylcellulose binder and water is clearly different from that for the clay plasticized body. The internal friction and ultimate shear strength of the alumina bodies appear to be dependent on the adhesion strength of the gelled organic film on the particles as well as the viscosity of the binder

*R. H. Weltmann and H. Green, *J. Appl. Physics.* **14**(11), 569–576 (1943).

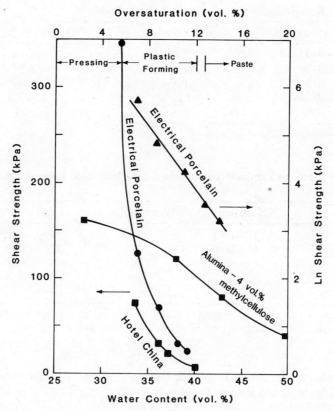

Fig. 14.23 Dependence of ultimate shear strength (direct shear test) on liquid content for several plastic bodies. The range of plastic forming is indicated for the electrical porcelain body.

solution. For each particle system, a particular concentration of plasticizer and liquid is required for plastic flow.

A variety of empirical tests for determining indices of plasticity have been devised and are discussed in the ceramics literature. Most involve imposing some simple or complex strain and measuring the resulting stress. These can be categorized as torque of mixing, indentation, tension, torsion, unconfined compression, and simple shear tests. The torque used to rotate mixing blades through a mixture is a quick, convenient way to note qualitative trends on varying the liquid content, chemical additives, colloidal particles, etc. (Fig. 14.24). Penetrometers indicate the resistance to deformation of a small volume of material. Simple torsion tests do not produce a uniform strain, and as normally practiced the specimen is not compressed. Tests in which the body is compressed during deformation, simulating industrial processing, should provide results that correlate with industrial workability. Studies of well-characterized systems using tests that provide a well-defined state of stress are needed to develop a science of plasticity.

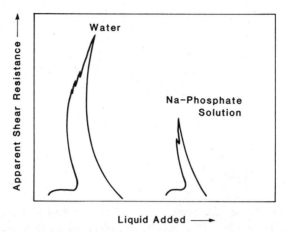

Fig. 14.24 Variation of torque for mixing with the content of water or deflocculant solution admixed into a plastic clay body.

The qualitative variation of the ultimate shear strength of a saturated plastic body with a change in composition and shear rate is often similar in behavior to that of a slurry of the materials. The rheology of saturated systems is discussed in Chapter 15.

SUMMARY

"Consistency" is a subjective term that describes several physical characteristics and rheological properties of the material. The consistency depends in a complex way on the physical nature, proportions, distribution, and association of the ingredients during deformation and flow. Common consistency states produced on adding a wetting liquid to a powder are an agglomerated powder, plastic material, a paste, and a slurry.

When a ceramic material is compressed, the pressure of fluid in the voids between particles reduces the effective stress carried by the particles. The contact stress between particles is high relative to the effective stress when the contact region between particles is small relative to the particle size. Interparticle sliding is restricted by friction at particle contacts and is very dependent on surface roughness and lubrication.

The resistance of a ceramic powder or particle system to shear is very dependent on the imposed compressive load which increases the effective stress and may increase the packing density. The compression of an unconsolidated ceramic particle system produces elastic and permanent volume strain which depends on particle parameters, packing, adsorbed additives, and the degree of pore saturation. Elastic springback on decompression is very dependent on elastic energy stored during compression.

The control of agglomerates is a common processing problem. Random agglomerates may produce inhomogeneity in a product, but controlled agglomeration is necessary to produce free-flowing powders and for cohesion and green strength. Both capillary liquid bonding and flocculation are important mechanisms for agglomeration.

Two mechanical properties of an unsaturated ceramic body are the cohesion and the internal friction. The ultimate shear strength of a saturated system is very dependent on the liquid content and shear conditions. Plastic behavior is commonly observed using empirical tests. The mechanical properties of plastic bodies depend on many variables and are incompletely understood.

SUGGESTED READING

1. G. Y. Onoda and M. A. Janney, Application of Soil Mechanics Concepts to Ceramic Particulate Processing, in *Advances in Powder Technology*, Gilbert Y. Chin (ed.), American Society for Metals, Metals Park, OH, 1982.
2. Soil Mechanics, in *Encyclopedia of Science and Technology*, Vol 12, 5th Ed, McGraw-Hill, New York, 1982.
3. H. Rumpf and H. Schubert, Adhesion Forces in Agglomeration Processes, in *Ceramic Processing Before Firing*, G. Y. Onoda Jr. and L. L. Hench (eds), Wiley-Interscience, New York, 1978.
4. A. Adamson, *Physical Chemistry of Surfaces*, Wiley-Interscience, New York, 1976.
5. T. William Lambe and Robert V. Whitman, *Soil Mechanics*, Wiley-Interscience, New York, 1969.
6. A. H. Cottrell, *The Mechanical Properties of Matter*, Wiley-Interscience, New York, 1964.
7. F. P. Bowden and D. Tabor, *The Friction and Lubrication of Solids*, Oxford University Press, London, 1950.
8. E. C. Bloor, Plasticity: A Critical Survey, *Trans. Br. Ceram. Soc.* **56**, 423–481 (1957).
9. George Y. Onoda Jr., Theoretical Strength of Dried Green Bodies with Organic Binder, *Am. Ceram. Soc. Bull.* **58**(5–6), 236–239 (178).

PROBLEMS

14.1 Illustrate the mean packing density as a function of the concentration of admixed liquid for bulky powder, agglomerate, plastic body, paste, and slurry states formed.

14.2 Illustrate the degree of pore saturation as a function of liquid content for the states in problem 14.1.

14.3 How do the packing factor and the uniformity of packing affect the liquid requirement to saturate the pores (DPS = 1) in the body?

14.4 Can a plastic clay body contain more water than the deflocculated slurry of the body? Why or why not?

14.5 A tank truck carrying relatively dry cement powder is loaded and unloaded pneumatically by pumping air into the powder. Explain the ease of flow in terms of Eq. 14.1.

14.6 An applied pressure of 10 MPa is applied to an unconsolidated concrete. After consolidation, the pore pressure is measured to be 3 MPa. What is the effective stress carried by the particles?

14.7 Calculate the mean contact stress in problem 14.6 if electrical interactions can be neglected and the contact area ratio A_c/A_T is 0.01. What is the contact stress when liquid drains from the pores?

14.8 Estimate the difference in contact stress for the two cases shown in Fig. 14.4. For the angular particles assume that the contact length is (0.1) particle size.

14.9 Explain why boundary lubricants are not low-molecular-weight surfactants.

14.10 Derive Eqs. 14.8, 14.9, and 14.10.

14.11 Illustrate an element of material with $\sigma_y = 300$ kPa, $\sigma_x = 100$ kPa, and $\tau_{xy} = 0$. What is $K_{h/v}$? Calculate the principal normal stresses, τ_{max}, and the angle of the plane for τ_{max}.

14.12 Construct a Mohr circle diagram for problem 14.11.

14.13 Construct a stress path diagram for isostatic pressing and dry pressing in a metal die. The range of applied pressure for each is 125 MPa. For dry pressing, assume that $K_{h/v} = 0.39$ and $\tau_{xy} = (0.21)\,\sigma_x$ for all pressures.

14.14 The theoretical value of $K_{h/v}$ for an isotropic material is $K_{h/v} = \nu/(1 - \nu)$ where ν is the Poisson's modulus of the material. Estimate $K_{h/v}$ for compressed alumina powder, assuming that ν does not vary with porosity. ($\nu = 0.23$.)

14.15 The diameter d of the contact area between two elastic spheres is

$$d = \left[\frac{12(1 - \nu^2)N}{E} \frac{R_1 R_2}{R_1 + R_2} \right]^{1/3}$$

where N is the normal load, E is Young's modulus, and R_1 and R_2 are the radii of the spheres. Derive an equation expressing the dependence of the ratio contact stress/applied stress as a function of ν, E, and σ_a. Assume $R_1 = R_2$ and simple cubic packing.

14.16 For a noncohesive powder, the shear stress at failure is one-half the normal stress. What is the angle of internal friction?

14.17 If the reversible compressibility of a powder is 0.02 MPa^{-1}, what is the linear springback on removing an applied stress of 5 MPa? Assume that the springback is isotropic.

14.18 Estimate the maximum strength of a compact due to capillary bonding of water at 20°C. Assume that the particles are randomly packed, the pore fraction is 0.4, and Eq. 14.14 is appropriate. Calculate the strength for particles of 5 and 0.5 μm in diameter.

14.19 Estimate the bond strength S_o for the highest concentration of binder in Fig. 14.18. Assume a pendular model and a porosity of 31%.

14.20 Graph the shear stress at failure as a function of applied normal stress at failure using the results in Table 14.3. Contrast the behavior for the clay body containing 34 vol % water and the alumina body containing 43 vol % water.

14.21 Calculate the relative difference in tensile strength for the pendular and coated cases when $\phi = 0.4$ and $V_b/V_p = 6\%$.

15

Rheological Behavior of
Slurries and Pastes

Rheology is the science of deformation and flow. Knowledge of rheological behavior is essential in designing or selecting equipment for storing, pumping, transporting, milling, mixing, atomizing, and forming a ceramic system. Rheological measurements are an integral part of the research and development of slurry systems, and rheological tests are used in programs for monitoring and controlling the consistency and behavior of slurries for casting, spray-drying, or glazing and pastes for surface printing and decorating.

Ceramic slurries and pastes are commonly multicomponent systems that are relatively complex in structure and often poorly characterized. Particles may range from granular sizes to colloids. Added electrolytes and polymers may significantly change interparticle forces and the state of dispersion. The interparticle spacing depends directly on the concentration of particles (solids loading), the state of dispersion, and the particle packing. Particle coagulation and polymer flocculation may produce a structural linkage, a gel structure, that varies with the mixing procedure and the shear history prior to measuring the rheological properties. Flow produced during a rheological measurement may alter the microstructure of the system.

In this chapter we will discuss fundamentals and techniques used to determine the rheological properties of these important systems. Rheological behavior is then described in terms of models for the effects of individual components and their mixtures.

15.1 EFFECTIVE STRESS AND SHEAR RESISTANCE IN A SATURATED SYSTEM

When a liquid suspension is compressed, the imposed pressure is supported by the liquid medium, and the system is relatively incompressible at common processing pressures if free from gas bubbles. The difference between the applied pressure and the liquid pressure, called the effective stress ($\bar{\sigma}$), was defined by Eq. 14.1. In an open packed suspension of completely dispersed particles (Fig. 15.1), the effective stress is

$$\bar{\sigma} = R - A \qquad (15.1)$$

where $(R - A)$ is the effective pressure due to repulsive and attractive electrical interactions between particles. The coagulation and flocculation of particles can alter $(R - A)$ and the resistance to shear. However, the effective stress is independent of the pressure imposed in processing, which simplifies the interpretation of the shear flow behavior. In a more crowded slurry, shear flow may be momentarily blocked by neighboring particles, and

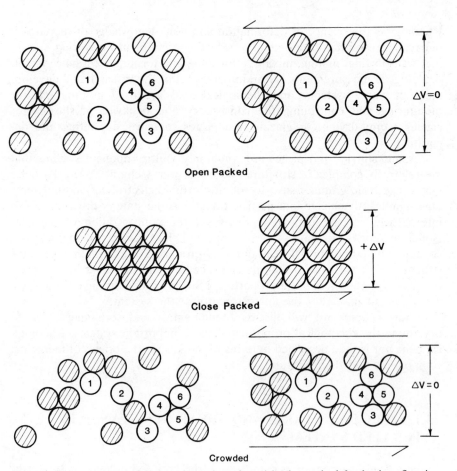

Fig. 15.1 Short-range translation and rotation of particles is required for laminar flow in a slurry, and laminar flow becomes more restricted when the interparticle separation decreases.

the resistance to shear flow is dependent on particle translation away from the direction of shear, which is time-dependent.

In a nearly close-packed system, initial flow produces significant contact stresses between particles, and a volume dilation of the system must occur to accommodate shear flow. For these systems the effective stress is described by Eq. 14.1 and the properties are not independent of the confining pressure.

15.2 RHEOLOGICAL PROPERTIES

A shear stress is required to initiate and maintain laminar flow in a simple liquid. When a shear stress τ is linearly dependent on the velocity gradient $-dv/dr$ (Fig. 15.2), the liquid is said to be Newtonian, and

$$\tau = \eta_L \frac{-dv}{dr} \tag{15.2}$$

The constant of proportionality is the coefficient of viscosity η_L. The coefficient of viscosity indicates the resistance to flow due to internal friction between the molecules of the liquid. The velocity gradient $-dv/dr$ is the shear rate $\dot{\gamma}$; for simplicity, $\dot{\gamma}$ will be used in equations that follow.

In liquids and solutions containing large molecules and suspensions containing nonattracting anisometric particles, laminar flow may partially orient the molecules or particles. When orientation reduces the resistance to shear, the apparent viscosity decreases with increasing shear rate, and the behavior is said to be pseudoplastic (Fig. 15.3). This behavior is often described by an empirical power law equation

$$\tau = K(\dot{\gamma})^n \tag{15.3}$$

where K is the consistency index and n is the shear thinning constant which indicates the departure from Newtonian behavior. The apparent viscosity of a power law liquid is

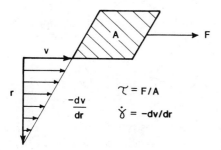

$$\tau = F/A$$
$$\dot{\gamma} = -dv/dr$$

Fig. 15.2 Shear stress F/A and shear rate $-dv/dr$ used in defining the coefficient of viscosity.

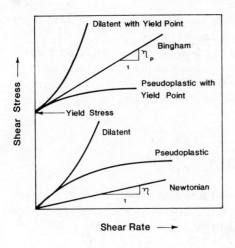

Fig. 15.3 Dependence of shear stress on shear rate for Newtonian, Bingham, and pseudoplastic and dilatant systems with and without a yield point.

$$\eta_a = K(\dot{\gamma})^{n-1} \tag{15.4}$$

For a pseudoplastic material, $n < 1$.

The power law with $n > 1$ may also approximate the flow behavior of moderately concentrated suspensions containing large agglomerates and concentrated, deflocculated slurries. Particle displacement may reduce interference at a low flow rate, but with an increase in the shear rate, the particle interference and the apparent viscosity increases. This dependence on shear rate is called dilatant behavior. Power law materials have no yield point.

Slurries containing attracting molecules or particles require a finite stress called the yield stress τ_y to initiate flow. A material is called a Bingham plastic when the flow behavior is described by the equation

$$\tau - \tau_y = \eta_p \dot{\gamma} \tag{15.5}$$

The plastic viscosity η_p indicates the stress in excess of τ_y causing a shear rate $\dot{\gamma}$, i.e.:

$$\eta_p = \frac{\tau - \tau_y}{\dot{\gamma}} \tag{15.6}$$

The total resistance to shear is expressed by the effective or apparent viscosity η_a which is higher when the yield stress is higher but decreases with increasing shear rate, i.e.,

$$\eta_a = \eta_p + \frac{\tau_y}{\dot{\gamma}} \tag{15.7}$$

For a Bingham plastic material, the apparent viscosity indicates the sum of the shear resistance of the liquid of Newtonian behavior and the shear

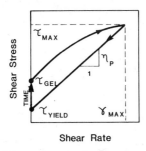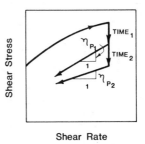

Fig. 15.4 Thixotropic behavior showing change of yield strength and plastic viscosity η_p after shear flow.

resistance of the particle structure which decreases with increasing shear rate. Note that for both a Bingham plastic and a pseudoplastic system, the apparent viscosity decreases with the shear rate, and two parameters are required to characterize the viscous behavior.

Liquids and slurries with a yield point may exhibit pseudoplastic or dilatant behavior. A general empirical equation describing the different modes of behavior shown in Fig. 15.3 is the Casson equation:

$$\eta_a^m = \eta_\infty^m + (\tau_y/\dot{\gamma})^m \tag{15.8}$$

where η_∞^m is the viscosity limit when structure is broken down at a very high shear rate, and m is a constant indicating the deviation from linearity. For materials with a yield point, the variation of log η_a with log $\dot{\gamma}$ is nonlinear.

The rheological behavior discussed above was assumed to be independent of the shear history and shearing time. For some materials the apparent viscosity at a particular strain rate may decrease with shearing time (Fig. 15.4). This behavior, called thixotropy, is commonly observed during the shearing of pseudoplastic and Bingham materials when the orientation and bonding of molecules or particles vary with the time of shear flow ($\dot{\gamma} \cdot t$). Thixotropic behavior in ceramic slurries is commonly reversible. For a Bingham material that is thixotropic, the apparent yield strength is higher after the suspension has been at rest and a particle structure has reformed. This higher apparent yield stress after a period of rest is often called the gel strength.

15.3 DETERMINATION OF VISCOSITY

Ranges of shear rate encountered in ceramic processing operations are indicated in Fig. 15.5. Viscosity data are more relevant when the viscosity is determined in the range of shear rate of the process.

The most widely used viscometer for ceramic suspensions and slurries is

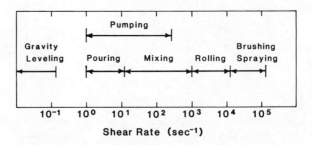

Fig. 15.5 Ranges of shear rate for different processing operations.

the variable-speed rotating cylinder viscometer. Depending on the design, either the inner or outer cylinder of length L rotates at an angular frequency ω, and the torque T produced by the viscous medium is recorded (Fig. 15.6). At a radial position r in the viscous material in the annulus, $\dot{\gamma} = -r(d\omega/dr)$, and

$$\tau = \frac{T/r}{2\pi rL} \tag{15.9}$$

The apparent viscosity is obtained using Eq. 15.2 and integrating between the radii of the two cylinders, a and b,

$$\eta_a = \frac{T}{\omega_a} \frac{b^2 - a^2}{a^2 b^2 4\pi L} \tag{15.10}$$

The length L in Eq. 15.10 may be increased to correct for the end effect. For suspensions, it is assumed in deriving Eq. 15.10 that the spacing $(b-a)$ of the annulus is much greater than the maximum particle size, the viscous

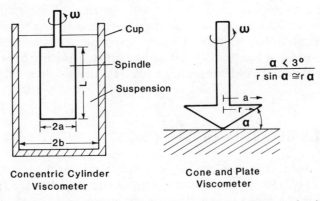

Fig. 15.6 Schematic of concentric cylinder and cone and plate viscometers showing parameters used in Eqs. 15.10 and 15.16.

material makes continuous contact with the surfaces of the cylinders, no slippage occurs, and laminar flow is fully developed. The shear rate varies with position in the annulus, and

$$\dot{\gamma} = \frac{2\omega_a}{r^2} \frac{a^2 b^2}{b^2 - a^2} \qquad (15.11)$$

The range of shear rates $(\dot{\gamma}_a - \dot{\gamma}_b)$ in the annulus is reduced by using a viscometer with a narrow annulus (<1 mm).

A concentric cylinder viscometer with a rotating cup is conventionally used for a Bingham plastic material. Slip at the wall of the inner cylinder may occur before the onset of flow of a Bingham material. Laminar flow begins at the inner cylinder when $T = T_a = \tau_y (2\pi a^2 L)$ and progresses outward to the point where $T = T_b = \tau_y 2\pi b^2 L$, as is shown in Fig. 15.7. Continued rotation at a constant speed causes the torque to increase to a value that is dependent on the rheological properties of the material. The plastic viscosity and yield strength are calculated using the equations

$$\eta_p = \frac{(T - T_o)}{\omega_b} \frac{(b^2 - a^2)}{b^2 a^2 4\pi L} \qquad (15.12)$$

and

$$\tau_y = T_o \frac{b^2 - a^2}{b^2 a^2 4\pi L \ln(b/a)} \qquad (15.13)$$

where T_o is the torque when ω is reduced to zero. The shear rate may be calculated from η_p and τ_y,

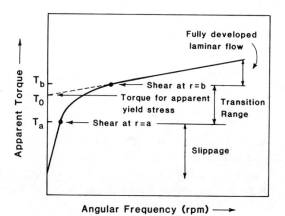

Fig. 15.7 Flow development in the annulus of a concentric cylinder viscometer for a Bingham material.

$$\dot{\gamma} = \frac{1}{\eta_p} \left(\frac{T}{2\pi r^2 L} \right) - \tau_y \qquad (15.14)$$

The apparent viscosity η_a is calculated using Eq. 15.7.

Rotating cylinder viscometers are widely used for determining the rheological properties of slips and slurries. An apparent viscosity at one shear rate is not sufficient for comparing the rheological behavior. The properly designed concentric cylinder viscometer provides a means for determining the rheological properties for a controlled range of shear rate. Rotational viscometers are commonly used at shear rates up to about 1200 s^{-1}; at higher rotational speeds, centrifugal effects become important.

The cone and plate viscometer shown in Fig. 15.6 is used for determining the viscosity of molecular liquids and colloidal suspensions. Interchangeable cones provide a selection of radius and cone angle α (from 0.3 to 3°). For a small α, the shear rate is independent of the radial position r, and

$$\dot{\gamma} = \frac{-dv}{dy} = \frac{-d(\omega s)}{d(r\alpha)} \cong \frac{\omega}{\alpha} \qquad (15.15)$$

The viscosity (no yield point) is calculated from the torque T using the equation

$$\eta = \frac{3\alpha T}{2\pi a^3 \omega} \qquad (15.16)$$

Slightly truncated cones are used for suspensions of colloidal particles.

The viscosity of a Newtonian liquid may be determined from the pressure drop ΔP during laminar flow at a volumetric rate Q in a capillary tube of radius R and length L. The viscosity is calculated from the Poiseville equation

$$\eta = \frac{\pi \Delta P R^4}{8QL} \qquad (15.17)$$

and

$$\dot{\gamma} = \frac{\Delta P r}{2L\eta} \qquad (15.18)$$

For a pseudoplastic liquid, the velocity profile in a capillary viscometer is somewhat flattened toward the center (Fig. 15.8), and the deviation of the shear rate from linearity increases when the flow constant n is lower. The laminar flow of a power law pseudoplastic liquid in a capillary tube is given by the equation

$$K = \frac{1}{2} \left[\frac{n}{3n+1} \right]^n \frac{R^{n+1} \Delta P}{\bar{v}^n L} \qquad (15.19)$$

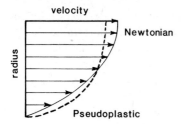

velocity

radius

Newtonian

Pseudoplastic

Fig. 15.8 The velocity profile of a pseudoplastic liquid is relatively flattened.

where $\bar{v}$ is the mean flow velocity. The flow behavior of a Bingham plastic fluid or a yield-pseudoplastic fluid in a capillary tube is much more complex. The analysis of the flow behavior of these systems is presented in references 2 and 5.

A capillary viscometer is commonly used for determining the viscosity of a Newtonian or pseudoplastic liquid at a shear rate beyond the limit for a rotational viscometer and for Bingham plastics of high-yield strength. Turbulent flow begins when a critical shear rate is exceeded and local fluctuations occur in the velocity and direction of flow. For flow in a pipe, the conditions causing a transition to turbulent flow can be predicted using the dimensionless Reynolds number (N_{Re}):

$$N_{Re} = \frac{2\bar{v}D_L R}{\eta_L} \tag{15.20}$$

where $\bar{v}$ is the average flow velocity of a liquid of density D_L and viscosity η_L in a tube of radius R. The transition to turbulent flow begins when N_{Re} is approximately 2100 for a Newtonian liquid; flow is fully turbulent when $N_{Re} > 3000$. The critical N_{Re} for the onset of turbulent flow is higher for pseudoplastic liquids and generally increases as the pseudoplastic constant n decreases. Turbulent flow is of importance in some mixing operations and atomization such as spray-drying.

The rate at which a heavy sphere falls through a suspension may also be used to determine the viscosity of a Newtonian liquid. Stokes' law (Eq. 7.2) is used for calculating the coefficient of viscosity from the time required to fall a fixed distance. The viscosity of very viscous liquids and paste can be determined by measuring the stress on a thin sheet or foil pulled through a slit filled with material. The viscosity may be calculated using Eq. 15.2. A variation of this is the axial displacement of a cylinder in a tube filled with material. If carefully controlled, the coefficient of viscosity of the material can be determined for a controlled displacement (shear) rate.

Other experimental techniques used to determine a relative viscosity include the viscous damping of a sonic probe, the torsional resistance to stirring, and mixing under conditions simulating industrial conditions. The time for flow through an orifice in a cup is also used as a quality control index of viscosity.

For a thixotropic slurry, the yield strength of undisturbed material, often called the gel strength τ_{gel}, is dependent on the time at rest following agitation. When using a concentric cylinder viscometer to determine the gel strength, the material is initially sheared for a few seconds in the viscometer at a high shear rate and then permitted to rest for a specified period of time (quiescent time). The gel strength is the maximum shear stress before the gel breaks, when the viscometer rotates at a specific slow rate (≈ 3 rpm). The relative gel strength may also be determined experimentally using a special torsion viscometer. An inner cylinder or disk of high mass initially twisted through some specific angle is immersed in the thixotropic material. After a particular rest period, the cylinder is released, and the angular rotation is measured as the torque in the spindle decreases. Less angular rotation occurs in a material with a higher gel strength.

Indices of thixotropic behavior may be conveniently determined using a concentric cylinder or cone and plate viscometer. An index of thixotropic structural buildup (B_{gel}) is the difference of the yield stress obtained after two quiescent times (such as 6 and 30 min):

$$B_{gel} = \frac{\tau_{y_{t2}} - \tau_{y_{t1}}}{\ln t_2/t_1} \qquad (15.21)$$

When the plastic viscosity is determined after shearing for two different times at a constant shear rate or two shear rates and a constant time, indices of structural breakdown can be calculated from the difference in plastic viscosity or extrapolated yield strength relative to the differential time or rate of shear flow (Fig. 15.4), i.e.:

$$B_{thix} = \frac{\eta_{P_1} - \eta_{P_2}}{\ln (t_1/t_2)} \qquad (15.22)$$

15.4 VISCOSITY OF LIQUIDS

The viscosity of simple liquids used in ceramic processing is Newtonian in behavior. For a series of liquids of similar type, the viscosity increases with molecular size as is indicated in Table 9.2. With only a few exceptions, heating increases the specific volume and reduces the viscosity of simple liquids; the variation of viscosity over a limited temperature range is approximated by the equation

$$\eta_L = A \exp (Q/RT) \qquad (15.23)$$

where A and Q are constants for a particular liquid. For water $A = 1.25 \times 10^{-3}$ mPa s and $Q/R = 1970°K$. The viscosities of several liquids at 0 to 70°C are listed in Table 9.3. In liquids containing large molecules, the molecules tend to be randomized by thermal vibration and oriented in laminar flow. Thermal energy increases the free volume and may facilitate orientation during flow, and the viscosity decreases with temperature and shear rate.

Most linear polymers such as polystyrene and polyethylene are non-Newtonian pseudoplastic materials. The viscosity increases with the mean molecular weight. A branched polymer that is more symmetrical in structure and orients less when sheared exhibits less pseudoplasticity when sheared. Polymers with a wider molecular-weight distribution generally exhibit a lower melt viscosity. Plasticizers may be either of the Newtonian or non-Newtonian type.

15.5 VISCOSITY OF BINDER SOLUTIONS

Binder molecules have a large sphere of influence relative to their size, and dissolved binder molecules may significantly increase the viscosity of a simple liquid. As shown in Fig. 15.9, binders with a higher average molecular weight are especially effective in increasing the viscosity. At low concentrations the dependence of viscosity of the solution η_s at low shear rates on the concentration C is often approximated by the equation

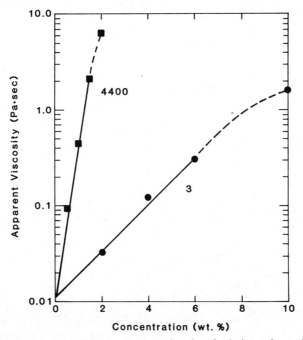

Fig. 15.9 Effect of concentration on the apparent viscosity of solutions of two viscosity grades of hydroxyethyl cellulose binder (Cellosize, Union Carbide Corp., New York).

$$\eta_S / \eta_L = \exp(SC) \tag{15.24}$$

where S is an effective sphere of influence for molecules (hydrated) of a particular molecular weight. Interaction between molecules that is reduced by solvation of the molecules causes a departure from Eq. 15.24 at concentration greater than about 2 to 10%.

The random orientation of dispersed binder molecules due to Brownian motion is offset by shear stresses in laminar flow, and binder solutions exhibit pseudoplastic behavior, as is shown in Fig. 15.10. The reduction in

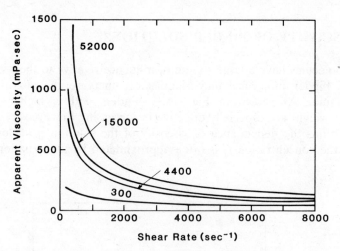

Fig. 15.10 Dependence of the pseudoplastic behavior of binder solutions on shear rate and the viscosity grade of the hydroxyethyl cellulose binder (Cellosize, Union Carbide Corp., New York).

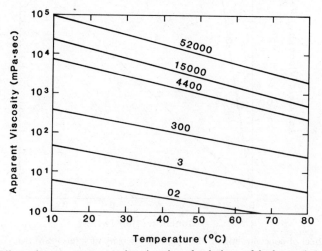

Fig. 15.11 Effect of temperature on the viscosity of solutions of hydroxyethyl cellulose of different viscosity grade (Cellosize, Union Carbide Corp., New York).

viscosity due to shear alignment is greater for solutions of higher-molecular-weight binders. When a binder is available in several viscosity grades, different grades can be blended to modify viscosity and pseudoplasticity.

The viscosity of solutions of dispersed binder molecules of different viscosity grade decreases with increasing temperature, as shown in Fig. 15.11. Temperature control is essential to obtain a reproducible viscosity. For binders that resist gelation, the variation of viscosity with temperature is often reversible between the freezing point and boiling point of the solvent.

15.6 VISCOSITY OF SUSPENSIONS OF DISPERSED COLLOIDAL PARTICLES

The viscosity of a suspension (η_S) is greater than the viscosity of the liquid medium (η_L) in the suspension, and the ratio is referred to as the relative viscosity (η_r). For a very dilute suspension of noninteracting spheres in a Newtonian liquid, the viscosity for laminar flow is described by the Einstein equation

$$\eta_r = \eta_s/\eta_L = 1 + 2.5 \, f_p^v \qquad (15.25)$$

where f_p^v is the volume fraction of dispersed spheres. The higher viscosity is caused by the dissipation of energy as liquid moves around the spherical particles. A more general equation for the viscosity of a suspension of nonspherical particles is

$$\eta_r = 1 + K_H f_p^v \qquad (15.26)$$

where K_H is the apparent hydrodynamic shape factor of the particles. In suspensions of particles with an aspect ratio greater than 1, particle rotation in the velocity gradient during flow produces a larger effective hydrodynamic volume (Fig. 15.12) and $K_H > 2.5$.

At a concentration above about 5–10 vol %, interaction between particles during flow causes the viscosity to increase at an increasing rate as the volume fraction of particles increases. Rheological data for suspensions of

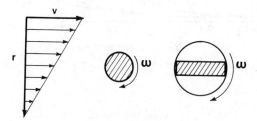

Fig. 15.12 The anisometric particle has a cross-sectional area equal to that of the sphere but a larger time average hydrodynamic shape factor K_H.

uniform spherical colloidal particles is approximated by the Dougherty-Kreiger equation*

$$\eta_r = \left[1 - \frac{f_p^v}{f_{cr}^v} \right]^{-K_H f_{cr}^v} \tag{15.27}$$

where f_{cr}^v is the packing factor at which flow is blocked. The variation of relative viscosity with volume fraction of dispersed particles, for different values of K_H, is shown in Fig. 15.13.

Woods and Kreiger* obtained values of K_H and f_{cr}^v indicated in Table 15.1 for aqueous suspensions of dispersed monosize latex spheres finer than 0.7 μm. The volume fraction of dispersed particles (f_p^v) was calculated from the volume fraction of spheres (f_s^v), their mean size ($\bar{a}_{V/A}$), and the effective thickness of the adsorbed surfactant Δ, using the equation

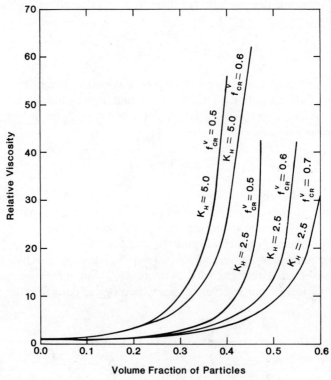

Fig. 15.13 Dependence of relative viscosity on solids concentration as indicated by Eq. 15.27.

*I. M. Krieger and T. J. Dougherty, *Trans. Soc. Rheol.* **3**, 137 (1959).
*M. E. Woods and I. M. Krieger, *J. Colloid Interfac. Sci.* **34**(1), 91–99 (1970).

Table 15.1 Dougherty-Kreiger Constants for Aqueous Suspensions of Latex Spheres

Shear Rate	K_H	f^v_{cr}
Low	2.7	0.57
High	2.7	0.68

Source: I. M. Krieger and T. J. Dougherty, *Trans. Soc. Rheol.* **3,** 137 (1959).

and

$$f^v_p = f^v_s \left[1 + \frac{6\Delta}{\bar{a}_{V/A}} \right] \tag{15.28}$$

$$\Delta = 4.5 \, \text{nm}$$

This correction becomes increasingly significant when the particle size decreases below 1 μm, as is seen in Fig. 15.14. Agglomeration reduces the effective f^v_{cr} (Table 15.1). Suspensions were observed to be pseudoplastic for $f^v_p < 0.50$, apparently because of the formation and persistence of agglomerates at low shear rates.

In suspensions of particles of anisometric shape both K_H and f^v_{cr} depend on particle orientation produced during laminar flow. Brownian motion somewhat randomizes the particles and increases the effective hydrodynamic shape factor.

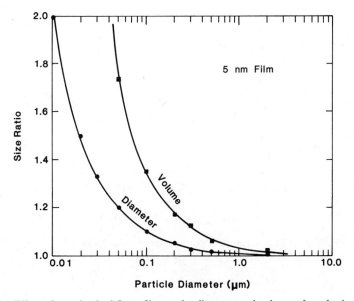

Fig. 15.14 Effect of an adsorbed 5-nm film on the diameter and volume of a spherical particle.

15.7 VISCOSITY OF SLURRIES OF DISPERSED POWDERS AND POWDER-COLLOID MIXTURES

Because of the complex interactions that occur during the flow of slurries containing coarse particles and mixed particle sizes, equations describing the dependence of the viscosity on parameters of the system are empirical. Farris* has proposed the equation

$$\eta_r = (1 - f_p^v)^{-K_F} \tag{15.29}$$

where K_F is a constant that ranges from about 3, for a continuous distribution of sizes that pack densely, to about 21, for coarse particles of uniform size. As seen in Fig. 15.15, the viscosity of defloculated suspensions of alumina is only roughly approximated by this equation. Equation 15.27, which is more general, provides a better empirical fit.

More efficient particle packing reduces the range of sizes of interstices among particles. For a particular solids loading, improved packing increases

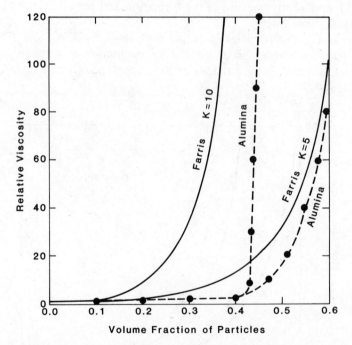

Fig. 15.15 Effect of solids concentration on the relative viscosity predicted by Eq. 15.29, and data for dispersed aluminas having a higher and lower f_{cr}^v.

*R. J. Farris, *Trans. Soc. Rheol.* **12**(2), 281–301 (1968).

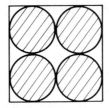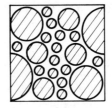

Fig. 15.16 A particle size distribution that packs more efficiently reduces the fraction of immobilized liquid. (The volume fractions of particles are equal in the two illustrations.)

f_{cr}^v and reduces the scale of segregation of the liquid medium, as is shown in Fig. 15.16. Slurries with relatively more mobilized liquid and a matrix of colloids may be relatively fluid and pseudoplastic over a wide range of shear rate. The general behavior for the two deflocculated suspensions of alumina differing in particle size distribution and f_{cr}^v is described by Eq. 15.27. The viscosity of the suspension having a wider size distribution and a higher f_{cr}^v rises less abruptly with increasing solids content.

Hindered rotation and mutual particle interference cause dilatant behavior above a particular shear rate (Fig. 15.17), and increasing the solids loading in the slurry generally decreases the shear rate at which dilatant behavior begins.* Dispersion of agglomerates or modification of the particle size distribution to produce additional fines and a higher f_{cr}^v may extend the shear range without dilatant flow, which can be very important in the pumping and spraying of slurries.

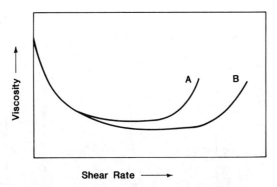

Fig. 15.17 General variation of viscosity with shear rate for two slurries indicating dilatant behavior at a high shear rate. The less well dispersed system (A) exhibits dilatant behavior at a lower shear rate.

*R. L. Hoffman, *J. Colloid Interfac. Sci.* **46**(3), 491–506 (1974).

15.8 RHEOLOGY OF COAGULATED SYSTEMS

Deflocculation disperses agglomerates of particles, and, as seen in Fig. 10.10, the viscosity of the slurry varies inversely with the zeta potential on adding a polyelectrolyte deflocculent. Coagulation can produce agglomerates or a continuous structure in the suspension. The structure may be partially broken down by shear forces, and the apparent viscosity is very dependent on the shear rate and time of shear flow.

In suspensions of oxides deflocculated using simple electrolytes, the viscosity also varies indirectly with the zeta potential. When the pH corresponds to the IEP of the particles, coagulation produces large, relatively porous agglomerates which may form a continuous gel structure if the solids content is sufficiently high. Smaller, less porous agglomerates form at a pH higher or lower than the IEP.

The viscosity of a slurry is very sensitive to the coagulated structure, as is shown in Fig. 15.18. A minimum in the apparent viscosity occurs when the

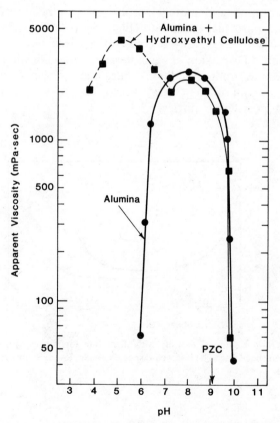

Fig. 15.18 Dependence of the viscosity of two slurries of a reactive alumina powder on pH adjusted using an addition of 0.2 M AlCl$_3$; an adsorbed binder alters the behavior at a low pH.

agglomerates are dispersed into small agglomerates or individual particles with a minimum hydrodynamic shape factor. Also, the coagulated structure of higher viscosity is commonly pseudoplastic and thixotropic, whereas the partially deflocculated slurry may be of lower viscosity but dilatant if sufficiently concentrated.

The maximum viscosity of a clay slip occurs when edge-face heterocoagulation produces large porous agglomerates, called flocs, or a continuous structure if sufficently concentrated. Increasing the pH reduces the floc size and eventually produces a dispersion of clay platelets with both negative edges and faces, which reduces the apparent viscosity, as is shown in Fig. 15.19. At low pH, the coagulated slip is not dispersed, because the faces of the clay particles remain negative. The relative maximum in the apparent viscosity and the yield stress are higher in a coagulated clay slip of finer particle size (Fig. 15.20); the dependence of the apparent yield stress on

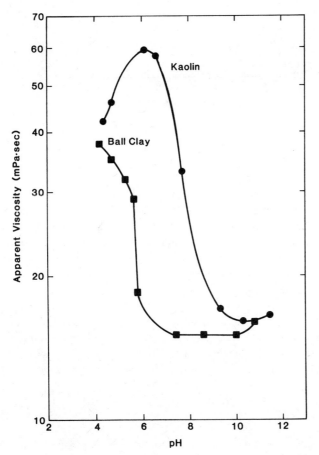

Fig. 15.19 Variation of viscosity of clay slurries with pH adjusted using HCl and NaOH (10 vol % clay).

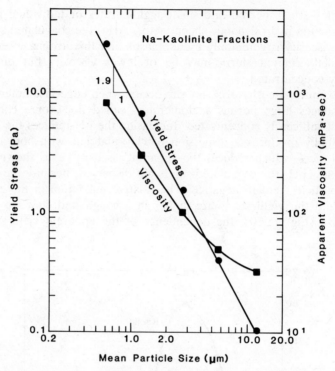

Fig. 15.20 Effect of particle size on the apparent viscosity and yield stress of coagulated suspension of kaolinite. (Data courtesy of W.G. Lawrence, Dean Emeritus of the College of Ceramics, Alfred University, Alfred, NY.)

particle size is approximated by Eq. 14.13. Substitution of clays containing more of the mineral montmorillonite and a higher clay loading can also produce a higher maximum apparent viscosity. The viscosity of a coagulated clay suspension may decrease or increase on heating. Heating reduces the viscosity of an aqueous medium but increases the electrochemical activity. The apparent yield strength may increase at temperatures approaching 100°C when lime is present.

The flow properties of thixotropic slurries and pastes are very dependent on the shear history. In the absence of an energy barrier, the coagulation rate is a function of the frequency of particle collisions. The number fraction of dispersed particles (N/N_o) remaining in a suspension of attracting particles after time t is given by the Von Smoluchowski equation*

$$N/N_o = [1 + t/\tau_c]^{-1} \tag{15.30}$$

*Paul Sennett and J. P. Oliver, Chapter 7 in *Chemistry and Physics of Interfaces*, D. E. Gushee (ed.), American Ceramic Society, Washington, D.C., 1965.

where τ_c is the period for $N/N_o = 0.5$. When t/τ_c can be expressed as $f/\dot{\gamma}$, where f is the coagulation frequency, a shear rate dependent coagulation index λ is

$$\lambda = \frac{c}{1 + \dot{\gamma}/f} \qquad (15.31)$$

where c is a constant. Pseudoplastic and thixotropic behavior is expected when $\dot{\gamma}/f > 0$. The constant f is dependent on the interparticle separation and attractive forces and is proportional to $N^2 T/\eta$, where N is the concentration of particles. An equation that is the combination of Eqs. 15.5 and 15.31 has been used by Worrall* to describe the viscosity of partially coagulated clay slurries which deviate from Bingham behavior:

$$\tau - \tau_y = \eta_p \dot{\gamma} + \lambda \dot{\gamma} \qquad (15.32)$$

Aging is the drift of the rheological properties of a suspension with time. Causes of the change in the coagulated structure are time-dependent dissolving, chemical reactions, and mechanical agitation. Aging is usually accelerated by increasing the temperature and the intensity of agitation. In clay suspensions, aging is expected and must be controlled to reproduce the flow behavior and thixotropic buildup.

15.9 RHEOLOGY OF SUSPENSIONS AND SLURRIES CONTAINING BINDERS

Systems containing dispersed particles, electrolytes, and an adsorbed binder are relatively complex and wide variations in rheological behavior can be effected by changing the proportions and types of components. When the binder flocculates particles in the suspension, the viscosity and yield strength may be increased in a range of pH where particle charging and dispersion would otherwise occur (Fig. 15.18). The effects of the binder may be profound at low shear rates, but with an increasing rate of shear, the viscosity behavior becomes less dependent on flocculation effects relative to effects of the particle size distribution, particle shape, and particle loading.

An ionic binder may act as either a flocculant or a deflocculant. A low-molecular-weight ionic binder that does not flocculate the particles may increase the viscosity of the liquid but decrease the viscosity of the slurry if particle dispersion is stabilized. Flocculation produced by an ionic binder increases the viscosity. Nonionic molecular binders of low molecular weight may act as a deflocculant at a low concentration and as a flocculant when the

---

*W. E. Worrall, *Clays and Ceramic Raw Materials*, Halsted Press, New York 1975.

concentration exceeds that for monolayer coverage. Nonionic binders of high molecular weight are usually flocculants. Flocculation and the presence of the pseudoplastic binder solution between the particles significantly alter the rheological properties. The viscosity of nonionic binder solutions is usually stable over a wider range of pH or concentration of highly charged ions than for solutions of ionic binders. On gelation of a binder, the gel strength increases with concentration and is higher for a binder of higher molecular weight (Fig. 15.21).

The viscous behavior shown in Fig. 15.22 is often preferred for slurries that must be mixed, stored, pumped, sprayed, etc.; examples are shown in chapters that follow. Steric hindrance of the binder resists the formation of hard agglomerates, and the yield stress provides a mechanism to prevent product separation during storage. However, shear thinning (pseudoplasticity) provides the necessary flow during mixing, pumping, and spraying. Similarly, pseudoplastic coatings with a yield stress resist running (creep flow) due to gravitational forces or mild mechanical vibrations.

Gelation of the binder solution may cause a significant increase in the viscosity of a slurry (Fig. 11.7). This characteristic can be valuable to obtain a strong coating without runs when a thermogelling slurry is applied on a hot surface. Gelation may also be produced by chemical means, as is indicated in Fig. 11.8.

A suspension of powder having a viscosity at a low shear rate less than about 1000 mPa s is commonly called a paste when the yield strength is in

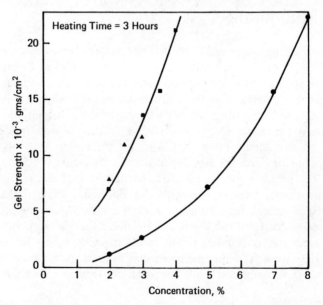

Fig. 15.21 Effect of binder concentration on gel strength of methylcellulose solutions of low (lower curve) and medium (upper curve) molecular weight (65°C). (Courtesy of Dow Chemical Co., Midland, MI.)

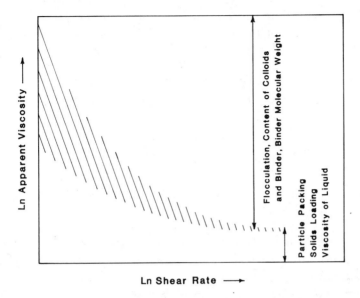

Fig. 15.22 Nominal rheological behavior of many industrial slurries at low to moderate shear rates.

the range of about 100–10,000 Pa. The suspension is described as being buttery, creamy, and watery as the yield strength decreases below 100 Pa. Intuitive descriptions of the general rheological behavior of a complex slurry may sometimes be insufficient or imprecise because of pseudoplastic or thixotropic behavior and variation in shear rate when the material is used.

One of the problems in interpreting the behavior of suspensions and slurries is determining the character of the system under flow conditions. Minimal characterization must include determining the characteristics of the ingredients and the state of dispersion. The agglomerate structure can be analyzed indirectly in a carefully diluted suspension using standard wet size analysis techniques. Agglomerates may also be separated by filtration or freeze-drying and examined microscopically. Coagulants and flocculants may be doped with a salt easily identified by microscale chemical analyses, prior to admixing into the system, to aid in analysis.

SUMMARY

The rheology of suspensions and slurries is a complex topic. Physical properties and interactions between the liquid, colloid, and particle components can vary widely. Colloids may be dispersed or agglomerated macromolecules and/or isometric and anisometric particles. Particles of variable character may be dispersed in a liquid-colloid matrix, agglomerated, or linked by coagulation and flocculation into a structure. The structure of the system may change with time, temperature, or shear flow. Rheological properties should be determined under conditions providing a controlled shear rate, as a function of shear rate, for characterized systems. In dispersed systems, the relative viscosity depends on the hydrodynamic shape factor and volume fraction of the dispersed particles with adsorbed surfactant. Coagulation and flocculation produce slurries having a yield strength and pseudoplastic behavior that is commonly preferred in industrial processing.

SUGGESTED READING

1. Temple C. Patton, *Paint Flow and Pigment Dispersion*, Wiley-Interscience, New York, 1979.
2. George R. Gray and H. C. H. Darley, *Composition and Properties of Oil Well Drilling Fluids*, Gulf Publishing, Houston, 1980.
3. W. E. Worrall, *Clays and Ceramic Raw Materials*, Halsted Press, New York, 1975.
4. Rex W. Grimshaw, *The Chemistry and Physics of Clays*, Wiley-Interscience, New York, 1971.
5. A. H. P. Skelland, *Non-Newtonian Flow and Heat Transfer*, *Wiley-Interscience, New York*, 1967.
6. Michael D. Sacks, Properties of Silicon Suspensions and Cast Bodies, *Am. Ceram. Soc. Bull.* **63**(12), 1510–1515 (1984).
7. R. L. Hoffman, Discontinuous and Dilatent Viscosity Behavior in Concentrated Suspensions. II. Theory and Experimental Tests, *J. Colloid. Interfac. Sci.* **46**(3), 491–506 (1974).
8. Martin E. Woods and Irvin M. Krieger, Rheological Studies on Dispersions of Uniform Colloidal Spheres, *J. Colloid Interfac. Sci.* **34**(1), 91–99 (1970).
9. R. J. Farris, Predictions of the Viscosity of Multimodal Suspensions from Unimodal Viscosity Data, *Trans. Soc. Rheol.* **12** (2), 281–301 (1968).

PROBLEMS

15.1 Write the flow equations for a pseudoplastic and dilatent material having a yield point.

15.2 What is the dependence of the viscosity of alcohols on molecular size and temperature?

15.3 Sketch the variation of apparent viscosity with shear rate for a pseudoplastic fluid for values of n between 0.1 and 1.0.

15.4 For a coagulated clay suspension exhibiting Bingham behavior, the yield stress is 2.0 Pa and the plastic viscosity is 10 mPa s. Calculate and graph the apparent viscosity with shear rate. Compare this to the behavior of a power law pseudoplastic material.

15.5 A concentric cylinder viscometer has a rotating inner cylinder of length 4.0 cm and radius 1.5 cm. The annulus $(b-a) = 0.1$ cm. Calculate the range of shear rate in the annulus during a test when $\omega_a = 60$ Hz. What is the range when $(b-a) = 1$ cm? Assume that the yield stress of the material is zero.

15.6 Calculate the viscosity for the material and geometrical parameters in problem 15.5 when $\omega_a = 100$ Hz and $T = 300,000$ N · m. What is the shear stress at the surface of the bob and the surface of the cup?

15.7 What is the range of shear rate $(\dot{\gamma}_a - \dot{\gamma}_b)$ in terms of ω_a for a concentric cylinder viscometer?

15.8 A concentric cylinder viscometer is used to determine the rheological properties of a Bingham plastic material ($L = 4.0$ cm, $a = 1.8$ cm, and $(b-a) = 0.1$ cm). The dial readings at $\omega = 300$ rpm and 600 rpm correspond to 5.0 and 8.0 N/M^2. Calculate τ_y, η_p, and η_a at 300 and 600 rpm. What is $\dot{\gamma}$ at 300 and at 600 rpm?

15.9 Contrast viscosity as a function of shear rate for a pseudoplastic suspension with $n = 0.25$ and a dilatent slurry with $n = 4$. Assume K is the same for each.

15.10 Show graphically why knowledge of the shearing stress at one shear rate is insufficient when comparing the rheological behavior of two Bingham slurries or a Bingham and a pseudoplastic slurry.

15.11 The relative velocity v/v_{max} of a power law pseudoplastic fluid as a function of the radial position r in a tube of radius R is $(v/v_{max}) = [(3n + 1)/(n + 1)] (1 - (r/R)^{(n+1)/n})$. Graph v/v_{max} as a function of r/R for a Newtonian liquid and a pseudoplastic liquid with $n = 0.15$.

15.12 Reduce the Casson equation to the simple logarithmic forms for Newtonian, pseudoplastic, dilatent, and Bingham materials, by the proper choice of m and τ_o.

15.13 The gel strengths obtained for a casting slip after quiescent times of 10 and 60 min were 3.5 and 8.0 Pa, respectively. Calculate the structural buildup index (B_{gel}) for this slip.

15.14 What is the value of the effective sphere of influence parameter S for the two grades of hydroxyethyl cellulose in Fig. 15.9?

15.15 The viscosity of a blend of two binder solutions can be approximated by the equation $\ln \eta = a \ln \eta_a + b \ln \eta_b$ where a and b are the weight fractions of each binder. Estimate the viscosity of a 50–50 blend of 2 wt % solutions of the binders in Fig. 15.9.

15.16 Calculate the hydrodynamic volume of a columnar-shaped particle 6 μm in length and the combined hydrodynamic volumes of two particles formed by breaking this particle at midlength. Assume that the particles may rotate in three dimensions.

15.17 Calculate and graph the variation of relative viscosity with solids loading predicted by the Dougherty–Kreiger equation when $f_{cr}^v = 0.6$ and $K_H = 3.5$.

15.18 Agglomeration occurs in a system with $f_{cr}^v = 0.6$ and $K_h = 3.5$, causing $f_{cr}^v = 0.45$ and $K_H = 4.5$. What is the change in relative viscosity for a suspension containing 40 vol % solids?

15.19 The maximum solids loading of a pourable suspension of zirconia particles is observed to be 40 vol % for particles of 1-μm diameter but only 20 vol % for particles of 0.02-μm diameter. The hull of polyelectrolyte deflocculent used is ≈ 5 nm in thickness. Can the adsorbed polyelectrolyte produce the lower loading limit?

15.20 Carry out the calculations in problem 15.17 for a system containing 1-μm particles and a second system containing 0.1-μm particles. Both systems are dispersed using a polyelectrolyte for which $\Delta = 5.0$ nm.

15.21 For the suspension of alumina in Fig. 15.18, sketch the expected variation of viscosity with shear rate for pH 8 and for pH 9.8.

15.22 A ceramic slurry that exhibits rheological behavior similar to that of house paint is described by the Casson equation with $m = 0.5$, $\eta_\infty = 200$ mPa s, and $\tau_y = 0.7$ Pa. Graph the variation of apparent viscosity with shear rate.

15.23 The yield stress required to prevent the gravitational settling of a particle of diameter $a \gg 20$ μm and density D_p in a slurry of density D_s is approximated by the equation $\tau_y \geq \frac{2}{3} a (D_p - D_s)g$. What is the maximum size that will remain suspended for the slurry in problem 15.22 if the volume fraction of solids is 0.50 and the particle density is $6.0 \, \text{Mg/m}^3$?

PART VI

BENEFICIATION PROCESS

Ceramic processing includes operations that modify the characteristics of the material system and the rheology for forming, to improve the microstructure in the ultimate product. These processes, called beneficiation processes, may be physical or chemical in nature. In this section we will consider processes for modifying the characteristics of batch components and their proportions and distribution in a multicomponent system. We will discuss the processes for size reduction called comminution processes in Chapter 16, and batching, dispersion, and mixing processes for dry and wet systems in Chapter 17. Beneficiation may include removing particles in a certain range of size and separating liquid from a slurry to concentrate the particles and remove soluble impurities, which is considered in Chapter 18. The preparation of powder agglomerates of controlled character is called granulation and is discussed in Chapter 19.

16

Communition

The comminution processes of crushing and milling are widely used in ceramic processing to reduce the average particle size of a material, to liberate impurities and reduce the porosity of particles, to modify the particle size distribution, to disperse agglomerates and aggregates, to reduce the maximum particle size, to increase the content of colloids, and to modify the shape of particles. Some milling processes also provide effective dispersion and mixing, which is discussed in Chapter 17, and are used to provide comminution and mixing simultaneously. The ceramic processing engineer must be familiar with these very important processes, which can have such a large impact on the rheology, fabrication behavior, sintering behavior, and ultimate microstructure of the product.

16.1 COMMINUTION EQUIPMENT

Primary crushers such as jaw crushers and cone crushers (Fig. 16.1), which produce compression and shear stresses (nipping), are commonly used individually or in series to reduce the size of coarse feed to an average size ranging down to about 5 mm or larger in size. Crushing rolls may be used to reduce less coarse feed to below 1 mm. In a hammer mill, rotating hammers pulverize particles of a brittle but relatively soft material and force the fines through the openings in a circular, wear-resistant screen (Fig. 16.2). A hammer mill is capable of producing a large reduction in size, down to about 0.1 mm. One or more of a variety of mills may then be used to further reduce the average particle size, as is indicated in Fig. 16.3. Common mills used for grinding ceramic materials are ball mills, vibratory mills, attrition mills, fluid energy mills, and roller mills.

A ball mill is a hollow rotating cylinder or conical cylinder partially filled with hard, wear-resistant media having the shape of rods, short cylinders,

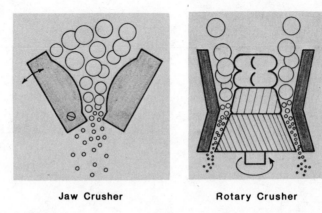

Jaw Crusher   **Rotary Crusher**

Fig. 16.1 Jaw and rotary crushers. (Illustrations courtesy of Sturtevant Inc., Boston.)

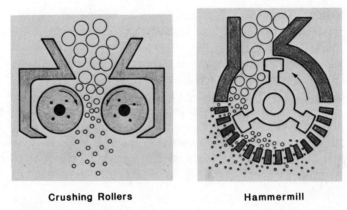

Crushing Rollers   **Hammermill**

Fig. 16.2 Schematic diagram of a roll crusher and a hammer mill. (Illustrations courtesy of Sturtevant Inc., Boston.)

balls, or pebbles, and granular feed material (Fig. 16.4). The mill may be as simple as a plastic bottle or porcelain jar, or a steel cylindrical shell with a hard, wear-resistant ceramic or hardened steel lining. Industrial mills range in diameter from several centimeters with a capacity of several grams to several meters in diameter with a capacity of several tons. The tumbling media in a rotating mill produce a grinding action by impacting and shearing the particles on their surfaces. Operating variables include the size and angular velocity of the mill, the size of the media relative to the size of the feed material, the loading of the mill, the relative volumes of media and feed material, the physical characteristics of the media, agglomeration of feed or product, and, in wet milling, the viscosity of the slurry during milling. In continuous ball milling, dry feed material is continually added, and the forced convection of air through the mill entrains and removes fine

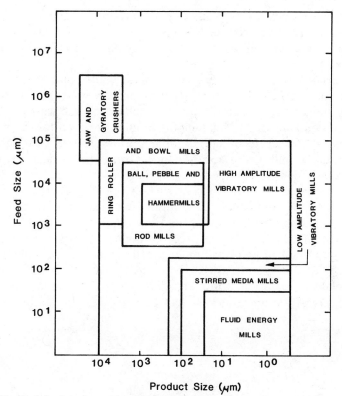

Fig. 16.3 Nominal feed and product mean size capabilities of industrial equipment. (From Wade Summers, *Am. Ceram. Soc. Bull.* **62**(2), 213 (1983).)

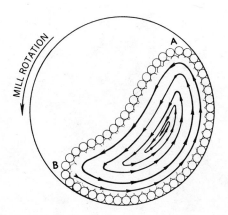

Fig. 16.4 Schematic diagram of a ball mill showing cascading from A to B and subsequent media movement.

particles. Ball mills are typically used to produce -200 and -325 mesh materials with a wide size distribution (more narrow distributions with a mean size of several microns) and to deagglomerate and mix slurries and powders.

Industrial vibratory mills are either of the horizontal tube type or the vertical torus type (Fig. 16.5). Low-amplitude mills are used for wet grinding submillimeter feed material to a submicron size with a minimum of wear and contamination. High-amplitude mills are used for the wet or dry grinding of coarser feed and to obtain a wider size distribution. The high-amplitude mill is also a satisfactory mixer. The vertical-type vibratory mill is nearly completely filled with cylindrical media and the feed slurry. Weights mounted eccentrically control the vibration pattern and reduce the power to maintain the vibration. Vibratory mills used in wet grinding usually have a rubber or ceramic lining and are supported on rubber or metal springs. Media acceleration produces an impact energy that is significantly greater than the energy in ball milling. Milling may be continuous in some models. With vibrations, discharge is quick even when the product is a pseudoplastic suspension. Vibratory mills are used industrially for milling a wide variety of ceramic materials to a particle size that is a few microns or

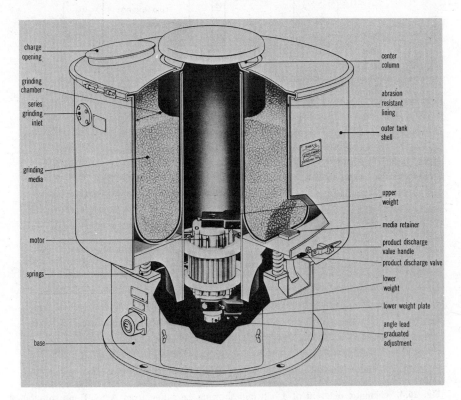

Fig. 16.5 Vibratory wet grinding mill. (Courtesy of SWECO Inc., Florence, KY.)

smaller. The grinding capacity of a large vibratory mill is limited to about 2 tons per hour, which is much smaller than can be achieved using a large ball mill.

A planetary attrition mill is a stirred media mill. A central shaft with arms rotating at 1–6 Hz continually stirs the slurry of particles and spherical media 0.3–1.0 cm and provides a means to vary the grinding energy (Fig. 16.6). Intense rolling and in-line impacts are produced by the differential velocity of media moving around the agitator arm into the cavity behind the trailing edge. The stirred media mill is used industrially for wet grinding to below 1 μm and for the dispersion of agglomerates of submicron particles. A pumping system in production models maintains circulation and uniformity of the slurry and is used for discharge. Oxidizable particles such as carbides and nitrides can be milled under an inert gas atmosphere, and a thermal jacket can be used to control the slurry temperature. Slurry capacities are quite large and continuous models are available. In a high-speed attrition mill, small grinding media and granular material are intensely agitated at 20–30 Hz to produce turbulence between a cylindrical rotor and a stator. The intense milling action produces considerable heat, and water cooling is required. Contamination due to wear of the mill lining and media is considerable and is commonly eliminated by chemical leaching, sedimentation, or magnetic separation. Attrition grinding is used for producing submicron powders of hard refractory oxides, carbides, nitrides, titania pigments, and paper-grade kaolin.

Fluid energy mills grind and classify in a single chamber. Feed material finer than about 5 mm may be either hard or soft in nature. Impacts and attrition between high-speed particles entrained in a fluid and moving in a circular orbit provide the grinding action. Centrifugal force causes oversize particles to remain in the peripheral grinding zone, but fines are drawn off in a central collector. The particle size and output of the product are controlled

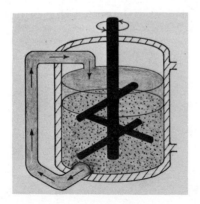

Attrition Mill

Fig. 16.6 Recirculating stirred media attrition mill. (Courtesy of Union Process Inc., Akron, OH.)

by the propellant pressure and the material feed rate. Mill capacities range from a few grams to several tons per hour. Mill linings may be wear-resistant hard materials or expendable plastic liners. Product heating in a fluid energy mill is lower than in other dry grinding operations, and an inert or nonoxidizing atmosphere such as N_2 may be used.

A roller mill in which feed material passes between a rotating table and arm supported rollers, and the disk mill in which material passes between a stationary and a rotating grinding disk are used for the comminution of moderately hard clays and porous calcined aggregates which break down under moderate stress.

Precious-metal inks of a relatively high viscosity used for electrodes and conduction paths in electronic ceramics and for decorating china and glass are commonly milled in a small, three-roller mill. The viscous paste adheres to the rolls. Differential roller speeds produce high shear stresses for dispersion and a precisely controlled particle size.

16.2 LOADING AND FRACTURE OF PARTICLES

Crushing and milling operations produce both compressive and shear loads on particles. Falling or vibrating media may produce a compressive in-line impact (Fig. 16.7). Shear is produced when a particle is seized between two surfaces moving with different velocities. Attrition is produced by frictional stresses. In milling situations using large media or rollers a combination of

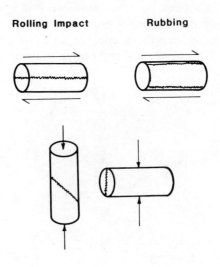

Fig. 16.7 Shear stresses in particles are produced by rolling and rubbing actions, and shear and tensile stresses by compresive loading.

in-line and rolling impacts can occur. The role of attrition increases as the media size and impact force decrease and as the frequency of rubbing contacts increases.

The grinding energy produced during milling is proportional to the mass (m) and the change in velocity (v) of the media on impact:

$$\text{Energy} = \Delta \left(\tfrac{1}{2} mv^2 \right) \tag{16.1}$$

The mass is increased by using media of a larger size or density (Table 16.1). Media with a high elastic modulus can produce a high dv/dt; media must also be hard, fine-grained, and nonporous to resist abrasive wear. Using the tetragonal to monoclinic phase transition in partially stabilized zirconia particles as an index of the grinding stress on milling, we see in Fig. 16.8 that vibratory milling with dense media produces a much greater stress than is produced in ball milling. Viscous flow and anelastic deformation reduce the grinding stress. When grinding an anelastic material, the impact load on the particle is lower, because the anelastic deformation reduces dv/dt. Deformable dense materials and porous agglomerates and aggregates are somewhat anelastic.

During milling, shear and tensile stresses are produced by in-line compressive loads; shear is also produced by rolling loads, and attrition is produced by the sliding and rubbing of particles between hard surfaces. The initial tensile and shear fracture of large particles is expected in microscopic regions of intensified stress produced by point loads and preexisting defects, as indicated in Fig. 16.9. The reduction of the fracture strength σ_f of a material due to the presence of stress-intensifying flaws of depth c is described by the equation

$$\sigma_f \sim \frac{K_{Ic}}{y\sqrt{c}} \tag{16.2}$$

where y is a constant that depends on a flaw geometry and K_{Ic} is the fracture toughness parameter that includes the work required to extend the crack.

Table 16.1 Density of Industrial Grinding Media

Material	Density (Mg/m^3)
Flint pebbles	2.4
Porcelain	2.3
Steatite porcelain	2.7
High-density alumina	3.6
Zircon	3.7
Zirconia	5.5
Steel (hardened or carburized)	7.8
Tungsten carbide	15.6

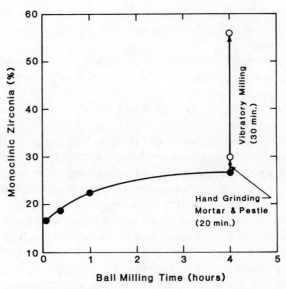

Fig. 16.8 Grinding stresses produced during ball milling with alumina media are considerably lower than those produced in high-amplitude vibratory milling using tungsten carbide media, as indicated by the stress-sensitive tetragonal to monoclinic phase transformation of a zirconia powder.

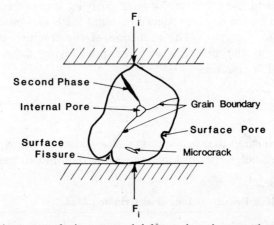

Fig. 16.9 Various types of microstructural defects reduce the strength of a particle.

Microfissures at the edges and surfaces of particles, surface pores, and internal microcracks and pores which increase c reduce the fracture strength of brittle particles. Fracture fragments containing defects that produce a smaller stress intensification are more resistant to grinding. Dense particles with a finer grain size and higher fracture toughness are more resistant to attrition. Abrasion grinding may increase the concentration of edge and surface flaws, which can aid in fracture. Particles exhibiting anelastic deformation are tougher particles. Attrition is more important for the grinding of anelastic particles and particles finer than a few microns. Particles having a glassy matrix are less tough below their glass transformation temperature.

In general, the mean size of fracture fragments is smaller when the impact force is higher, but the reduction ratio is also very dependent on the microstructure of the particle. Microstructures, integranular fracture paths, and material defects that cause crack branching may increse the apparent reduction ratio. As seen in Fig. 16.10, the milling rate of coarse-grinding tabular alumina is midway between that of single crystal fused alumina particles and a calcined, porous, finer-grained alumina aggregate.

Milling also produces atomic-scale lattice defects in particles, as revealed by an observed increase in the dislocation density, a change in the index of

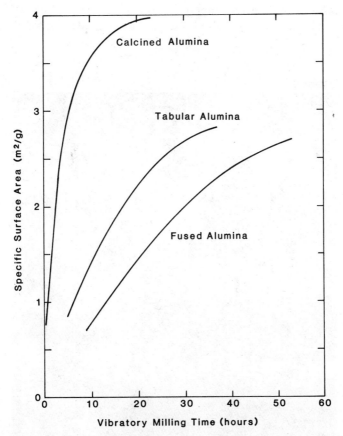

Fig. 16.10 Alumina particles having a higher density and strength have a lower grinding rate. (Data courtesy of SWECO Inc., Florence, KY.)

refraction, a phase transition, and a reduction of the coercive force of hard ferrite particles. These effects are usually less pronounced in wet ball milling, where the impact stresses are lower. When grinding calcined aggregates that contain fine crystals dispersed in a matrix phase, chemical leaching of a less resistant matrix phase by the milling liquid may significantly reduce the milling time and lattice damage.

Chemical mechanisms other than leaching may improve or retard comminution in a particular mill. Water and other chemical species may be adsorbed in cracks and reduce the strength of the material, which should aid grinding. However, water vapor often causes agglomeration in dry milling. Agglomerates absorb some of the impact energy, reducing the energy for particle fracture, and a "hard pack" of agglomerates on the lining of the mill isolates particles from the milling process. Surfactants such as alcohols, oleic acid, glycols, silicones, etc. added at a concentration less than 1% can minimize agglomeration and powder packing. Control of agglomeration is required for efficient and reproducible grinding. In wet milling, a deflocculent is added to disperse agglomerates and to increase the solids loading in the slurry. The deflocculant used should not degrade during the milling process.

Fig. 16.11 Silica glass fibers and fracture fragments produced after ball milling for 0.5 h using 1 cm media.

Fig. 16.12 Stabilized refractory-grade zirconia after ball milling for 3 h; angular particles produced by fracture are rounded by attrition.

It is generally observed that mineral particles from compression crushers and hammer mills are more angular and anisometric than particles produced by grinding. Particles with a low aspect ratio are produced rapidly on ball milling very uniform glass fibers, as is shown in Fig. 16.11. After 1 h no fibers are observed and the preponderance of particle surfaces are fracture surfaces. Both fresh angular fracture surfaces and rounded surfaces due to attrition are observed in ball-milled, stabilized refractory zirconia with a feed size of 200–500 μm (Fig. 16.12).

16.3 MILLING PERFORMANCE

Although the choice of a mill may depend on the ultimate capital cost, capacity, period of a milling cycle, particle size distribution, and material factors, the mill selected must be used efficiently commensurate with wear and maintenance expense. We can gain insight into factors effecting milling performance by examining Eq. 16.3.

$$\frac{\text{Particles generated}}{\text{Time}} = \frac{\text{Media collisions}}{\text{Time}} \frac{\text{Particle impacts}}{\text{Collision}} \frac{\text{Particles}}{\text{Impact}} \qquad (16.3)$$

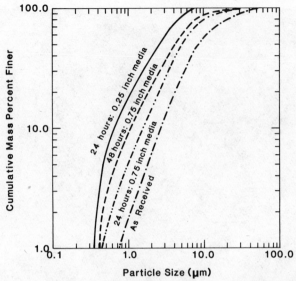

Fig. 16.13 The size reduction of talc (Mohs hardness = 1) during wet ball milling occurs more rapidly using finer media. (From W. Milani, M.S. thesis, Alfred University, Alfred, NY.)

Media Collision Frequency

The probability of fracture is higher statistically when the frequency of collisions is higher (Fig. 16.13). The collision frequency per unit volume of the mill increases rapidly with a decrease in the size of the media (Eq. 13.2) and an increase in the media velocity, which is higher in attrition mills and vibratory mills. In ball mills, the frequency of impacts is limited by the tumbling velocity and the one-dimensional nature of impacts, and because the mill is only partially filled. During ball milling, collisions causing grinding occur primarily in the tumbling layers near the center. In vibratory and attrition mills, collisions causing grinding occur between a relatively larger fraction of the media at some instant of time.

Particle Impact Frequency

The probability of hitting a particle during a collision is higher for cylindrical media, which pack more densely with finer interstices. In vibratory mills, both area and line impact zones occur, as shown in Fig. 16.14. The probability of an impact also increases as the concentration of particles on the surface of the media increases and when the particles are more uniform in size. Particles become more numerous as the particle size decreases. Agglomeration reduces the frequency of impacts. In wet milling, the slurry viscosity should be high enough to reduce the particle mobility and maintain an uniform coating on the media and retain particles in the impact zone.

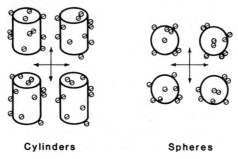

Cylinders **Spheres**

Fig. 16.14 Line and area impact zones between cylindrical media produce a higher frequency of particle impacts.

Particle Fracture

The probability of fracture is high when the grinding stress is large and the strength of the particles is low. Agglomerates may disperse on impact but absorb energy needed for fracture or attrition. Collision forces are dissipated when the slurry viscosity at the shear rate of milling is too high. A combination of shear and compression stresses may enhance the fracture of tough particles of a micron size. Crack branching during fracture produces finer fragments (i.e., the reduction ratio is larger); this is enhanced in crystals of low symmetry, in flaw-loaded particles, and when the fracture velocity is higher (higher impact force, chemical interaction at the crack

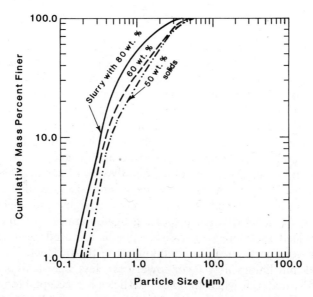

Fig. 16.15 The grinding rate of 95% alumina body for alumina substrates is increased on increasing the solids content and viscosity of the deflocculated slurry.

tip). The relative reduction ratio generally decreases as the particle size becomes smaller.

The factors in Eq. 16.3 indicate that the rate of size reduction decreases as the particles become finer and when no deflocculant is present to prevent the formation of agglomerates. For efficient milling of a slurry, the slurry should be deflocculated to disperse agglomerates, and the solids content should be increased to develop a coating of an adequate viscosity on the media to prevent the escape of particles from the grinding zone but not completely dissipate the grinding stress (Fig. 16.15); this will also minimize media wear and slurry contamination. Larger media are used for larger and stronger feed particles. Smaller media, ranging down to about 1 mm in diameter, are used for milling micron-size particles or dispersing agglomerates, because the rate of fine grinding is very dependent on the frequency of collisions. A grinding action that increases both the frequency and energy of the collisions increases the rate of milling.

16.4 MILLING PRACTICE

In ball milling, impact occurs between tumbling media; attrition also occurs between these media and other rotating media near the bottom. The height of the media before cascading is a function of the media charge, solids loading, angular speed, and the viscosity of the suspension in wet milling. The critical angular frequency ω_{cr} (Hz) causing centrifuging is

$$\omega_{cr} = 0.5R^{-1/2} \qquad (16.4)$$

where R (meters) is the radius of the mill. Adhesion produced by a viscous slurry that wets the mill effectively reduces ω_{cr} and the mill is normally operated at $(0.65-0.85)\,\omega_{cr}$ to produce a maximum media height of 50–60° measured from the horizontal through the center of rotation. A lining with baffle bars may be used to minimize media slippage. Typically, the media charge is about 50% of the volume of the mill, and powder or slurry is added to slightly exceed the void spaces in the media; this combination provides a good compromise between grinding efficiency and rate of wear of the media and lining. Wear increases rapidly when the material does not fill the interstices between the media. The wear resistance of the media and lining should be matched. Smaller media charges and higher material loading are often used when dispersing agglomerates and mixing are the primary objectives. More wear-resistant media are used for tough materials and when wear particles are a contaminant in the product. The net cost of media and mill linings may favor lower-cost, less wear-resistant media for easily milled material when the wear particles are acceptable in the milled product. Media exceeding 8 cm in diameter are available. Media 1.3 cm in diameter are very popular, and 0.6 cm media are extensively used for fine

Rolling Impact   **Rubbing**

grinding. Media of mixed size may wear more rapidly unless the feed particles are soft. In general, the ratio of media size to feed size should exceed 25 to 1. Higher-density and/or larger media are used when the viscosity of the suspension is high. In wet milling the viscosity of the deflocculated slurry is in the range 500–2000 mPa s at the milling temperature. The viscosity should be sufficient to form a film of slurry on the media, which holds particles in the impact zone and protects media from wear, and to minimize slippage between the media and the wall of the mill. Results indicate that when milling to a micron-size chemical deflocculation is critical,

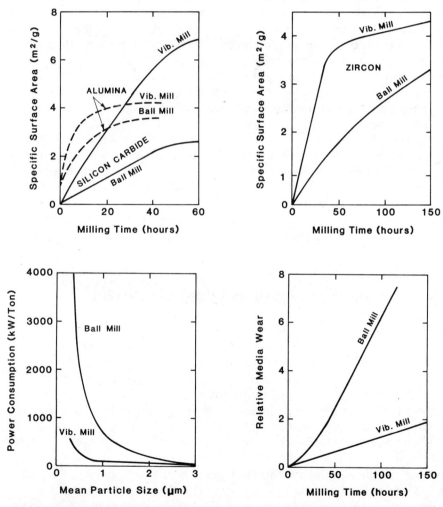

Fig. 16.16 Milling results for calcined alumina, silicon carbide, and zircon and the power consumption and media wear for the vibratory and ball milling of zircon under optimal conditions. (Courtesy of SWECO Inc., Florence, KY.)

the viscosity should not be increased by partial flocculation. When milling a deflocculated slurry, the viscosity may decrease with milling time as more colloids are produced or as the packing efficiency increases.

Low-amplitude wet vibratory milling is normally limited to feed sizes smaller than 250 μm, but feed sizes may range up to about 1.25 cm in high-amplitude mills. Cylindrical media 1.3 cm in diameter are recommended for low-amplitude milling to a fine size, because smaller media do not produce a uniform distribution of vibration energy. Larger media are used in high-amplitude mills. Grinding rates in terms of the increase in the specific surface area of the product are reported to be 10–50% greater than for ball milling when the eccentric weights are properly adjusted. The viscosity of the feed suspension is normally the maximum that can be discharged from the mill with vibration. In low-amplitude milling, the feed suspension should be premixed and fill the void space between the media. Comparative results for the milling of silicon carbide and zircon sand to a submicron size, as shown in Fig. 16.16, indicate that power consumption and media wear are about seven times lower for vibratory milling.

Feed material in attrition mills is usually finer than about 50 μm, and the solids content of slurries can range from 30%–70%. The size of the spherical grinding media is in the range of 0.5–5 mm. Attrition mills are operated at a velocity that produces turbulence and an increase in the apparent volume of the fluidized charge of slurry and media. Very rapid attrition is produced by the intense combined compression and shearing stresses, and the frequency of collisions is very high.

Compressed air, superheated steam, or compressed inert gases may be used as the dynamic medium in fluid energy milling. Feed material may range up to several millimeters in size.

16.5 PARTICLE SIZE DISTRIBUTIONS

The particle size distribution of the milled product will vary with the grain size and fracture behavior of particles of the feed material and the type of mill and its operating conditions. The grain size distribution in many calcined materials is approximately log-normal if regular grain growth has occurred, and the particle size distribution of the milled product is often approximately log-normal if milling disperses the grains. On grinding materials containing particles that are amorphous or single crystals, a somewhat different size distribution that is a function of their different fracture behavior may be produced.

The particle size distribution after the industrial ball milling of a -200 mesh calcined alumina composed of particles that are aggregates of grains finer than 5 μm is shown in Fig. 16.17. The finest particle size is about 0.1 μm, and the distributions are log-normal in form. Milling preferentially reduces particles coarser than the geometric mean $\bar{a}_{gM}$ and adds to the

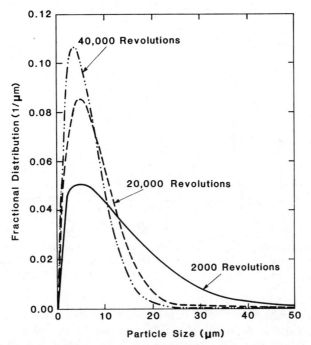

Fig. 16.17 Change in size distribution $f_M(a)$ with mill revolutions on wet ball milling a -200 mesh calcined alumina.

concentration of fines smaller than $\bar{a}_{gM}$. The maximum particle size decreases slowly, and the distribution remains skewed to larger sizes. Particles as coarse as 30 μm remain after milling for 20,000 revolutions. The general effect of increasing the material loading of the mill is to decrease the rate of reduction of the maximum size, and a very broad size range is produced. Dry ball milling typically produces fewer submicron particles because of agglomeration.

The comparative log-normal distribution of sizes produced on ball milling and vibratory milling a calcined, aggregated alumina is shown in Fig. 16.18. A smaller maximum size and a narrower particle size distribution are obtained after vibratory milling, because larger particles are less likely to escape impact in three-dimensional vibratory milling using cylindrical media. Vibratory milling is typically used to grind refractory oxides such as alumina and zirconia to a size finer than 10 μm and titanates and ferrites for electroceramics and zircon opacifiers with a size range of about 0.1–1 μm.

Attrition milling produces powders having a relatively narrow size distribution with a mean size ranging from a few microns to less than 0.1 μm. Fluid energy milling is used to obtain a narrow particle size distribution with an average particle size below 10 μm and relatively little submicron material.

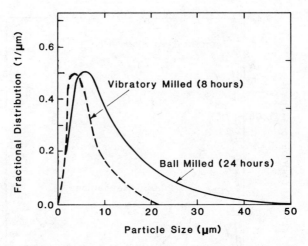

Fig. 16.18 Size distributions f_M (ln a) of -60 mesh calcined alumina after vibratory and ball milling under optimal conditions. Log normal distribution (ball mill $\bar{a}_{gM} = 5.6$ μm, $\sigma_g = 2.24$; vibratory $\bar{a}_{gM} = 2.4$, $\sigma_g = 1.97$). (Data courtesy of SWECO Inc., Florence, KY.)

The rate of size reduction of dense hard particles such as zircon, alumina, and silicon carbide decreases rapidly when the mean size $\bar{a}_{V/A}$ is less than about 1 μm in ball milling, less than 0.5 μm in vibratory milling, and less than about 0.1 μm in attrition milling. For each material and mill, there is a practical grinding limit that is a function of the grinding energy and the strength and toughness of the particles.

16.6 MILLING EFFICIENCY

The history of comminution is replete with equations attempting to describe energy consumption during milling. A quantitative description of milling can aid in selecting comminution equipment, in optimizing the design and operation of a particular mill, and in understanding the mechanics of material breakdown for different milling conditions. In industrial milling situations, the kinetic impact energy is only a fraction of the total energy input. Considerably more energy is consumed in accelerating the container and media, the elastic compression of media and lining materials, shearing viscous liquid, and as heat.

Charles* studied the fracture of glass cylinders of size a_o using a dropped weight crusher of constant mass and observed that the characteristic mean size $\bar{a}$ of the crushed produce ($\bar{a} \ll a_o$) could be correlated to the kinetic energy input U_K:

$$U_K = A(1/\bar{a}^n) \tag{16.5}$$

*R. J. Charles, *Trans. AIME* **208**, 80–88 (1957).

The parameters A and n ($n > 1$) are empirical constants for a particular material and constant mode of fracture. This relation was later confirmed for the compressive fracture of glass spheres. An implication of Eq. 16.5 is that the energy for size reduction increases rapidly as the product size decreases.

In milling, the total energy input in producing a unit of product, U_T, is commonly monitored. A general empirical equation which often approximates the dependence of U_T on a characteristic mean particle size during milling is:

$$U_T = A_c[1/\bar{a}^m - 1/\bar{a}_o^m] \tag{16.6}$$

where A_c is an efficiency constant for a particular milling system and m is a fracture constant for a particular material. Equation 16.6 may be expected to describe the dependence of the energy consumption on particle size during milling when the fracture mode and general form of the size distribution remain constant, so that a single characteristic size can be used to represent the size distribution. The constant m, indicating the particle size dependence, is listed in Table 16.2 for several different materials. Higher values are observed for porous aggregates of fine grains, as would be expected from fracture theory. As is indicated in Table 16.3 for the milling of zircon to a submicron mean size, the lower value of A_c for vibratory milling indicates that the specific energy consumption is less than for ball milling. For the stirred media, mill, the exponent m is constant for both materials, but the apparent efficiency is much higher for the softer limestone.

Results such as those presented in Table 16.3 may be used to evaluate the effect of changing milling conditions or to ascertain the relative efficiency of mills of different capacity or type. In terms of the new surface energy relative to energy input in milling processes, milling is less than 1% efficient.

Table 16.2 Size Dependence Index m in Eq. 16.6 for Several Industrial Raw Materials[a]

Material	m
Fused alumina	1.1
Silicon carbide	1.3
Quartz	1.4
Tabular alumina	1.6
Bauxite	2.4
Calcined alumina (porous aggregate)	4.0[b]
Calcined titania (porous aggregate)	4.4[b]

[a]ball milling and low-amplitude vibratory milling to a submicron mean size. (Data courtesy of Sweco Inc., Florence, KY.)
[b]Varies with calcination temperature.

Table 16.3 Values of m and A_c for Milling to a Submicron Mean Size

Mill	Material	m	$A_c(KW(\mu m)^m)$ (T)	$A_c(KW(\mu m)^m)$ (m^3)
Ball	Zircon[a]	1.8	650	3350
Vibratory	Zircon[a]	1.8	100	520
Attrition	Quartz[b]	1.8	920	2680
Attrition	Limestone[b]	1.8	500	1500

[a]Data courtesy of Sweco Inc., Florence, KY.
[b]From J. A. Herbst and J. L. Sepulveda, *Proceedings of Powder and Bulk Solids Handling Conference*, Chicago, 1978.

A small improvement in efficiency may be a very significant factor in reducing the processing cost.

SUMMARY

Crushing and grinding are important processes used to change the particle size distribution and disperse agglomerates in materials. Feed material finer than about 5 mm may be ground in a variety of mills that differ in capacity, frequency of impacts for each mill, and impact stress. The feed material must be of a specific size range and consistency. Control of the mass and size of the grinding media and the solids concentration and dispersion of the feed slurry are important to maximize the rate of grinding to a particular size with a minimum of contamination. Particle fracture is controlled by the frequency and energy of particle impacts in the mill and the fracture resistance of the particles. Relative to ball milling, vibratory milling and attrition milling produce finer particles and a narrower size distribution at a faster rate. Ball milling is used for high-capacity grinding and when a wide size distribution or mixing and dispersion of agglomerates with a minimum of damage to particle or additive phases is preferred. Control of the mill materials, mill loading, and feed consistency is required to minimize wear and chemical contamination during milling.

SUGGESTED READING

1. Errol G. Kelly and David J. Spottiswood, *Introduction to Mineral Processing*, Wiley-Interscience, New York, 1982.
2. Temple C. Patton, *Paint Flow and Pigment Dispersion*, Wiley-Interscience, New York, 1979.
3. P. Somasundaran, Theories of Grinding, in *Ceramic Processing Before Firing*, G. Y. Onoda and L. L. Hench (eds). Wiley-Interscience, New York, 1978.

4. Wade Summers, Broad Scope Particle Size Reduction by Means of Vibratory Grinding, *Am. Ceram. Soc. Bull.* **62**(2), 212–215 (1983).

5. S. M. Wiederhorn and H. Johnson, Effect of Zeta Potential on Crack Propagation in Glass in Aqueous Solutions, *J. Am. Ceram. Soc.* **58**(7–8), 342 (1975).

6. D. A. Stanley et al., Attrition Milling of Ceramic Oxides, *Am. Ceram. Soc. Bull.* **53**(11), 813–815, 829 (1974).

7. B. Dobson and E. Rothwell, Particle Size Reduction in a Fluid Energy Mill, *Powder Tech.* **3**, 213–217 (1969/70).

8. B. Steverding, Brittleness and Impact Resistance, *J. Am. Ceram. Soc.* **52**(3), 133–136 (1969).

9. L. D. Hart and L. K. Hudson, Grinding Low Soda Alumina, *Am. Ceram. Soc. Bull.* **43**, 1–17 (1964).

10. R. J. Charles, Energy-Size Reduction Relationships in Comminution, *Trans. AIME* **208**, 80–88 (1957).

PROBLEMS

16.1 Define grinding rate, grinding efficiency, and contamination rate.

16.2 What is the approximate volume fraction of particles in a ball mill when milling a slurry containing 40 vol % solids using spherical media?

16.3 Calculate the critical angular frequency of a ball mill 100 cm in diameter and compare this to the 20-Hz motor speeds used in vibratory milling and the 1- to 6-Hz speed of a stirred media mill.

16.4 Calculate and compare the number of contacts per cubic meter in a mill containing spherical media of 3, 6, 13, and 25 mm in diameter. Assume simple cubic packing.

16.5 Calculate and compare the relative mass of spherical media of 3- and 13-mm diameter if composed of alumina, zirconia, steel, or tungsten carbide. What is the mass ratio for spherical and cylindrical media (aspect ratio = 1) of the same nominal size?

16.6 How similar are the Young's moduli E of the materials listed in problem 16.5 and lining materials such as polyurethane, polypropylene, and neopreme rubber? What is the effect of the E of media in milling?

16.7 What are possible causes of the differences in the grinding behavior of the three materials in Fig. 16.10? Compare differences for vibratory relative to ball milling for each material, and explain the results.

16.8 Calculate the relative surface area/bulk packing volume for 3-mm spherical and cylindrical media (aspect ratio = 1) assuming a simple cubic packing configuration for each.

16.9 Chemical additives called grinding aids may be added for a variety of reasons. State three reasons for using additives. Consider both wet and dry milling.

16.10 For dispersion, the frequency of shearing a particular number of particles per unit volume of the mill should be a maximum. How is this effected?

16.11 How should milling to a micron size be conducted to minimize contamination? Consider both the material to be milled and the materials used in the mill.

16.12 Explain why the temperature of the slurry should be monitored and controlled during milling.

16.13 At what shear rate should the slurry viscosity be measured for ball milling and vibratory milling?

16.14 Explain the three m values for alumina materials in Table 16.2 in terms of the microstructure and Griffith fracture theory. (Consult micrographs of calcined and tabular alumina in Chapters 3 and 8.)

16.15 What is the ranking of the grinding parameter A_c for the three materials in Fig. 16.10?

16.16 Consider a cylindrical particle of diameter a in contact with crushing rolls of diameter D and a gap x. The angle of nip is defined by the intersection of tangents at the particle-roll contact. Draw a free body diagram indicating the compressive and shear stresses at the contacts. If the coefficient of friction between the particle and the steel roll is 0.4, what is the critical roll diameter D to reduce particles of $a = 1$ cm to <0.5 cm?

16.17 What is the dependence of the milling rate $df_M(a)/dt$ on $\bar{a}_{gM}$, $f_M(a)$, and $(a_i/\bar{a}_{gM})$?

17

Batching and Mixing

Batch additives may be in the form of a granular material, a powder, a liquid, a chemical solution, an emulsion, or a slurry. Stored raw materials are transported for feeding, proportioned gravimetrically or volumetrically, and then charged into a mixer to form a batch. Mixing combines, distributes, disperses, and intermingles the batch materials differing in chemical and/or physical form.

Mixing must be well controlled to reproducibly wet the particles, disperse additives, and produce a homogeneous batch of the proper consistency. Improper or nonreproducible batching and mixing is often the root cause of microscopic defects in sintered ceramics. In this chapter we will consider batching and mixing equipment and processes, and the analysis of the mixedness of a material system.

17.1 BULK SOLIDS TRANSPORT AND BATCHING

Ceramic processing operations involve the transport of materials that may range widely in consistency. Bins or silos are commonly used for storing incoming dry materials and to provide "surge capacity" when there is a change in the input and output rates. The conical or wedge-shaped portion at the bottom of the bin is called a hopper. Dry solids are conveyed using a continuous belt, bucket, chain, screw, vibratory slide, or pneumatic slide type of device (Fig. 17.1). Slurries are commonly stored in agitated tanks and pumped through pipes using a positive displacement or centrifugal pump to motivate flow. A very viscous slurry or paste may be pumped using a reciprocating, progressing cavity pump.

Bin and hopper designs that produce first-in, first-out flow called mass flow produce less material segregation and supply material that is more uniform in bulk density and flow rate (Fig. 17.2). A decrease in the slope of

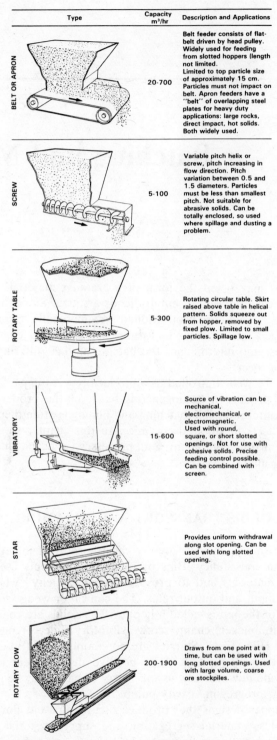

Type	Capacity m³/hr	Description and Applications
BELT OR APRON	20-700	Belt feeder consists of flat-belt driven by head pulley. Widely used for feeding from slotted hoppers (length not limited. Limited to top particle size of approximately 15 cm. Particles must not impact on belt. Apron feeders have a "belt" of overlapping steel plates for heavy duty applications: large rocks, direct impact, hot solids. Both widely used.
SCREW	5-100	Variable pitch helix or screw, pitch increasing in flow direction. Pitch variation between 0.5 and 1.5 diameters. Particles must be less than smallest pitch. Not suitable for abrasive solids. Can be totally enclosed, so used where spillage and dusting a problem.
ROTARY TABLE	5-300	Rotating circular table. Skirt raised above table in helical pattern. Solids squeeze out from hopper, removed by fixed plow. Limited to small particles. Spillage low.
VIBRATORY	15-600	Source of vibration can be mechanical, electromechanical, or electromagnetic. Used with round, square, or short slotted openings. Not for use with cohesive solids. Precise feeding control possible. Can be combined with screen.
STAR		Provides uniform withdrawal along slot opening. Can be used with long slotted opening.
ROTARY PLOW	200-1900	Draws from one point at a time, but can be used with long slotted openings. Used with large volume, coarse ore stockpiles.

Fig. 17.1 Feeding devices for dry solids. (Reproduced with permission from *Introduction to Mineral Processing*, Wiley-Interscience, New York, 1982.)

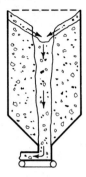 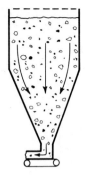

Funnel – Flow Bin   Mass – Flow Bin   **Fig. 17.2** Funnel and mass flow from a bin.

the hopper or an increase in the wall friction of the hopper or internal friction of the material increases the tendency for first-in, last-out flow called funnel flow or "rat holing" (Fig. 17.3). Funnel flow is usually acceptable only for rather coarse materials that do not segregate. An internal baffle or "insert" may decrease the central flow velocity and the tendency for funnel flow.

Powders and granular materials stored in bins and silos are subjected to the compressive load of the overlying material. The internal friction angle of the material can be determined as a function of compressive stress using the direct shear test (Eq. 14.11). A related flow property is the dynamic angle of repose, which is the slope of the cone of material formed when the material is poured and flows freely onto a horizontal surface. Flow also depends on the compressibility and cohesion of the material, as is indicated in Table

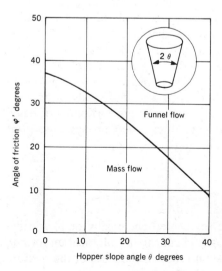

Fig. 17.3 Dependence of type of flow on the slope and kinetic friction angle between the material and wall of a conical hopper. (Courtesy of Jenike and Johanson Inc., Winchester, MA.)

Table 17.1 Effect of Compressibility on Flow of Powders

Compressibility (%)	Unloaded	Flow	Angle of Repose
5–18	Free-flowing granules	Good to excellent	25–35
18–22	Powder and granules	Fair	35–45
22–28	Very fluid powder	Floodable[a]	33–40
>28	Cohesive powder	Very poor	>60

Source: Ralph L. Carr Jr., *Ceram. Age* **11**, 21–28 (1970).
[a] Unstable, gushing, discontinuous flow.

17.1. Cohesive materials gain strength when compressed. Adsorbed moisture tends to increase the cohesion and formation of lumps in common materials, and the flowability decreases with an increase in the moisture content. Cyclic freezing and thawing may produce agglomerates or lumps that impede flow. Compaction and cohesion increase the tendency for blocked flow caused by arching and may cause the flow induced pneumatically to be nonuniform floodable flow, due to variations in the air permeability.

The flow of a powder through an opening decreases significantly when the size of the flow unit is larger than about 0.15 of the opening, because of arching. The free flow of particles and agglomerates finer than about 44 μm is retarded by particle adhesion. Ceramic particles with an adsorbed processing additive may exhibit poorer flowability when the temperature is increased and especially when the melting point is approached or the glass transition temperature of the surface film is exceeded.

Abrasion is a problem when transporting ceramic slurries through pipes. Valves should not restrict flow when open, and elbows in pipe systems should be of a large radius to minimize wear and contamination. Positive displacement pumps provide a volumetric flow that is less dependent on the suction or discharge pressure and are commonly used where a large pressure head is needed. Centrifugal pumps are used to produce a high flow rate.

Materials handling systems for batching ceramics are now commonly automated. A computer may be used to open and close valves, activate feeding systems, verify the gravimetric or volumetric proportioning of batch ingredients to specified tolerances, activate dust collection systems, print batching records, and maintain an inventory of raw materials.

17.2 MIXING AND MIXEDNESS

In the initial batch, raw materials are segregated chemically or physically. Mixing is the process used to improve the chemical and physical uniformity of the mixture. The mixedness of a system refers to the state of the mixture that can be described after chemical or physical analysis.

The length, area, or volume of the largest region of each component in the mixture is referred to as the scale of segregation of that component. In a chemical solution the minimum scale of segregation is limited by the size of the largest molecules, but in a particle system it is limited by the largest particle size. A mixture is homogeneous when the composition does not vary with position. All mixtures are inhomogeneous when examined on a scale that is smaller than the size of the components. The requisite "scale of scrutiny" is the sample size requisite for determining unacceptable variations in the composition or structure of the ultimate product. During mixing the minimum scale of segregation and maximum homogeneity are obtained only when agglomerates and viscous additives are well dispersed.

When two components are initially combined, an unmixed arrangement exists. Mixing caused by random flow will produce, at best, a fully randomized mixture containing incompletely dispersed components. Although randomly mixed, the degree of segregation is larger than for the fully dispersed (nonrandom) mixture, as is shown in Fig. 17.4. The intensity of segregation indicates the relative deviation of a chemical or physical characteristic when a component of one type has been intermingled with components of another type. Idealized mixtures differing in both the scale and intensity of segregation are shown in Fig. 17.5.

The degree of mixedness must be determined quantitatively to assess the homogeneity of the mix and the efficiency of a mixing device. Numerous statistical indices have been proposed to indicate the departure of a mixture from a randomly mixed state. A random mixture of particles differing only in composition but not physical form can be defined in statistical terms as one in which the probability that a particle drawn at random is of a

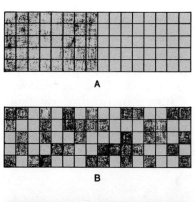

A

B

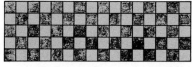

C

Fig. 17.4 Particle arrangements in idealized two-dimensional mixtures: (A) completely segregated, (B) completely random, and (C) completely dispersed.

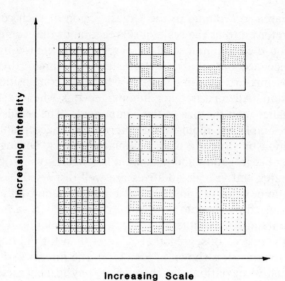

Increasing Scale

Fig. 17.5 Diagrammatic representation of mixtures varying in both intensity and scale of segregation.

particular type depends only on the proportion of that type. According to probability theory, if n particles are withdrawn from a binary random mixture one at a time and each is replaced before the next is withdrawn, the probability $p(r/n)$ that some number r would be of the first component in n trials is given by the binomial theorem

$$p\left(\frac{r}{n}\right) = \frac{n!}{(n-r)!r!}\, p^r q^{n-r} \tag{17.1}$$

The probability of failure $q = 1 - p$. If a sample containing n particles is withdrawn as is common in sampling, the probability given by Eq. 17.1 is correct when n is a very small fraction of the total number of particles in the mixture. If n is large and neither p nor q is close to zero, the binomial distribution is closely approximated by the Gaussian (normal) distribution.

The variance s^2 and its square root, the standard deviation s

$$s = \left[\frac{\sum_{0}^{N}(C_i - \bar{C})^2}{N-1}\right]^{1/2} \tag{17.2}$$

where N is the number of samples analyzed, C_i is the fractional concentration of the component in a sample, and $\bar{C}$ is the mean fractional concentration of that component, are used as indices of mixedness but are dependent on the mean concentration of the component analyzed. Many different indices of mixedness that range from 0 to 1 have been defined in terms of the standard deviation of the original segregated mixture (σ_0), the standard deviation after mixing (s), and the ultimate standard deviation for a completely random mixture (σ_r). The index M, where

$$M = \frac{\log \sigma_0 - \log s}{\log \sigma_0 - \log \sigma_r} \tag{17.3}$$

has been recommended because it provides a clear distinction between mixtures approaching the theoretical optimum.[*] For a completely random two-component mixture of particles of separate identity but otherwise physically uniform, the standard deviation σ_r decreases as the number of particles in the sample n (i.e., the sample size) increases;

$$\sigma_r = \left(\frac{C_1(1 - C_1)}{n} \right)^{1/2} \tag{17.4}$$

In a mixture of two different materials varying in particle size, the standard deviation of a completely random mixture may be calculated using the equation

$$\sigma_r = \left[C_1 C_2 \left(\frac{C_2 \left(\sum fW \right)_1 + C_1 \left(\sum fW \right)_2}{M_s} \right) \right]^{1/2} \tag{17.5}$$

where M_s is the mass of a sample taken and $\sum fW$ is the sum of the product of the weight fraction f of particles in each size class and the mean particle weight W in the class. As defined by Eq. 17.5, the standard deviation of a statistically random mixture σ_r decreases as the proportions of the two components becomes less equal and the mean particle size of the components decreases. The standard deviation of completely segregated material is[*]

$$\sigma_0 = \left[\frac{C_1}{1 - C_1} \right]^{1/2} \tag{17.6}$$

A multicomponent system may be treated as a binary system by considering one component dispersed in a mixture of the other components.

Indices such as s and M are used to assess the spatial uniformity of bulk material—i.e., the macroscale mixedness—as a function of parameters of a mixing operation and to compare the effects of different mixing techniques.

Analyses used to determine macroscale mixedness include chemical, mineral, and phase analysis; colormetric analysis; and the analysis of phosphorescent or radioactive tracers. Samples should be removed from

---

[*] F. H. H. Valentin, *Chem. Eng.* **208**(5), 99–104 (1967).

diverse regions of the mixture following a scheme that will reveal segregation and inhomogeneity. The sampling technique used should not alter the mixedness in any significant way.

Sampling and analysis errors must be kept to a minimum if the calculated variance is to provide an accurate assessment of the mixedness. The uncertainty in the standard deviation calculated for a mixture is low only when a large number of samples are taken (Table 17.2). A comparison of

Table 17.2 Uncertainty in the Ratio of Estimated s to True s_t Standard Deviation

| | Range of s/s_t | |
N	95% Confidence	99% Confidence
5	0.60–2.87	0.52–4.39
10	0.69–1.82	0.62–2.29
15	0.73–1.58	0.67–1.85
20	0.76–1.46	0.70–1.67
50	0.84–1.24	0.79–1.34
100	0.88–1.16	0.85–1.22
200	0.91–1.11	0.88–1.14
500	0.94–1.07	0.91–1.09

Source: F. H. H. Valentin, *Chem. Eng.* **5**, 99–104 (1967).

Fig. 17.6 Organic binder segregated in barium titanate appears dark in (A) secondary electron emission and (B) backscattered electron emission. (Courtesy of Ward Votava, Alfred University, Alfred, NY.)

the mixedness of two systems may be meaningless when only a few samples have been taken.

Microscopic analyses and microscale chemical analyses may be used to determine the scale of segregation in an individual sample—i.e., the microscale mixedness; examples of microscale inhomogeneities are shown in Fig. 17.6. Microscopic examination and chemical reactivity that depends on the diffusion path between components may be used to study dispersion and microscale mixedness. Agglomerates may increase the diffusion path and the time or temperature to complete a chemical reaction. Organic inhomogeneities of low mean atomic number may be identified conveniently using the backscattered electron detector of the scanning electron microscope.

17.3 MIXING MECHANISMS

Mixing is produced by the mechanisms of convection, shear, and diffusion (Fig. 17.7). Convection transfers components from one region to another. Shear increases the interface between components by deforming their shape. Diffusion interchanges molecules and particles randomly between neighboring regions of the mixture.

The relative importance of each mixing mechanism depends on the design of the mixer, the consistency and rheology of the material, and the energy input in mixing. Flow produced by feeding devices and stirring and baffling contribute importantly to convection. Mixer surfaces moving at different velocities will produce a shear stress gradient in intervening material if slippage does not occur. Intense shear stresses can disperse agglomerates and mix viscous materials on a microscale. Eddy currents, cavitation, impact, and sonic vibration can produce turbulence, which disperses agglomerates and accelerates diffusive mixing.

The relative contribution of the mixing mechanisms is dependent on the design of the mixer and the Reynolds number during material flow. Bulky

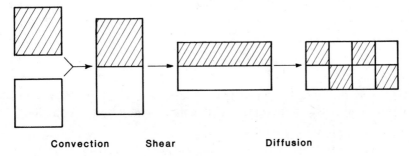

Convection   Shear   Diffusion

Fig. 17.7 Flow of two materials during mixing and mixing mechanisms.

powders and liquids and suspensions of low viscosity may be mixed by convection, shear, and turbulence. In the mixing of very viscous liquids, pastes, slurries, and plastic bodies, only laminar flow commonly occurs, and the mixing mechanisms are convection and shear.

17.4 MIXING EQUIPMENT AND PRACTICE

Each material system has a particular consistency and flow resistance that may vary with the time of mixing and flow rate. Agglomerates in the mixture have a range of strengths, and a mixing intensity in excess of that required for bulk flow is normally required to disperse the agglomerates. Mixing at a low shear rate but a high shear flow, the product of shear rate and time is often used for the bulk mixing of low-viscosity suspensions. The shearing stress developed can be increased by initially mixing at a higher solids content and viscosity and then admixing additional liquid. Relatively high shear stresses produced at a high shear rate are usually required to completely disperse agglomerates in a low-viscosity suspension. Many commercial mixers are designed with two or more mixing elements to produce both high shear mixing in a local region of the mixture and low shear bulk mixing. When mixing a system of high viscosity, considerable torque and mixing energy are required for both the dispersion of agglomerates and bulk flow. High-torque mixers for ceramic systems must be rugged in design, abrasion-resistant, and relatively expensive. Material systems of high viscosity such as an extrusion body may be mixed as a slurry with excess liquid which is removed later by filter pressing or spray-drying.

Slurries and Pastes

Each type of mixer is effective over a particular range of viscosity, as is shown in Fig. 17.8. Impeller mixers that produce both circulation and circumferential flow (Fig. 17.9) are used for mixing a wide variety of suspensions and slurries. In blungers and storage tanks, agitation and flow are produced by rotating paddles, a marine propeller, or a turbine impeller. The flow type in an impeller mixer can be characterized using a Reynolds number (Re),* where

$$\text{Re} = \frac{ND_L(\text{Dia})^2}{\eta} \tag{17.7}$$

where N is the impeller speed (rev/time) and Dia is the diameter of the agitator. The power requirement P is approximated by the equation*

---

* E. J. Kelly and D. J. Spottiswood, *Introduction to Minerals Processing*, Wiley-Interscience, New York, 1982.

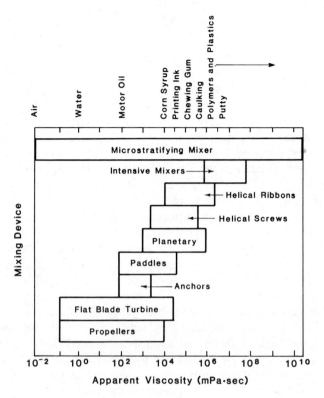

Fig. 17.8 Working ranges of mixers for mixing slurries, pastes, and plastic bodies. (Courtesy of TAH Industries Inc., Imlaystown, NJ.)

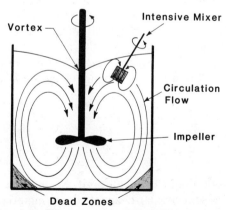

Fig. 17.9 An impeller mixer with a secondary intensive mixing device.

$$P = C_d N^3 D_L (\text{Dia})^5 \qquad (17.8)$$

where C_d is the drag coefficient for the impeller. An increase in those factors that increase Re also increase P, and a single impeller is not satisfactory for creating bulk volumetric flow and turbulence in a slurry of high viscosity. Small, high-speed (several 1000 rpm), serrated, or vaned impellers or an ultrasonic vibratory probe may be introduced to produce cavitation and turbulence for dispersing agglomerates and friable particles in a local region. The high-intensity mixing device may also be installed in line to continuous-ly disperse material entering or leaving the tank. The flow of a suspension through an orifice can produce intense microscale mixing if turbulent flow occurs. Ultrasonic tanks, probes, and throats provide a controlled, intense dispersion of suspensions by creating cavitation. Slurries of moderate vis-cosity are also mixed using anchors, paddles, and planetary impellers. Mixing under a vacuum to remove air from the slurry is accomplished by using a vacuum tight cover; the elimination of bubbles may also aid in dispersing agglomerates. These systems are also commonly mixed in ball mills and high-amplitude vibratory mills. Media impacts disperse agglomer-ates, and the tumbling media provide convection and shear for bulk mixing.

More viscous slurries and pseudoplastic pastes may be mixed in a planetary mixer, helical mixer, or intensive mixers (Fig. 17.10). Mixing elements are relatively large and strong to provide the high torque needed for shear flow in bulk. These mixers are sometimes provided with a heating or cooling jacket to improve wetting, control the viscosity more precisely, or to prevent overheating.

A microstratifying mixer produces a controlled stratification in the ma-terial rather than random flow. Static mixing elements fixed in a conduit divide, invert, and radially mix material received from the preceding element with a programmed precision and efficiency, as is indicated in Fig. 17.11. Two inlet streams become four alternating streams after the first element, eight after the second element, etc. The number of strata increases exponentially with the number of mixing elements n and the striation thickness X_s is

$$X_s = \frac{\text{Dia}}{2^n} \qquad (17.9)$$

where Dia is the inside diameter of the unit. Shear force produced during circulation flow in the semicircular channel causes material to migrate from the center of the pipe to the inside wall every 5 elements. The mixer may be jacketed for control of temperature. Some axial mixing is provided by the constantly changing velocity profile. Microstratification may be used for mixing liquids, suspensions, and pastes ranging widely in viscosity (Fig. 17.8).

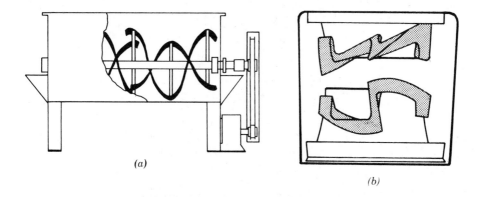

(a)

(b)

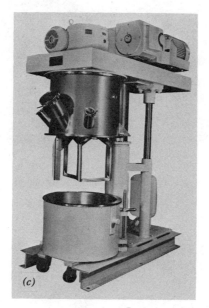

(c)

Fig. 17.10 Intensive mixers of the (a) helical ribbon type and (b) sigma type. (c) A double planetary mixer

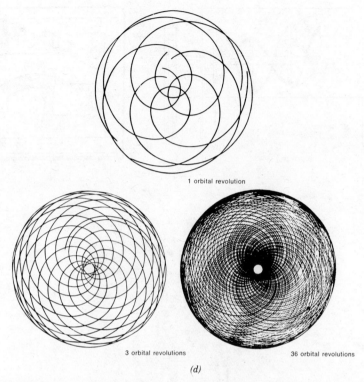

1 orbital revolution

3 orbital revolutions

36 orbital revolutions

(d)

Fig. 17.10 (d) Its mixing action. (Photos c and d courtesy of Chas. Ross and Son Co., Hauppauge, NY.)

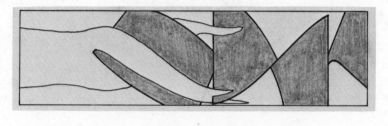

Element Number

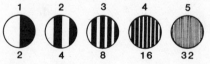

Number of Striations

Fig. 17.11 Mixing produced by flow through a microstratifying mixer. (Courtesy of TAH Industries Inc., Imlaystown, NJ.)

Plastic Bodies

High-energy mixers constructed from wear-resistant materials are required to directly mix ceramic bodies having a plastic consistency. Cylindrical pan mixers, which may vary somewhat in design, are used for mixing plastic material. Blades or plows direct the flow of material, and heavy Muller wheels rolling over material or rapidly rotating blades or vanes may be used to provide local high-shear energy to break up agglomerates (Fig. 17.12). Liquids are commonly introduced as a spray after dry solids have been mixed. Batching, mixing, and discharge can be automatically controlled, and the temperature of the pans may be controlled. Pan-mixed material is often mixed further in a pug mill. In the pug mill, angled paddle-type knives cut and shear material and then consolidate it under pressure and extrude it onto a carrier. Large de-airing pug mills can continuously process tons of material per hour. Plastic filter cake is often preblended in a version of a pug mill called a wad mill. Intensive mixers that typically have very heavy duty sigma blades and heavy-duty double-planetary mixers (Fig. 17.10) are also used widely for mixing plastic bodies in relatively small batches.

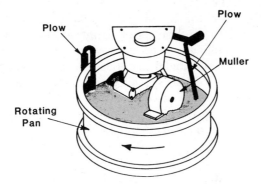

Fig. 17.12 Cylindrical pan mixer. (Adapted from Lancaster Mixers Lebanon, PA.)

Granular Materials and Powders

Granular materials and powders are typically mixed with a small amount of liquid, binder solution, etc. using a pan mixer, conical blender, ribbon mixers, or rotating drum mixers. Muller-type wheels or rotating beaters are required to disperse agglomerates and more uniformly distribute the liquid or solution introduced using a spray nozzle. Wetted powders are not free-flowing unless they become granulated after distribution of the liquid additives. The Muller wheel or rotary disperser may be elevated to allow for granulation. Conical blenders are also widely used for mixing free-flowing powders and liquid additives introduced as a spray (Fig. 17.13). Tumbling

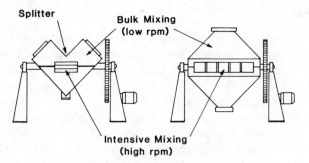

Fig. 17.13 V-type and conical mixers.

flow provides convection and shear mixing; in the double-cone blender, the intersection of the cones splits the flow once every revolution. Models with an internal intensive mixing bar provide additional energy for comminuting agglomerates in dry feed powders and those formed on admixing liquid. Tumbling without intensive mixing produces granulation. Construction may be of either plastic or stainless steel, and some models are available with a thermostatically controlled jacket.

When mixing slurries, the liquid and soluble processing additives are normally introduced into the tank prior to introducing particle materials. Automatic material handling systems convey, meter, and charge materials under computer control. Colloidal materials and recycled scrap is usually slurried separately at a particular specific gravity and added prior to coarser materials. When mixing more viscous slurries and pastes, solids may be blended dry before adding to the liquid or a colloidal suspension. Intense dispersers may be incorporated in the mixing tank or in the piping system following discharge.

In mixing granular materials and powders to produce pressing material or a plastic body, the extreme fines tend to adsorb the liquid and form granules. Prewetting of the coarser particles, a fine liquid spray, and the use of Muller wheels or rotary beaters to disperse agglomerates aid in producing coarse particles coated with wetted fines. Systems with porous particles or extremely fine colloids may absorb considerable liquid. Powders are usually dry-blended and deagglomerated before adding liquids. Humidified aging may improve the moisture uniformity. A viscous binder solution is not easily admixed uniformly into fine ceramic powder. Pressing powders for high-performance technical ceramic are commonly mixed as a slurry and spray-dried into pressing granules to achieve a higher degree of homogeneity.

17.5 MIXING PERFORMANCE

Mixing is a complex process, and the performance of a mixing system is usually evaluated experimentally. However, a simple model can aid in visualizing the process. A general mixing law for the rate of mixing dM/dt produced in a particular mixer is

$$\frac{dM}{dt} = A(1 - M) - B\phi_u \qquad (17.10)$$

where M is the mixedness parameter, A and B are mixing and unmixing constants, respectively, for a particular mixing system, and ϕ_u is the unmixing potential of the material. The constant A is larger when the mixing action is intense relative to the cohesive resistance of the material. The unmixing potential ϕ_u indicates the tendency of materials to demix owing to differences in the particle size, shape, density, and surface attraction, and B is a function of the dynamics of the mixer.

The mixedness initially increases with the number of revolutions, but on nearing a maximum, the rate levels off (Fig. 17.14). The maximum mixedness achieved depends on the mixing and unmixing rates for the system, and the optimum mixing time depends on the unmixing potential. A change that reduces the unmixing potential such as an increase in viscosity may reduce the mixing constant A as well.

The increase in mixedness with number of revolutions for the mixing of initially segregated 5-mm spherical particles in a two-dimensional double-cone blender is shown in Fig. 17.15. The rate of mixing dM/dt is dependent on both the number of rotations and the mixer loading. There is an optimum residence time for each loading condition, and the number of revolutions required to achieve a nearly random mix increases as the loading increases.

The correlation of the microstructure of a ceramic material with mixing parameters has been reported in relatively few studies. Macroscale nonuniformity arises from nonflow, biased flow, and segregation. Relatively little mixing occurs in stagnant regions, where convection is poor. The ability of a small, high-speed impeller to distribute batch components in the bulk and disperse all agglomerates decreases as the apparent viscosity increases as is shown in Fig. 17.16. Settling of larger or denser particles is a problem when mixing a low-viscosity suspension. Chemical changes during

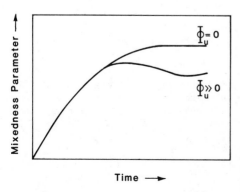

Fig. 17.14 The general dependence of mixedness on mixing time described by Eq. 17.10.

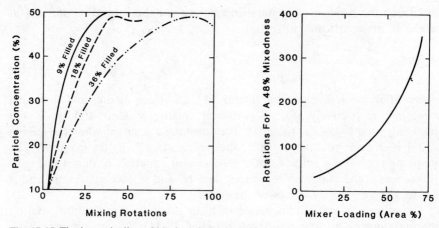

Fig. 17.15 The intermingling of black and white spheres is a function of the mixing revolutions and the loading of a two-dimensional version of a v-blender. The initial feed was totally segregated, and the relative concentration of each type was 0.5.

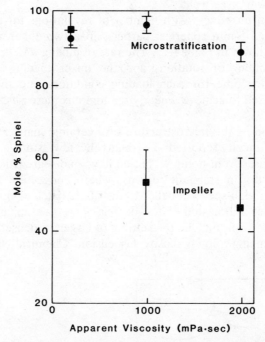

Fig. 17.16 The average amount of $ZnAl_2O_4$ spinel formed on reacting samples of mixed ZnO and Al_2O_3 powders (in slurry form) depends on the homogeneity of the mixture; at a higher viscosity, the dispersion of ZnO aggregates, and bulk convection is reduced using a small impeller.

294

mixing that cause agglomeration may also produce segregation. Fibrous particles or colloidal particles may adhere more strongly to mixer surfaces. When mixing dry powders ranging widely in size, vibrations may cause fine particles to percolate down through pores in the coarse particles. Corners, surfaces of mixing hardware, and wear pockets may collect poorly mixed material which may later escape into the bulk. Close tolerances between moving and static portions of the mixing hardware aid in minimizing stagnant zones.

Difficult mixing situations include the admixing of a high-viscosity solution into a fine powder or low-viscosity liquid, dispersing agglomerates in a high-viscosity matrix, and dry-mixing a minor additive uniformly in a powder or granular material. A sensitive additive is sometimes premixed with a component of the batch and its state checked before adding to the total mixture.

SUMMARY

Raw-material handling systems and batching have become highly mechanized and automated, and factors affecting the flow behavior of materials must be recognized to ensure proper flow and to minimize segregation. Mixing is used to reduce the scale of segregation of the components and improve the homogeneity of the mixture. The degree of mixedness is determined by analyses and is specified using a parameter such as the standard deviation or mixedness index, which indicates the variation of the concentration of a component from the mean for a set of samples. Analysis is also required to determine the dispersion of agglomerates and mixedness on a microscale.

The choice of a mixer depends on the consistency and flow properties of the material, the homogeneity required, and the batch capacity. More energetic mixing is required when the flow resistance is high. Demixing tendencies reduce the degree of mixedness achievable and may be influenced by reducing the particle size or by chemical coagulation and flocculation. The rate of mixing and homogeneity achieved depend on the mechanics of the mixing device, the degree of loading, and the chemical and physical properties of the mixture.

SUGGESTED READING

1. Errol G. Kelly and David J. Spottswood, *Introduction to Mineral Processing*, John Wiley and Sons, New York, 1982.
2. V. W. Uhl and J. B. Gray, *Mixing: Theory and Practice*. Vol. 1, Academic Press, New York, 1966.
3. Milton N. Kraus, *Pneumatic Converying of Bulk Materials*, McGraw-Hill, New York, 1980.
4. Richard Hogg, Grinding and Mixing of Nonmetallic Powders, *Am. Ceram. Soc. Bull.* **60**(2), 206–211, 220 (1981).

5. Thomas D. Croft, James S. Reed, and Robert L. Snyder, Micros-tratified Mixing of Ceramic Systems: Viscosity Dependence, *Am. Ceram. Soc. Bull.* **57**(12), 1111–1115 (1978).

6. G. Daniel Lapp, A Numerical Index of Stirrer Performance in Continu-ous Flow, *Am. Ceram. Soc. Bull.* **56**(2), 186–188, 193 (1977).

7. H. M. Sutton, Flow Properties of Powders and the Role of Surface Character, in *Characterization of Powder Surfaces*, G. D. Parfitt and K. S. W. Sing (eds.), Academic Press, New York, 1976.

8. J. M. Pope and K. C. Radford, Blending of UO_2–ThO_2 Powders, *Am. Ceram. Soc. Bull.* **53**(8), 574–578 (1974).

9. M. Cable and J. Hakim, A Quantitative Study of the Homogeneity of Laboratory Melts Using Simple Stirrers, *Glass Tech.* **14**(4), 90–100 (1973).

10. H. Ries, The Importance of Mixing Technology for the Preparation of Ceramic Bodies, *Interceram.* **3**, 224–228 (1971).

11. Jerry R. Johanson, Working Guide: Feeding and Storage Methods, *Ceram. Age* **10**, 25–30 (1970).

12. Ralph L. Carr Jr., Properties of Solids: How They Affect Handling, *Ceram. Age* **11**, 21–28 (1970).

13. R. C. Rossi, R. M. Fulrath, and D. W. Fuerstenau, Quantitative Analyses of the Mixing of Fine Powders, *Am. Ceram. Soc. Bull.* **49**(3), 289–292 (1970).

14. F. H. H. Valentin, The Mixing of Powders and Pastes: Some Basic Concepts, *Chem. Eng.* **208**(5), 99–104 (1967).

15. K. Stange, Degree of Mixing in a Random Mixture as a Basis for Evaluating Mixing Experiments, *Chem.-Eng. Tech.* **26**(3), 331–337 (1954).

16. P. M. C. Lacey, Developments in the Theory of Particle Mixing, *J. Appl. Chem.* **4**(5), 257–268 (1954).

PROBLEMS

17.1 Explain why vibrations might cause a fine powder stored in bin to flow more poorly.

17.2 Should the outlet dimension X_B of a bin depend on the moisture content of the material in the bin? Explain, using the relationship $X_{arch} < 2\sigma_y/D_b g$, where X_{arch} is diameter for which arching is expec-ted, σ_y is the yield strength of consolidated material, D_b is the bulk density of the material, and g is the acceleration of gravity, in composing your answer.

17.3 When a material varying widely in particle size distribution is charged into the center of a bin, the coarser particles tend to segregate at a greater radius (nearer the wall). Explain.

17.4 When the bin of material in problem 17.3 is unloaded, describe the particle size distribution of the material as a function of unloading time for the cases of mass flow and funnel flow.

17.5 Explain pneumatic-induced flow using the concept of effective stress. Why may air pressure reduce the tendency for arching?

17.6 The appropriate propeller size for mixing a slurry depends on the size of the mixer and the viscosity of the slurry. Explain why it is difficult to produce both bulk flow in the most remote parts of the tank and local turbulent flow using a single propeller when the viscosity of the slurry increases.

17.7 Large shear stress gradients may aid in dispersing agglomerates in viscous systems. Under what conditions are large gradients produced in mixing? Illustrate a minor dispersed phase of both a small and large scale of segregation dispersed homogeneously and inhomogeneously. Illustrate mixtures that are homogeneous with a nonuniform scale of segregation and inhomogeneous but with a uniform scale of segregation.

17.8 Prepare two grids containing 90 squares numbered 10–99. On one grid shade one-half of the squares corresponding to a microstratified dispersion. Generate random numbers and shade one-half of the other grid corresponding to a random mixture. Contrast the scale of segregation of each and the scale of scrutiny that can reveal each to be inhomogeneous.

17.9 The weight percent concentrations of an ingredient in 10 samples removed at random from a mixture after 10 min of mixing are 20.2, 21.1, 21.0, 16.4, 20.3, 17.1, 22.0, 20.4, 23.4, and 20.1 wt%, and after 30 min, 19.5, 21.0, 23.0, 18.6, 20.1, 18.2, 23.0, 19.1, 21.2, and 19.3 wt%. Calculate the standard deviation for each data set. Did the additional 20 min of mixing improve the mixedness (use a 95% confidence interval)? Explain.

17.10 If the standard deviations calculated in problem 17.8 had been obtained for 20 samples, would the conclusion about the effect of additional mixing time be different?

17.11 Show that the standard deviation for a completely random mixture of black-and-white particles, otherwise identical, increases as the fraction of the minor component increases (assume an invariant sample size).

17.12 Calculate the standard deviation for a completely random mixture of equal proportions of alumina and quartz (Table 7.6). The sample size is 5 g.

17.13 If the experimental standard deviation of the mixture in problem 17.11 is 0.09 after mixing for 20 min, calculate the mixedness index (M).

18

Particle Separation, Concentration, and Washing Processes

The microstructure of the final ceramic product is very dependent on the range of particle sizes and impurities in the processing system. The separation of particles by size is a rather common operation in ceramic processing. Impurity particles in raw materials and those introduced during processing such as from the abrasion of materials handling and processing machinery are sometimes removed based on a difference in size, density, surface behavior, or electromagnetic properties. Liquid separation processes are used to remove soluble impurity salts and to reduce the concentration of ions that interfere with deflocculation or coagulation processes. Separation processes are also used to remove liquid to concentrate particles and change the consistency of the system and to remove particles from recycled waste liquids.

Separation and concentration processes have long been used on an industrial scale for beneficiating industrial minerals and in the processing of traditional ceramics. These beneficiation operations should also be considered in developing processes for producing many of the newer ceramics.

18.1 PARTICLE SIZING

Continuous screening and batch-sieving operations are used to directly separate rather coarse particles. Finer particles are separated by differences in their settling rate, which is called classification. Sizing may involve the separation of particles into a relatively narrow range of size, or discrete size ranges, or it may eliminate extremely coarse or fine sizes. Cyclone and centrifugation techniques are commonly used for particle sizing in the subsieve range of size.

298

Screening Techniques

Screening apparatus is commonly used for the continuous separation of particles ranging from about 200 mm down to about 37 μm in size. A vibratory continuous wet-screening apparatus used for multiple separations is shown in Fig. 18.1. Particle motion is assisted by mechanical vibrations which cause shear flow over the mesh and by entrainment in the liquid passing through apertures in the screen. Standard sieves ranging up to about 2 m in diameter are used, and screening rates of several thousand gallons per hour for slurries with a nominal viscosity ranging up to 5000 mPa s are attainable.

Both dry forced-air screening and sonic sieving are used for the dry separation in the range of 850 to about 37 μm. Production models may be configured in series to obtain multiple size cuts. Disk and rotary drum sieves are used for high-capacity dry sieving.

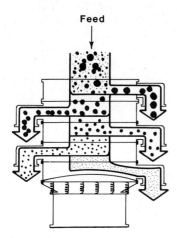

Feed

Vibratory Sieving

Fig. 18.1 Vibratory continuous wet screening separation. (Courtesy of Sweco Inc., Florence, KY.)

Classification Techniques

The purpose of classification is generally separation by size, but other factors, such as particle density and shape, fluid flow, and hindered particle flow, affect the separation. Classification by differential sedimentation is widely used in minerals processing, and a wide variety of equipment is available. A detailed discussion of these operations may be found in E. G. Kelly and D. J. Spottiswood, *Introduction to Mineral Processing* (Wiley-Interscience, New York, 1982).

Cyclone separators, which are relatively simple in construction with no moving parts, are widely used for particle separation in ceramic processing.

Cyclones are inertial separators. The feed suspension enters the cyclone tangentially under pressure and circulates in a spiral path, as shown in Fig. 18.2. The centrifugal force on a particle is opposed by the inward-directed drag force of the fluid. Particles coarser than the cut point migrate downward in the primary spiral along the wall and are discharged. Finer particles migrate into the secondary vortex moving upward and are discharged with the majority of the fluid. The mechanics of flow is rather complex, and design is based on empirical correlations of the parameters listed above for classifiers and the feed pressure and size of the apparatus. The cut size is a function of the settling rate of particles suspended in the slurry, the gradient of the spiral, the fluid velocity, and the size of the classifying chamber. The sharpness of the cut is high when the particle density and shape and fluid currents are very uniform. Industrial models may accommodate a feed rate ranging up to about 1 ton/h and a cut size (separation) in the range of 0.5–50 μm. Laboratory models are available for classifying a few grams of material. Hydroclones are cyclones used for separating particles in liquid slurries and suspensions. Industrial models provide a cut in the range of

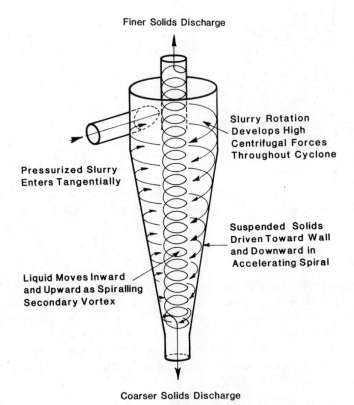

Fig. 18.2 Hydroclone separation indicating fluid flow.

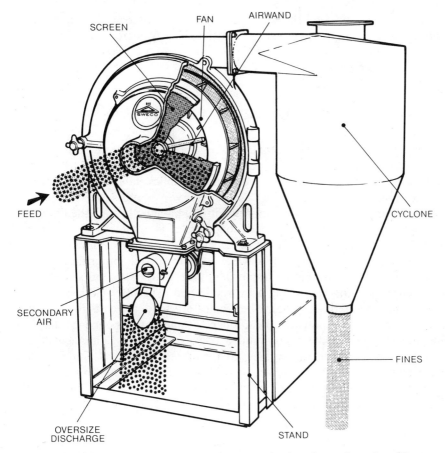

Fig. 18.3. Continuous disk screening and cyclone separation in series configuration. (Courtesy of Sweco Inc., Florence, KY.)

2–100 μm. Hydroclones are available with wear-resistant or expendable polymer linings. A progressing cavity pump is used to provide feed slurry under a controlled pressure. Industrial units processing up to several hundred liters per minute may be connected in parallel to a manifold to increase the capacity and efficiency. Models that process a few liters per minute are available for laboratory samples.

Two widely different size cuts may be produced by combining a screening device in series with a cyclone or a hydroclone (Fig. 18.3). Coarse particles are separated first by a screen. Finer particles entrained in the fluid are fed into a cyclone for a second size cut.

Centrifuging

Centrifuges are used for separate particles that settle under conditions described by Stokes' law (Eq. 7.1). At particle volume concentrations higher than about 5%, the average density and viscosity of the medium through which the particle settles are significantly increased, and particle settling is hindered. Centrifugal classification is used for separating particles by size, as was discussed in Chapter 7.

18.2 FILTRATION AND WASHING PROCESSES

Filtration is widely used in ceramic processing to concentrate particles from a suspension or slurry. The plate and frame filter press (Fig. 18.4) is the most common filtering device used in the ceramics industry for reasons of low maintenance and flexibility. The press receives pressurized suspension and distributes it to the filter surfaces. Filtrate passing through the filter exits through an outlet duct. A solid concentrate is retained in the frames between the filter surfaces. Some models (Fig. 18.5) have separate ducts for introducing and removing wash water with dissolved impurities. Modern filter presses have alternating frames and filter plates hung on a rack which are compressed together using a hydraulic ram. The filter cloth is nylon or some other synthetic material. The frames are made of cast iron, stainless steel, high-strength polymer, or rubber. Frames are opened by hand or hydraulically, and cake drops onto a conveyor below. Equipment size is directly dependent on the surface area required to form the product at a particular rate. A common practice is filtration at a constant rate followed by constant pressure filtration to minimize the initial loss of fine particles and maintain a constant liquid constant.

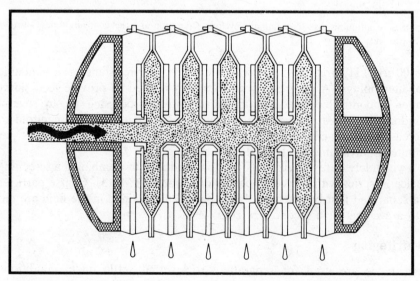

Fig. 18.4 Center feed industrial filter press. (Courtesy of Netzsch Inc., Exton, PA.)

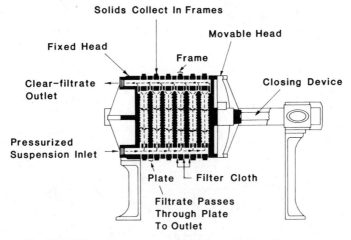

Fig. 18.5 Filter press for concentrating and washing filter cake.

Rotary disk, rotary drum, and horizontal vacuum filters are used for the continuous filtering of high-permeability precipitates, slurries, etc. Vacuum is used to remove water, and filter cake adhering to a belt, rotating disk, or drum is removed after rising above the slurry.

Liquid Permeability

Filtration depends directly on the permeability of the filter cake, and the coefficient of permeability is an important property of any porous system. A model for the permeability of a packed bed of particles was introduced by D'Arcy in 1856 to explain permeation through a sand filter. He observed that the uniaxial volume flow rate Q through the filter (Fig. 18.6) was proportional to the differential pressure ΔP across the filter of thickness L and cross-sectional area A. The D'Arcy equation is

$$Q = \frac{K_{D} A \, \Delta P}{L} \tag{18.1}$$

where K_{D} is the D'Arcy permeability coefficient. Normalizing Eq. 18.1 for the viscosity of the fluid η_{f}, the coefficient of specific permeability K_{P} is

$$K_{P} = \frac{\eta_{f} Q L}{A \, \Delta P} \tag{18.2}$$

Equation 18.2 is valid when the fluid flow through the pores is laminar, which may be estimated from the equation

$$Re = \frac{D_{f} \bar{v} a_{V/A}}{\eta_{f}} < 1 \tag{18.3}$$

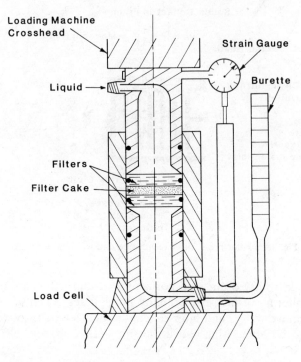

Fig. 18.6 Apparatus for determining the coefficient of permeability and compressibility of filter cake.

where $\bar{v} = Q/A(1 - \phi)$ is the apparent velocity of the fluid of density D_f in the pore channels among particles with a mean size $\bar{a}_{V/A}$.

Capillary flow models assume that the connecting interstices among particles can be represented as a network of closed channels. Kozeny considered the porous system to be a bundle of parallel, uniform, but tortuous channels of total length (L_T) and hydraulic radius (R_H). The hydraulic radius (R_H) is defined as the cross-sectional area divided by the wetted perimeter of the channel and is calculated from the specific surface area per unit volume of the particles S_S:

$$R_H = \frac{\phi}{S_S(1 - \phi)} \tag{18.4}$$

Carmen showed that the apparent liquid velocity was also dependent on the tortuosity of the pores. The Carmen-Kozeny* equation is

$$K_P = \frac{\phi R_H^2}{K_0} \cdot \tag{18.5}$$

---

* P. C. Carmen, *Trans. Inst. Chem. Eng.* (Lond.) **16**, 168–188 (1938)

where K_0 is a constant representing the shape and tortuosity of the pores and has a value of 5 for a cylindrical capillary.

Interconnected interstices among packed particles have a varying cross-sectional area, and the permeability is very dependent on the minimum pore dimension. Lukasiewicz† has shown that the specific permeability of interstitial pores is better represented by the equation

$$K_P = \frac{\phi \bar{R}_L^2}{K_L} \tag{18.6}$$

where $\bar{R}_L$ is the volume average linear mean pore radius determined by mercury penetration porosimetry and K_L is a constant for a particular pore network. For a cylindrical pore, $K_L = 8$.

Fluid migration is higher when the pressure differential and permeability are high and the viscosity of the fluid is low. The permeability is relatively more dependent on the effective pore size than the porosity (Fig. 18.7). In compacts of dry powders, agglomeration can cause a significant increase in the effective $\bar{R}_L$ and K_P, although the porosity is only slightly changed.

The permeability of packings of particles with a continuous distribution of particle sizes depends significantly on the form of the size distribution. Figure 18.8 illustrates four different distributions with the same range of sizes; the permeability decreases for the sequence A through D as a result of more fines in the system packing to give smaller pore interstices. For

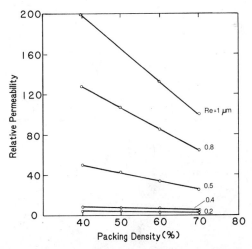

Fig. 18.7 Dependence of relative permeability coefficient on packing density and pore radius (calculated from Eq. 18.6, base condition is Re = 0.1 μm and PF = 0.70).

† S. Lukasiewicz, Ph.D. thesis, Alfred University, Alfred, NY, 1983.

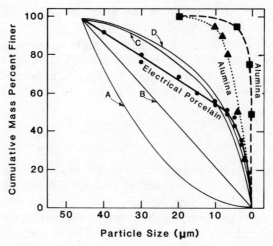

Fig. 18.8 Cumulative size distributions having the same maximum and minimum particle size which vary greatly in packing and permeability and curves for a reactive and a calcined alumina. (After M. K. Bo et al., *Trans. Inst. Chem. Engr.* **43**, 228–232 (1965).)

distributions of type A and B, increasing the size range will increase K_P, but for those of type C and D, an increase in the size range will decrease the permeability. The specific permeability is sensitive to the pore structure and size influenced by the particle packing, as is indicated in Table 18.1. The presence of an organic binder in the liquid may drastically reduce K_P. The adsorbed polymer chains may cause a slight increase in ϕ, but the bridging of polymers across pores reduces the effective hydraulic radius and the permeability.

Table 18.1 Permeability Coefficients

System	Porosity (%)	$K_P (10^{-18} \text{ m}^2)$
Gypsum mold	40	28,600
Porcelain casting body (deflocculated)	35	30
Hole china body (flocculated)		170
Electrical porcelain body (50% clay)		
Filter cake (flocculated)	43	85
From pug mill	29	12
Compressed at 2.5 MPa	25	5
Alumina powder ($<5 \mu$m)		
Flocculated cast	56	1,100
Deflocculated cast	30	0.8
Deflocculated cast (0.5% binder)	36	0.2
Zirconia power ($<0.05 \mu$m)		
Deflocculated cast	59	6

Filtration

In uniaxial filter pressing, the increase in thickness dL of the cake with time dt can be related to the volume of filtrate dV_f using a materials balance:

$$dL = \frac{W \, dV_f}{A(D_s - \phi D_s - W\phi)} = \frac{J \, dV_f}{A} \tag{18.7}$$

where W is the mass of solids in suspension per unit volume of liquid, ϕ is the fraction of interstices among the particles in a cake of cross-sectional area A, and D_s is the density of the solids in the cake. The parameter J is the volume of cake per volume of filtrate and is a contraction for $W/(D_s - \phi D_s - W\phi)$. Substituting Eq. 18.7 into Eq. 18.2 for liquid permeability and expressing the volume flow rate in differential form,

$$\frac{1}{A} \frac{dV_f}{dt} = \frac{K_p \, \Delta P_c \, A}{\eta_L J V_f} \tag{18.8}$$

where ΔP_c is the pressure drop across the cake of thickness dL. In practice, the total pressure drop ΔP_T across the cake and filter medium is measured. Assuming the filter medium has a constant resistance R_m, and expressing the specific resistivity α_c of the cake as $\alpha_c = K_P^{-1}$,

$$\frac{1}{A} \frac{dV_f}{dt} = \frac{\Delta P_T}{\eta_L \left(\dfrac{J V_f}{A} \alpha_c + R_m \right)} \tag{18.9}$$

For an incompressible cake, $J\alpha$ is constant, and on integrating

$$L = \left[\frac{2J \, \Delta P_T \, t}{\alpha_c \eta_L} + \frac{R_m^2}{\alpha_c^2} \right]^{1/2} - \frac{R_m}{\alpha_c} \tag{18.10}$$

When $R_m/\alpha_c \ll 1$, Eq. 18.10 may often be expressed as

$$\frac{L^2}{t} = \frac{2J K_p \, \Delta P_T}{\eta_L} \tag{18.11}$$

The pressure drop ΔP_c across the cake produces a gradient in the effective stress. For a compressible cake, J and K_P decrease with pressure, and J and K_P in Eq. 18.11 are apparent values.

Filtration data graphed as shown in Fig. 18.9 for a well deflocculated suspension of alumina may be analysed using the reciprocal of Eq. 18.9. The filtration behavior of a partially deflocculated clay porcelain body where $R_m/\alpha_c \ll 1$ is described by Eq. 18.11, as is shown in Fig. 18.9. The parameter L^2/t is termed the filtration rate. The filtration rate is significantly dependent on the viscosity of the filtrate, and heating of the slurry or substitution of a

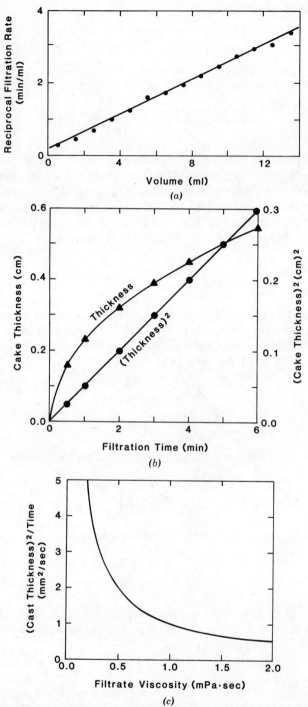

Fig. 18.9 (a) Filtration results for deflocculated alumina slurry plotted in form described by Eq. 18.9. (b) Filtration results for a deflocculated semivitreous whiteware slip. (c) Dependence of the filtration behaviour of a deflocculated alumina slurry on filtrate viscosity calculated using Eq. 18.9.

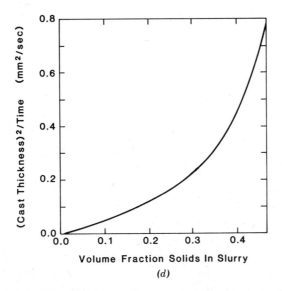

Fig. 18.9 (d) Calculated dependence of filtration rate on solids concentration in a deflocculated alumina slurry; solid fraction of the filter cake is 0.60.

liquid of lower viscosity may reduce the filtration time. A slurry temperature of 35–40°C is commonly used in filtering slurries in the production of electrical porcelain. Filtration also occurs more rapidly when the solids loading in the slurry and J is higher (Fig. 18.9). Filtration is very sensitive to both dispersion and agglomeration, which affect K_p. As is shown in Fig. 18.10, the filtration rate of an alumina slurry is significantly reduced by the presence of an organic binder of a higher viscosity grade (molecular weight)

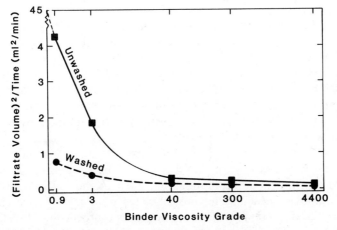

Fig. 18.10 Dependence of the filtration rate of a slurry of unwashed and washed Bayer process alumina on the viscosity grade of the admixed hydroxyethyl cellulose binder (0.5 wt % binder in solution; 7 vol % solids in slurry, $V_f^2/t = 45 \text{ ml}^2/\text{min}$ with no binder addition).

Table 18.2 Impurity Removed by Washing

Material	Concentration (wt %)	
	As Received	Washed
Bayer process alumina powder (<5 μm)		
(Na$_2$O impurity)	0.04	0.003
Zirconia powder from chloride precursor (<0.05 μm)		
(Cl impurity)	0.79	0.15

and by removing coagulating ions from the surfaces of the particles prior to filtration.

Washing operations may be conducted after filtering with the filter cake in place. Washing rates are of the order of the final filtration rate dV_f/dt when wash liquid passes through the thickness of the final cake. Batch or continuous centrifugation may also be used for concentrating solids and for washing operations. Washing may significantly reduce the concentration of soluble impurities in industrial minerals, inorganic chemicals, precipitates, etc., as indicated in Table 18.2. Manganese zinc ferrite powders are washed during industrial processing to reduce the alkali content and thus to obtain a much more uniform grain size on sintering and improved magnetic properties. Other industrial chemicals are washed during processing. Filter pressing is widely used in preparing plastic clay bodies and removes significant quantities of soluble salts.

18.3 OTHER PARTICLE CONCENTRATION PROCESSES

Particles may be concentrated using a cyclone and by gravity sedimentation, flotation, magnetic separation, and electrical precipitation. Sedimentation and flotation are widely used in processing industrial minerals. In froth flotation, surfactants are used to produce a stable froth containing suspended particles having a hydrophobic surface that has a greater affinity for air bubbles than for water; surface-specific surfactants may provide separation by mineral type. Particles differing in density may be separated by sink/float behavior in a liquid or suspension of intermediate density.

Magnetic separation is used to remove magnetic minerals and magnetic iron particles produced by the abrasive wear of processing machinery. Superconducting magnetic separations are used in the bulk processing of kaolins to remove mica and ferrous titanates of a micron size. Smaller magnets are used to remove tramp iron in wet processing. In electrical precipitation operations, fine particles charged on adsorbing gas ions are removed when they migrate to the collecting electrode.

SUMMARY

Particle separation and classification processes are used to modify the range
of particle size, to remove impurities, and to concentrate particles and
remove liquids. Screening and cyclone-type separators are widely used to
remove extremely coarse or fine sizes and to separate particles by size in the
range of 0.5 μm to several 100 mm. Hydrocloning, filtration, and centrifuging are used to concentrate particles in a slurry. The filtration rate is very
dependent on the solids content of the slurry and the permeability of the
filter cake. The permeability of a system of packed particles is an important
property influencing the processing of ceramic systems and can range over
several orders of magnitude for similar systems differing in agglomeration
and pore size. Magnetic separation and other separation processes are used
to remove impurity particles and to concentrate particles having a different
density or surface behavior. Washing processes are used to remove soluble
substances.

SUGGESTED READING

1. Errol G. Kelly and David J. Spottiswood, *Introduction to Minerals
   Processing*, Wiley-Interscience, New York, 1982.
2. F. A. L. Dullien, *Porous Media*: *Fluid Transport and Pore Structure*,
   Academic Press, New York, 1979.
3. J. Bear, *Dynamics of Fluids in Porous Media*, American Elsevier, New
   York, 1972.
4. Terrence J. Fennelly and James Reed, Mechanics of Pressure Slip
   Casting, *Am. Ceram. Soc. Bull.* **55**(5), 264–268 (1972).
5. D. S. Adcock and I. C. McDowall, Mechanism of Filter Pressing and Slip
   Casting, *Am. Ceram. Soc. Bull.* **40**(10), 355–362 (1957).
6. P.M. Heertjes, Studies in Filtration, *Chem. Eng. Sci.* **6** 190–208 (1957).
7. H. P. Grace, Resistance and Compressibility of Filter Cakes, *Chem. Eng.
   Prog.* **49**(6), 303–318 (1953).
8. David Cornell and D. L. Katz, Flow of Gases Through Consolidated
   Porous Media, *Ind. Eng. Chem.* **45**(10), 2145–2152 (1953).
9. P. C. Carmen, Some Physical Aspects of Water Flow in Porous Media,
   Disc. Faraday Soc. **3**, 72–77 (1948).

PROBLEMS

18.1 Does the efficiency of a screening operation for a particular size
depend on the frequency, amplitude of vibration, and material loading? Explain the dependence on each paramerer.

18.2 Using Stokes' law as a qualitative description of the balance between the centrifugal force on the particles and the drag force of liquid flowing inward in a hydroclone, predict the change in "cut size" on increasing the liquid viscosity, particle density, and tangential velocity of flow.

18.3 Calculate the time to centrifuge zirconia particles larger than 1 μm in diameter from a column of an aqueous suspension 6 cm in height using a 1000-rpm centrifuge. The initial radial position is 12 cm (25°C, $D_p = 6.02 \, \text{Mg/m}^3$).

18.4 The volume flow rate of water (20°C) through a filter that is 60% porous, 0.6 cm thick, and 3.5 cm in diameter is 3.7 ml/s at a pressure head of 420 kPa. Calculate the specific permeability of the filter assuming laminar flow.

18.5 What is the apparent axial flow velocity v in problem 18.4? Calculate Re to check for laminar flow.

18.6 Estimate the change in permeability of a filter cake when the cylindrical pore diameter is decreased from 1.0 to 0.1 μm and the pore fraction increases from 0.35 to 0.5.

18.7 The permeability constant for a filter cake is $10 \times 10^{-18} \, \text{m}^2$ and the pore fraction is $\phi = 0.45$. Assuming cylindrical pores, estimate the apparent pore radius (R) and the hydraulic radius (R_H).

18.8 Do the results in Fig. 18.10 suggest that the binder molecules flow freely through interstices in the filter cake? Explain your answer.

18.9 Calculate (J = volume of cast/volume of filtrate) for filter pressing a slurry containing 300 g alumina ($D_p = 3.98 \, \text{Mg/m}^3$) and 100 ml water at 25°C when the pore fraction of the cake is 0.40.

18.10 Calculate the permeability of the filter cake formed from the slurry in problem 18.9 when the filtrate flow rate $\Delta V_f^2/\Delta t$ is uniformly 5 ml²/s and $R_m \ll \alpha_c$. Calculate the casting rate L^2/t (mm²/min) and the thickness after 1 h.

18.11 For the slurry in problem 18.9, calculate and graph the curve for cake thickness L (mm) as a function of time (min) for the following conditions ($K_P = 10^{-17} \, \text{m}^2$, ϕ cast = 0.40, $R_m \ll \alpha_c$):

    (a) $\Delta P = 0.14 \, \text{mPa}$, $T = 20°C$, $W = 2 \, \text{g/ml}$
    (b) $\Delta P = 0.14 \, \text{mPa}$, $T = 20°C$, $W = 4 \, \text{g/ml}$
    (c) $\Delta P = 1.40 \, \text{mPa}$, $T = 20°C$, $W = 4 \, \text{g/ml}$
    (d) $\Delta P = 1.40 \, \text{mPa}$, $T = 40°C$, $W = 4 \, \text{g/ml}$

Which parameter change produced the largest increase in L^2/t?

19

Granulation

Bulk materials containing an admixed binder and liquid and semidry powder containing agglomerates varying in size, shape, and density do not flow well, pack densely, or compact into a uniform microstructure. Granulation processes are used to produce a satisfactory feed material consisting of controlled agglomerates called granules. Granulated material is used extensively for modern pressing operations, as a feed material for calcining and melting processes, and as a catalyst support.

Granules may be formed directly by the proper admixing of a liquid or a solution of the binder and other additives into a stirred powder or, indirectly, by spray-drying the system in slurry form. Spray-drying may also be used to reduce the liquid content of a slurry, which would form a filter cake of extremely low permeability, such as an extrusion body containing a high-molecular-weight binder.

19.1 DIRECT GRANULATION

Powder granules may be produced directly by compaction, extrusion, and spray granulation (pelletizing) processes. Materials prepared in this way include technical aluminas, ferrites, clays, tile bodies, porcelain bodies, conventional refractory compositions, catalyst supports, and feed materials for glass melting and metal refining.

Fine powders premixed with a few percent of a wetting liquid or a binder solution may be compacted in a tableting die or between briquetting rolls or belts. A mixer with a roller and a perforated bottom plate is also used to produce a granulated powder. Granules produced by compaction are commonly dense, hard, and strong if compacted above about 10 MPa. An auger extruder or a kneader may be used to produce a granular product from premixed material with a plastic consistency. Spaghetti-shaped material exiting from an orifice plate is cut with rotary knives. In roll extrusion, moist feed is forced through perforated rolls or through a perforated plate. Dried material is crushed to reduce the particle size.

Spray granulation is the formation of granules when a liquid or a binder solution is sprayed into a continually agitated powder. The formation of the granules may be considered to occur in two stages: the nucleation of primary agglomerates at random, and growth by the addition of particles or small agglomerates to the primary agglomerate.

During mixing, particles roll and slide, and fines become airborne. A minute agglomerate called a nucleus or seed forms initially when a droplet of liquid hits and is adsorbed on the surface of a group of particles. Capillary forces and binder flocculation give the nuclei strength. Nuclei are more numerous when the liquid is introduced rapidly as a fine mist and the powder is agitated vigorously. Capillary forces may cause particle sliding and rearrangement, densification, and the migration of liquid to the surface of the agglomerate nuclei.

The growth of granules by layering occurs by the contact and agglomeration of particles to the nuclei. Agglomeration of young nuclei and fracture fragments can also produce a granule. The two growth processes depend on the liquid feed rate, adsorption of liquid into the agglomerate, and the mixing action. Rubbing between granules during tumbling may cause particle transfer at the surfaces and surface smoothing. The fracture of granules and attrition are higher when the forces during mixing are higher and the granules lack strength. The size, shape, and surface appearance of granules formed by spray granulation are shown in Fig. 19.1.

The liquid requirement for granulation is greater when the specific surface area of the particles and the adsorption of any additive phases are high. For each material system, there is a critical moisture range for granule formation. At the low end of the range, granules are eventually formed if the processing energy is sufficiently high. The addition of a greater amount of liquid reduces the granulation time but generally increases the size distribution and the porosity in the granules. Liquid contents used in granulating tile and high-alumina bodies containing clay are commonly in the range of 20–36 vol% (8–14 wt%), which indicates that the pores are incompletely saturated.

Rotating pan-type mixers, ribbon mixers, double planetary mixers, v-blenders, and continuous rotating drum and disk pelletizers are used for directly granulating ceramic powders. Rotating drums and disks are commonly used for producing rather coarse granules 1–10 mm in size as a feed for melting operations. Mixers with a more intense mixing action produce smaller granules larger than 44 μm for pressing. A colloidal or polymer binder aids in forming granules and increases the strength of granules of coarser powders. Increasing the temperature of the powder tends to produce smaller granules. Granules produced by spray granulation are more nearly spherical, in contrast to granules produced by compaction and extrusion.

Fig. 19.1 Scanning electron micrographs of granules formed by spray granulation: (a) general view of size and (b) shape and surface character of granules formed by layering.

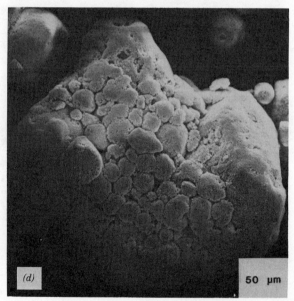

Fig. 19.1 (c) general view and (d) surface structure of granule formed by agglomerating nuclei and surface smoothing.

Capacities of commercial pelletizers may range up to several tons per hour. Spray-granulated material must usually be partially dried to produce a powder feed that is satisfactory for pressing operations.

19.2 SPRAY-DRYING

Spray-drying is the process of spraying a slurry into a warm drying medium to produce nearly spherical powder granules that are relatively homogeneous (Fig. 19.2). The high specific surface area of the droplets and the dispersion of the droplets in the drying air contribute to a relatively high drying efficiency. Spray-drying is used widely for preparing granulated pressing feed from powders of ferrites, titanates, other electrical ceramic compositions, alumina, carbides, nitrides, and porcelain bodies. Batch mixing in a slurry form and atomizing facilitate mixing and the achievement of mixedness. When properly controlled, spray-drying produces nearly spherical, relatively dense granules larger than 20 μm in size which flow and compact well in pressing operations. Capacities of industrial spray-dryers range from less than 10 to several 100 kg/h.

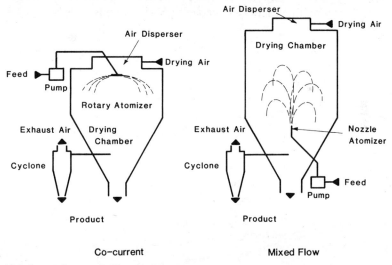

Fig. 19.2 Spray dryers: (left) cocurrent spray-dryer with centrifugal atomizer, and (right) mixed-flow spray-dryer with a nozzle atomizer.

The Spray-Drying Process

The batch components are first mixed to form a deflocculated pseudoplastic slurry. Consideration must be given to soluble inorganic impurities, retained in spray-drying, and the effect of the drying temperature on the degradation or evaporation of the organic binders and plasticizers. Aqueous systems are commonly used, but recycled organic liquids are sometimes used when the chemical or thermal reactivity of the powder or binder precludes an aqueous system. The slurry should be of relatively high solids content and free of air bubbles. Air bubbles are undesirable, because they reduce the density of the slurry and the resultant granules. The viscosity must be determined over a wide range of shear rate to check for reproducibility and the absence of dilatent behavior at the shear rate which may exceed $10^4\,s^{-1}$ during atomization. The viscosity behavior of alumina slurries for spray-drying is shown in Fig. 19.3.

Slurry that has been sieved to remove coarse agglomerates is commonly atomized using a nozzle or rotary atomizer. In a pressure nozzle, slurry accelerating through small channels breaks up into droplets. Pressure nozzles are relatively inexpensive but are subject to blocked flow, and the high operational pressures ranging up to 10 MPa increase abrasion rates. In a two-fluid nozzle (Fig. 19.4), a high-velocity gas is mixed with the slurry to produce turbulence and droplets; operating pressures may be reduced to less than 1 MPa; external mixing produces a smaller droplet size. Centrifugal energy from a disk or wheel rotating at several thousand rpm produces droplets in a rotary atomizer; the lower operating pressures and larger flow channels increase the feed rate capacity and reliability. However, spray-

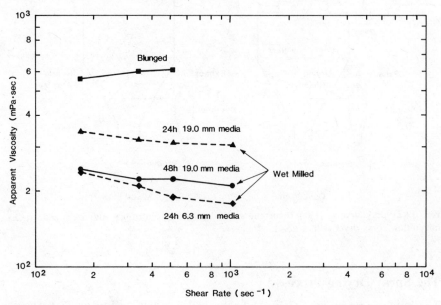

Fig. 19.3 Viscosity behavior of alumina slurries for spray-drying determined using a concentric cylinder viscometer.

Two Fluid Nozzle

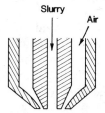

Slurry

Air

Fig. 19.4 Two-fluid nozzle (external mixing) for atomization of a slurry.

dryers using radial atomization must be relatively large in diameter because of the relatively large radial/axial flow pattern of the spray.

The droplet size in the spray depends on turbulence produced in the nozzle (Eq. 15.20) and varies directly with the viscosity and solids content of the slurry and inversely with the flow velocity. Characteristics of spray-dried alumina granules indicated in Table 19.1 are indicative of these trends. Granules ranging up to about 400 μm may be produced using pressure nozzles, and sizes up to about 150 μm using a rotary atomizer. Relative to a pressure nozzle, a two-fluid nozzle usually produces a greater fraction of sizes<44 μm.

A cocurrent dryer design positions the atomizer and the inlet for the drying air at the top of the dryer; atomized slurry is of maximum liquid content when it encounters the laminar flow of hot incoming air. The maximum product temperature is relatively low, and the evaporation time is relatively short. However, the product exits with moist air, and the exit temperature must be relatively high to obtain a dry product. In a counter-current design, slurry and drying air inlets are at opposite ends of the drying chamber. Partially dried droplets encounter the incoming hot air, and the

Table 19.1 Characteristics of Spray-Dried Alumina Granules[a]

Slurry Variables		Granule Size[c] (μm)				
Solids (vol%)	Binder (Wt%)[b]	90%<	50%<	10%<	Granule Density(%)	Fill[d] Density(%)
35	1.0	172	92	40	54	30
35	0.5	144	74	27	55	32
50	0.5	202	118	54	58	34
52	0.5	216	113	41	57	34

[a] Two-fluid nozzle, utility dryer.
[b] Fully hydrolyzed polyvinyl alcohol; powder basis.
[c] Sieving data.
[d] Bulk density after pouring into a graduated cylinder.

product heating is greater. The mixed flow design (Fig. 19.2), which increases the residence time of the spray and provides a means for reducing the size of the chamber needed for drying, is a convenient compromise and is characteristic of many small industrial and laboratory dryers.

Considerable evaporation must occur during the first few milliseconds of drying to prevent the formation of agglomerated granules and a coating of sticky agglomerates on the wall of the dryer. Filtered drying air is commonly heated using natural gas or electrically to a temperature in the range of 300–600°C. The inlet air temperature depends directly on the dryer design, the product residence time, and the thermal limitations of the product. The product temperature is normally lower than the outlet air temperature and is usually maintained below 100°C when using organic binders and additives.

The evaporation rate increases initially as the droplet is heated, and a skin is formed. The drying rate remains high as long as the surface is saturated with liquid. Evaporation retards the heating of the product. Rapid capillary flow (Eq. 2.16) and slower evaporation (i.e., a relatively low inlet air temperature and/or a relatively high relative humidity) extend the duration of evaporation from the surface. Capillary flow is reduced by parameters that reduce the permeability such as binder molecules and the migration of fine particles or precipitated salts into pores near the surface. Rapid heating may cause liquid to evaporate within the droplet forming a vapor bubble, which may expand the agglomerate. When the bubble breaks through the skin and collapses and the agglomerate is plastic, a dimpled or donut shape is formed. The evaporation rate decreases when the evaporation front recedes below the surface, and the agglomerate becomes less plastic. The granule is nearly in equilibrium with the warm, moist air leaving the dryer, and the outlet gas temperatures is controlled to maintain a constant moisture content.

The majority of the spray-dried product is discharged through a rotary valve into interchangeable containers attached at the base. Product fines entrained in the exhaust air are separated using a cyclone or bag filters. Product fines may be used in some pressing operations but are often recycled into the feed slurry. Periodic cleaning of the dryer using a liquid spray is required to remove the granulate crust; this is facilitated when the dried binder system is readily redissolved. Abnormally large granules and flakes of granulate crust from the wall of the dryer are undesirable and are commonly removed by screening.

Granule Character

Surface tension produces spherical droplets in the initial spray. Colliding droplets may coalesce into larger agglomerates, some of which are less nearly spherical in shape, or a granule that is a cluster of smaller agglomerates, as is shown in Fig. 19.5. Air bubbles present in the feed slurry or

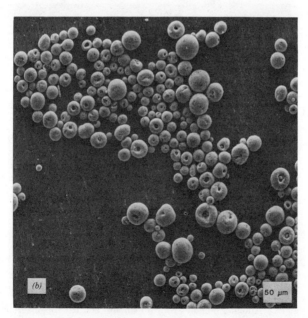

Fig. 19.5 Scanning electron micrographs of spray-dried granules: manganese zinc ferrite in form of (a) dense granules and (b) hollow granules.

Fig. 19.5 Alumina granules with (c) smooth and (d) rough surfaces.

occluded during atomization using a two-fluid nozzle may persist as relatively large pores in the ultimate granule.

Granules with a large crater or a donut shape are often observed in industrial spray-dried material (Fig. 19.5), regardless of the atomization method used. The tendency for these shapes is higher when the inlet temperature is relatively high, the binder content or molecular weight is relatively high, and the solids loading in the slurry is relatively low. Rapid surface drying and the formation of surface material of low permeability produce a greater initial heating rate. The momentary internal evaporation and bursting of the granule due to gas pressure, and subsequent flow induced by surface tension, can produce a surface crater that may persist when exposed material is partially dried and more viscous. Granules containing an abnormally large internal pore may be produced when the migration of finer particles, salts, or binder during drying produces a dense surface layer that shrinks relatively less than the interior (Fig. 19.6).

Comparative characteristics of whiteware granules prepared by spray-drying and spray granulation are listed in Table 19.2. The spray-dried granules are somewhat finer, of comparable bulk density, and slightly faster

250 μm

Fig. 19.6 Segregation of fines at the surface and large internal pores in a spray-dried granule containing particles of very different size. (Courtesy of J. Cline, PhD Thesis, Alfred University, Alfred, NY.)

Table 19.2 Characteristic of Granulated Whiteware Bodies[a]

Preparation	Wall Tile Spray-Drying	Hotel China Spray-Drying	Hotel China Spray Granulation
Moisture Adsorption (wt%)			
33% RH	2.7	3.5	2.4
92% RH	4.8	4.9	3.5
Granule size (μm)[b]			
90%	300	446	992
50%<	151	148	403
10%<	55	99	122
Flow rate (s)[c]			
33% RH		7	10
92% RH		8	12
Fill density (%)	38	40	41
Tapped density/fill density	1.15	1.13	1.20

[a] Granules larger than 1000 μm removed.
[b] Sieving data.
[c] Funnel test.

in flow rate owing to their more nearly spherical shape. The ratio of tap density to fill density is an index of hindered packing during gravity settling after pouring and is slightly higher for the spray-dried granules.

SUMMARY

Powders for pressing, calcining, and melting operations are commonly granulated. Compaction and extrusion processes are used to produce relatively large granules of higher density than the bulk powder. Spray granulation is used to produce granules that are satisfactory for use as a pressing powder for traditional ceramics and refractories and some fine-grained technical ceramics. The process must be controlled well to produce granules that are approximately spherical and reproducible in size and density. Granules produced by spray-drying are commonly more homogeneous; the formulation of the precursor slurry and drying process must be properly designed to avoid the formation of granules with large internal pores and irregular shapes. Spray-drying is relatively efficient, because the product is well dispersed in the drying medium and has a high specific surface area.

SUGGESTED READING

1. P. J. Sherrington and R. Oliver, *Granulation*, Heyden and Sons, London, 1981.
2. K. Masters, *Spray Drying*, Wiley-Interscience, New York, 1976.

3. H. J. Helsing, Advanced Processing of Oxide Ceramics by Spray Drying, *Powder Metal* **1**(2), (1969).

4. W. H. Engelleitner, Pelletizing as Applied to the Ceramic Field, *Am. Ceram. Soc. Bull.* **54**(2), 206–207 (1975).

5. H. B. Ries, Methods of Preparing Bodies for the Preparation of Oxidic and Non-Oxidic Ceramics, *Keramische Zeitschrift* **35**(2), 67–71 (1983).

6. David A. Lee, Comparison of Centrifugal and Nozzle Atomization is Spray Dryers, *Am. Ceram. Soc. Bull.* **53**(3), 232–233 (1974).

7. K. V. S. Sastry and D. W. Fuerstenau, Mechanisms of Agglomerate Growth in Green Pelletization, *Powder Tech.* **7**, 97–105 (1973).

8. S. Katta and W. H. Gauvin, Some Fundamental Aspects of Spray Drying, *J. AIChE* **21**, 1 (1975).

9. W. R. Marshall, Atomization and Spray Drying, *Chem. Eng. Prog.* monograph No. 2, Vol. 50 (1954).

PROBLEMS

19.1 Pressing granules of 10 wt% liquid are prepared by partially drying a spray-granulated tile body containing 14 wt% liquid. What is the weight percent of liquid removed relative to the weight of product (assume that the pycnometer density of the solids is $2.6 \, Mg/m^3$). Compare this to spray-drying when the feed slurry is 40 vol% solids.

19.2 A barium titanate powder was both spray-granulated and spray-dried. The granule size distribution of each determined by SEM analysis is log-normal: spray-dried $\bar{a}_{G_N} = 75 \, \mu m$, $\sigma_g = 1.47$, spray-granulated $\bar{a}_{g_N} = 120 \, \mu m$, $\sigma_g = 1.69$. Calculate the 90%<, 50%<, and 10%< sizes expected when sieving these powders, and compare the size distributions obtained for the two processes.

19.3 The sieving data in Table 19.1 are log-normal. Calculate and compare $\bar{a}_{g_M}$ and σ_g for these powders.

19.4 The sieving data for the product from the spray-drying of two slurries of alumina (69 wt% solids), for mesh sizes of 80, 120, 170, 200, 230, and 325, are as follows: two-fluid atomization 15, 45, 65, 75, 83, 90% retained; centrifugal atomization 29, 50, 68, 75, 85, 94% retained. Are the distributions log-normal? Compare them.

19.5 Alumina slurries of 60, 70, and 80 wt% solids are spray-dried. Assuming that the evaporation rate (mass water/time) is a constant during drying, calculate the relative product formation rate if the residual moisture content in the product is 0.1 wt%. Also, contrast the relative mass of liquid removed per mass of product for the three slurries.

19.6 Slurries with solid volume fractions 0.3, 0.4, and 0.5 are spray-dried producing granules with a solid fraction of 0.53, 0.54, and 0.58, respectively. Calculate the average radial shrinkage for each.

19.7 Assuming that in spray granulation the liquid requirement is directly proportional to the specific surface area of the raw materials, compare the liquid requirements for two powders with $\bar{a}_{V/A}$ of 5.0 and 0.5 μm and a 2/3–1/3 (by weight) mixture of each (assume that the shape factor and particle densities are equal).

19.8 Explain the results in Table 19.1 in terms of the expected effect of liquid and binder content on slurry viscosity, K_P, and evaporation during drying.

19.9 Alumina granules with a solids fraction of 0.40 are produced on spray-drying an aqueous slurry containing 80 wt% alumina. Did the droplets shrink in size during drying?

19.10 Sketch the formation of the granules during spray granulation (consider both layering and agglomeration processes).

19.11 Compare a granule density of 55% with the bulk density of the barium titanate and alumina powders in Chapter 5.

PART VII

FORMING PROCESSES

Forming transforms the unconsolidated system of feed material into a coherent, consolidated body having a particular geometry and microstructure. The selection of a ceramic-forming operation for a particular product is very dependent on the size, shape, and dimensional tolerances of the product, the requisite microstructure characteristics and levels of reproducibility required, and capital investment and productivity considerations. Other considerations include the surface character after forming, die or mold requirements, energy requirements, and safety.

The mechanics of forming are very dependent on the consistency and rheological response of the material. Coarse granular materials and granulated powders are commonly formed by uniaxial dry-pressing in a hard die or isostatic pressing in a flexible mold. Because the degree of saturation DOS of the press feed is in the range of 0.01–0.5, the applied stress in forming is relatively high, about 20–200 MPa. Feed materials having a plastic consistency and a DOS > 0.9 are formed by extrusion, pressure molding, injection molding, and jiggering operations. In conventional plastic forming, the degree of saturation of the feed material increases when the material is compressed; pressurized liquid in pores reduces the effective stress, and moderate forming pressures of 1–20 MPa are commonly used. Feed material used in ordinary injection molding is saturated (DOS = 1) with a thermoplastic or thermosetting polymer resin; temperature gradients are controlled during forming to facilitate flow through small channels and subsequently to increase the rigidity of the formed part. A pourable or pumpable slurry (DOS ≥ 1) is used in conventional casting. Forming pressures are < 1 MPa, because the slurry or paste is saturated with a viscous matrix. New molecular forming techniques are being developed for applying thin films and to produce bulk shapes.

External effects such as the surface roughness and adhesion or lubrication of the wall of the die or mold and the thermal and fluid transport into or from the die or mold are important considerations. Most ceramic products continue to be formed at temperatures less than a few hundred degrees centigrade using relatively moderate pressures. Significant margins exist for improving products by optimizing the forming process and the feed material. We will consider these important topics in Chapters 20–24.

20

Pressing

Pressing is the simultaneous compaction and shaping of a powder or granular material confined in a rigid die or a flexible mold. Powder feed for industrial pressing is in the form of controlled granules containing processing additives produced by spray-drying or spray granulation. Coarse granular compositions containing a binder are commonly in the form of a poorly flowing, semicohesive mass. For reasons of productivity and the ability to produce parts ranging widely in size and shape to close tolerances with essentially no drying shrinkage, pressing is the most widely practiced forming process. Products produced by pressing include a wide variety of magnetic and dielectric ceramics, various fine-grained technical aluminas including clip carriers and spark plugs, engineering ceramics such as cutting tools and refractory sensors, ceramic tile and porcelain products, and coarse-grained refractories, grinding wheels, and structural clay products.

Pressing by means of punches in hardened metal dies, commonly called dry-pressing, is commonly used for pressing parts thicker than 0.5 mm and parts with surface relief in the pressing direction. Isostatic pressing in flexible rubber molds, commonly called isopressing, is used for producing shapes with relief in two or three dimensions, shapes with one elongated dimension such as rods and tubes, and very massive products with a thick cross section. Pressed products are also produced using a combination of metal die and isostatic pressing and by roll pressing.

20.1 PROCESS VARIABLES IN DRY-PRESSING

Stages in dry pressing include (1) the filling of the die, (2) compaction and shaping, and (3) ejection, as is shown in Fig. 20.1. Free-flowing granules are fed to the die by means of a sliding feed shoe and are metered volumetrically. Poorly flowing feed is usually preweighed and fed by hand or by

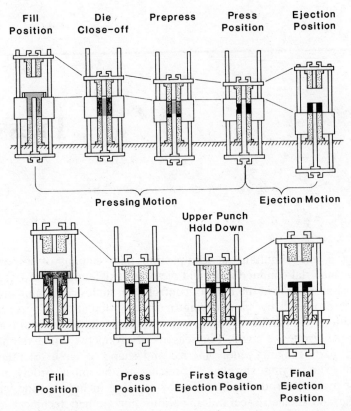

Fig. 20.1 Pressing and ejection motions using a floating die for (top) a one-level part and (bottom) a two-level part. (Courtesy of Dorst-Maschinen und Anlagenbau, Kochel A. See, West Germany.)

mechanically induced flow. Punch and die motions may be coordinated to produce a vacuum assisting settling into the cavity of the die.

Pressing modes defined in terms of the motion of the punches and die are listed in Table 20.1. Movement of the punches and die is synchronized, as is shown in Fig. 20.2. Punch stops are set at a particular displacement using mechanical control and a particular pressure using hydraulic control. Vibrat-

Table 20.1 Dry-Press Modes

Type	Die	Top Punch[a]	Bottom Punch[a]
Single action	Fixed	Motion	Fixed
Double action	Fixed	Motion	Motion
Floating die	Moves	Motion	Fixed

[a] Simple or composite.

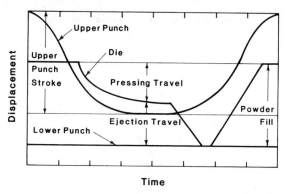

Fig. 20.2 Synchronized punch and die motions during a pressing cycle.

ing punches are used in pressing some coarse granular products such as refractories and grinding wheels. Tooling for dies and punches is commonly constructed of hardened steel, but special steels and carbide or ceramic inserts are used in high-wear areas. The clearance between the die and punch is about 10–25 μm when pressing micron-size powders and up to 100 μm when pressing granular particles. Punch motions are commonly contolled in relation to the centroid of the pressed part. The die wall is sometimes tapered (< 10 μm/cm) to facilitate ejection. Pressed parts may be ejected with or without contact pressure from the top punch. Small ejected pieces are commonly displaced to a conveyer by the leading edge of the feed show. Larger pieces are lifted and moved mechanically often using a vacuum assist, or by hand.

Pressing rates range from a fraction of a second for small parts to several minutes for large parts, using a single-action press, and rates exceeding 5000 parts per minute are achieved using a multistation rotary press. Press capacities range up to several hundred tons. The maximum pressing pressure used in dry-pressing is commonly in the range of 20–100 MPa; higher pressures are used for technical ceramics than for clay-based materials. Die-set life may range up to several hundred thousand pieces for a simple die and a low pressing pressure. Advertised tolerances are now better than ±1% in mass and ±0.02 mm in thickness. Dimensional tolerances and the uniformity of the microstructure are more difficult to control when the part has several surface levels.

Systems used to prepare granulated pressing powders by spray-drying commonly contain a deflocculant, binder, placticizer, and sometimes a lubricant, wetting agent, and a defoamer; representative examples of the organic systems used are listed in Table 20.2. A deflocculant is used to reduce the liquid requirement to form granules. The binder content is usually in the range of 2–12 vol %. The plasticizer decreases the moisture sensitivity of the binder; moisture in the powder commonly acts as a plasticizer and should be controlled. Lubricants may be introduced before

Table 20.2 Processing Additives Used in Industrial Pressing Powders

Product	Binder	Plasticizer	Lubricant
Alumina	Polyvinyl alcohol	Polyethylene glycol	Aluminum stearate
MnZn ferrite	Polyvinyl alcohol	Polyethylene glycol	Zinc stearate
Barium titanate	Polyvinyl alcohol	Polyethylene glycol	Stearic acid
Alumina substrates/ spark plug insul.	Microcrystalline wax emulsion	KOH: tannic acid	Wax, talc, and clay
Steatite insulator	Microcrystalline wax and clay	Water	Colloidal talc and wax
Refractories	Ca/Na ligno sulfonate	Water	Stearic acid
Tile	Clay	Water	Colloidal talc and clay
Hotel china	Clay and polysaccharides	Water	Colloidal clay

granulation or admixed to coat the granules in a separate operation. Lubricants can reduce die wear and ejection pressures and improve density uniformity in the pressed part. Soft wax binders and properly plasticized binder systems provide some lubricating action. A list of additives for a slurry for spray drying, containing a polyvinyl alcohol binder, are listed in Table 20.3. Pressing powders should be free-flowing, of a high bulk density, composed of deformable granules, and stable under ambient conditions, and should cause a minimum of die wear and not adhere to the punch. Pressed parts must be sufficiently strong to survive ejection and subsequent handling. The binder content is usually as low as practicable to minimize both the higher binder cost and the amount of gas produced during binder burnout.

Table 20.3 Organic Additive System for a Pressing Powder Produced by Spray-Drying

Ingredient	Concentration (wt%)
Polyvinyl alcohol	83.1
Glycerol	8.3
Ethylene glycol	4.1
Deflocculant	4.1
Nonionic surfactant	0.2
Defoamer	0.2

Source: Du Pont Inc., Wilmington, DE.

20.2 POWER FLOW AND DIE FILLING

Good powder flow is essential for reproducible volumetric filling, a uniform density of filling, and rapid pressing rates. Flow rates are often determined by measuring the time for a specific mass or volume of powder to flow through a standard funnel or inferred from the angle of repose of powder poured onto a flat surface.

Dense, nearly spherical particles or granules with smooth, nonsticky surfaces that are coarser than about 40 μm have good flow behavior and are preferred for pressing (Fig. 20.3). The presence of more than 5% of a finer size may sometimes stop flow altogether, and fines may enter the annulus between the punch and die wall, causing friction and reducing the escape of air. Extremely large granules are usually irregular in shape, and the bridging action between coarse granules may impede flow and the achievement of a uniform bulk density when powder flows into the die. They should be removed by sieving. When spaces are small between core rods or surfaces of a composite die, the maximum granule size must be a fraction of the cross

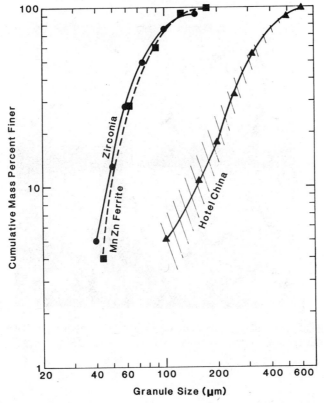

Fig. 20.3 Granule size distributions of spray-dried feed used is industrial pressing.

Fig. 20.4 Industrial pressing granules with (a) a relatively spherical shape and a smooth surface and (b) an external wax lubricant.

section available for flow. The surface of granules is smoother when composed of a finer particle size and when small granules do not adhere to larger ones (Fig. 19.5). Binders become stickier when the temperature approaches the melting point or exceeds the glass transition temperature of the binder system, and control of the concentration of the binder system and plasticizer (and temperature) is important for controlled flow. The relative humidity of the storage environment may have a significant effect on flow behavior of granules containing a water-soluble binder system (Fig. 20.5).

Pressing problems are generally minimal when the bulk density of the feed in the die, called the fill density, is high. A higher fill density reduces both the content of air in the powder and the punch travel. Powders composed of high-density granules that pack efficiently on flowing into the die have a higher fill density. Powder characteristics that retard flow may hinder packing and produce nonuniformities in the fill density. Because mechanical vibration may cause rearrangements into a higher packing density, the ratio of the vibrated bulk density to the poured density is used as an index of hindered flow and filling.

A typical fill density (D_f) for a granulated powder is in the range of 25–35%, as seen in Table 20.4. The granule size distribution for these spray-dried powders is approximately log-normal, and the fill bulk density is affected by the granule density and packing behavior. Powders composed of granules containing large pores, donut-shaped granules, and granules with rough surfaces have a relatively low packing density.

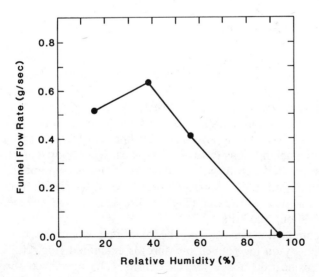

Fig. 20.5 Dependence of the flow rate of pressing granules containing a polyvinyl alcohol binder on the ambient (storage) relative humidity (funnel flow test).

Table 20.4 Characteristics of Several Industrial Spray-Dried Pressing Powders

	Particle Size[a] (μm)	Organics (Vol %)	Granule Size[b] (μm)	Granule Density (%)	Fill Density (%)
Alumina substrate	0.7	3.6	92	54	32
Spark plug alumina	2.0	13.3	186	55	34
Zirconia sensor	1.0	10.5	75	55	37
MnZn ferrite[c]	0.7	10.0	53	55[d]	32
MnZn ferrite[e]	0.2	8.5	56	31	18
Silicon carbide	0.3	20.4	174	45	29
		Water (Vol %)			
Wall tile	9	13–21	134		38

[a] Volume mean.
[b] Volume geometric mean (50%).
[c] Conventional powder.
[d] Donut shapes (low value).
[e] Coprecipitated powder.

20.3 COMPACTION BEHAVIOR

In dry-pressing, pressure produced by the moving punches compacts the dispersed powder into a cohesive unit with a particular shape and microstructure. The rate of densification is high initially but then decreases rapidly for pressures above 5–10 MPa (Fig. 20.6). The initial stress is transmitted by means of contacts between particles coated with the binder system. Granule deformation occurs by the sliding and rearrangement of particles within granules, and their deformation reduces the porosity and increases the number of intergranular contacts. Air compressed in pores migrates and partially exhausts between the punch and die. Relatively little densification occurs above 50 MPa, but the die wear is higher at these pressures when pressing hard ceramic materials. Industrial pressing pressures are commonly less than 100 MPa for high-performance technical ceramics and less than 40 MPa for whiteware compositions. The final compact density is less than the ultimate packing density of the particles. The compaction ratio is the ratio of the pressed density to the fill density and is commonly less than 2. Powders composed of ductile particles such as Al, Cu, and KBr have a much higher compaction ratio and may be pressed to $\approx 100\%$ density at a pressure of about 500 MPa.

Three compaction stages may be identified when the dependence of compact density on log (punch pressure) is examined (Fig. 20.7).* In stage I, slight densification above the fill density (D_f) may occur owing to the

* R. A. DiMilia and J. S. Reed, *Am. Ceram. Soc. Bull.* **62**(4), 484–488 (1963).

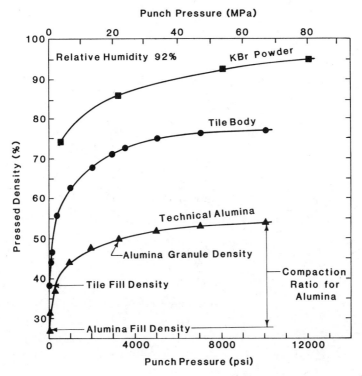

Fig. 20.6 Compaction behavior of spray-dried alumina and tile body granules and a KBr powder as a function of punch pressure (density calculated on inorganic basis).

sliding and rearrangement of granules. Interstices among granules are much larger than the average pores within the granules. In stage II, granules deform or fracture, reducing the volume of the relatively large interstices (Fig. 20.8). The apparent yield pressure (P_y) of the granules is less than 1 MPa when the binder system is soft and ductile. Denser granules with a higher binder content and/or a poorly plasticized binder system resist deformation, and a higher pressure is needed to compact these to an equivalent density. The densification in stage II may be approximated by the equation

$$D_C = D_f + m \ln(P_a/P_y) \tag{20.1}$$

where D_C is the compact density at an applied pressure P_a and m is a compaction constant that depends on the deformability and packing of the granules. Stage III begins when most of the large pores among deformed granules have disappeared, and the higher applied pressure causes the sliding and rearrangement of particles or fracture fragments within granules

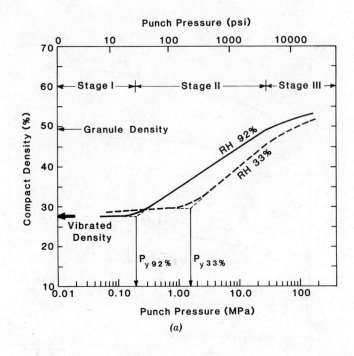

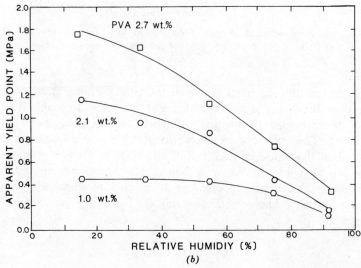

Fig. 20.7 Compaction diagrams indicating (a) stages of compaction and dependence on granule deformability and (b) dependence of granule hardness on the binder content and aqueous plasticizing of the binder.

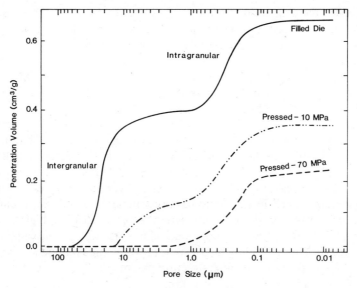

Fig. 20.8 Cumulative pore size distribution indicating intergranular and intragranular porosity of spray-dried alumina pressing granules on filling the die and after pressing (Hg-porosimetry technique).

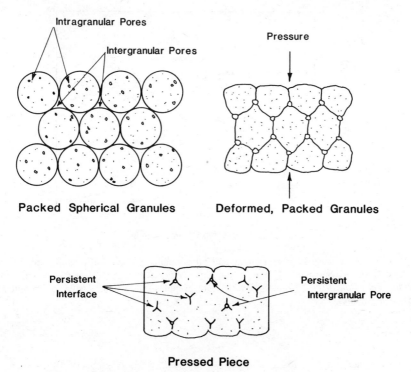

Fig. 20.9 Change in granule shape and bimodal pore size distribution during compaction.

into a denser packing configuration (Fig. 20.9). Interfaces between smaller and softer granules begin to be eliminated in stage II, and groups merge into a homogeneous mass. Interstices between large granules or harder granules and within large granules often persist into stage III (Fig. 20.10). The roughness of pressed surfaces and persistence of large pores and interfaces is increased when using hard granules larger than 100 μm, as shown in the microstructures in Fig. 20.11.

The compaction of a bulky powder and granules with the binder removed proceeds somewhat differently,* as is shown in Fig. 20.12. The fill density of a bulky powder may be quite similar to that of the granulated powder, but the pore size distribution is broad and monomodal. Pressing causes particle sliding and rearrangement, which decreases the volume of pores larger than the particle size. Particle translation is predominantly axial. Eventually the compaction imitates stage III compaction for granules. A high stress at interparticle contracts may fracture accicular particles, porous aggregates, and large particles; in a highly aggregated powder, the compaction is typical of aggregates in a matrix of finer fracture fragments.

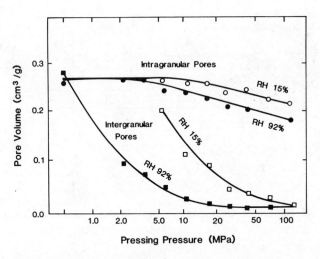

Fig. 20.10 Dependence of intergranular and intragranular porosity on pressing pressure and granule hardness controlled by relative humidity. (After R. A. DiMilia and J. S. Reed, *Am. Ceram. Soc. Bull.* **62**(4), 484–488 (1983), P_y of granules is 1.2 and 0.2 MPa for 15 and 92% RH, respectively.

* S. J. Lukasiewicz and J. S. Reed, *Am. Ceram. Soc. Bull.* **57**(9) 798–801 (1978).

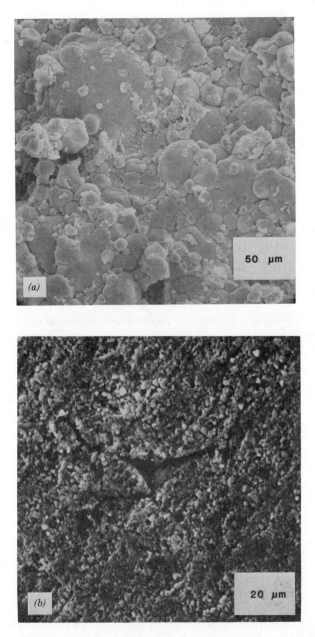

Fig. 20.11 Fracture surfaces of compacts of granules pressed to stage III indicate that (a) relatively hard granules are poorly joined and (b) interstices among large granules persist.

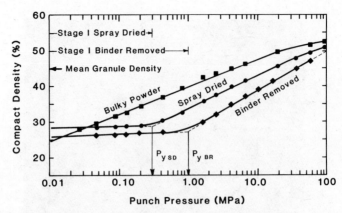

Fig. 20.12 Compaction behavior of MnZn ferrite feed in granular form with and without polyvinyl alcohol binder phase and comminuted granules (bulky powder) containing binder. (After S. J. Lukasiewicz and J. S. Reed, *Am. Ceram. Soc. Bull.* **57**(9), 798–801,805 (1978)).

20.4 EJECTION AND TRANSFER

On pressing a granulated powder, elastic compression of the granules begins in stage II and increases into stage III of compaction. Stored elastic energy produces an increase in the dimensions of the pressed part on ejection, called springback. Some differential springback between the compact and hardened steel punch is necessary to cause the compact to separate from the punch, and a linear springback less than about 0.75% is desirable. Excessive springback and differential springback due to stress gradients may cause compact defects on ejection from the die.

Springback is generally higher when the concentration of organics is higher and the pressing temperature is below the glass transition temperature of the binder system, as is shown in Table 20.5. The presence of an additive such as gum arabic, which increases the moisture adsorption, increases the plasticity of the binder system, and reduces the springback in granules stored at a lower relative humidity (Fig. 20.13). For the higher moisture content, the greater springback with gum arabic present is apparently due to its greater molecular size. Springback is also higher when the pressing pressure is higher.

The force for ejection depends on the taper and surface condition of the die, elastic springback of the compact, lubrication of the die wall, and ejection rate. Lubrication provided by a plasticized binder system may drop the ejection pressure by a factor of 5–10. The use of a die wall lubricant can reduce the ejection force still further, as is shown in Fig. 20.14; lubrication reduces die wear and the tendency for slip-stick sliding (Fig. 20.15). The ejection pressure is also larger for a die with a rough surface.

Transfer of the ejected part to a conveyer is accomplished by pushing with the front edge of the filling show or other mechanical means, a vacuum

Table 20.5 Mean Diametral Springback on Ejecting Compacts of Barium Titanate Granules[a]

T_g (°C)	P_y (MPa)	Initial Springback Pressure (MPa)	Pressed at 400 MPa Springback (%)
<10	0.3	22	0.31
22	0.5	—	0.65
42	1.0	7	1.6

[a] 2 wt % polyvinyl alcohol binder, 24°C.

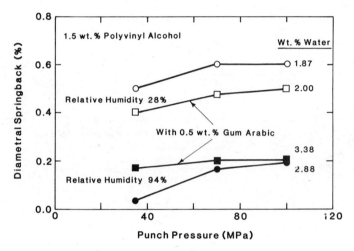

Fig. 20.13 Dependence of springback of compacts of alumina powder pressed at 25°C on pressing pressure and adsorbed aqueous plasticizer in binder.

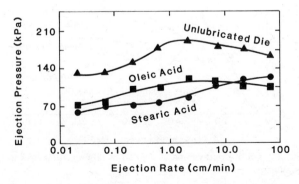

Fig. 20.14 Effect of lubrication on ejection pressure for an alumina compact with a polyvinyl alcohol binder (23°C).

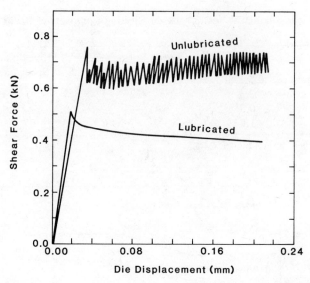

Fig. 20.15 Regular and slip-stick shear produced on displacing the die while the compact is compressed. (From R. A. DiMilia and J. S. Reed, *J. Am. Ceram. Soc.* **66**(9), 667–672 (1983).

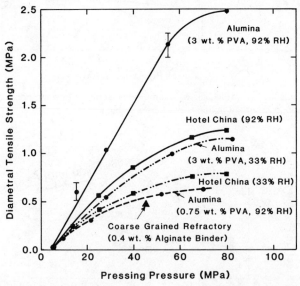

Fig. 20.16 Tensile strength of dried compacts pressed at different pressures using poorly and well-plasticized granules.

lifting device, or gravity fall when using a horizontal pressing action. Pressed parts must absorb the impact energy without failure. A small amount of plasticizer in the binder reduces the tensile strength but increases the impact strength of the binder system and aids in bonding between granules. Laminations between poorly joined granules and large pores reduce the strength. Compacts pressed at a higher pressure using plasticized granules are mechanically stronger, as is shown in Fig. 20.16; a diametral tensile strength above about 0.5 MPa is needed to obtain compacts with sharp edges.

20.5 DIE WALL EFFECTS

A portion of the applied load is transferred to the die wall during compaction. The force produced by the die wall friction causes pressure gradients and density gradients in the compact. For uniaxial compaction (Fig. 20.17), the mean shearing stress at the wall ($\bar{\tau}_w$) is related to the mean axial pressure $\bar{P}$ by the equation

$$\bar{\tau}_w = fK_{h/v}\bar{P} + A_w \tag{20.2}$$

where f and $K_{h/v}$ are defined in Chapter 14 and A_w is the adhesion at the wall. The shear stress increases with pressing pressure but is reduced by lubrication of the die (Fig. 20.18). The horizontal/vertical stress ratio parameter $K_{h/v}$ is higher when the yield strength of the granules is lower and when the pressing speed is slower.

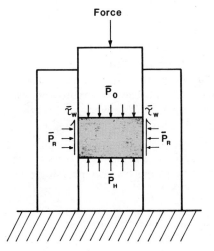

Fig. 20.17 Mean stresses on the compact during unidirectional compaction.

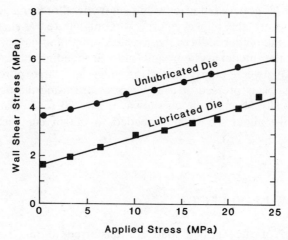

Fig. 20.18 Effect of pressing pressure on shear stress at the die wall for an alumina compact with a polyvinyl alcohol binder.

For one-directional pressing, the axial pressure transmission through the compact is approximated by the equation

$$\frac{\bar{P}_H}{\bar{P}_a} = \exp\left(\frac{-4fK_{h/v}H}{\text{Dia}}\right) \qquad (20.3)$$

where $\bar{P}_H/\bar{P}_a$ is the fraction of mean applied pressure $(\bar{P}_a)$ transmitted at a depth (H) for a cylindrical compact of diameter (Dia). Stress transmission for a granulated, micron-size alumina powder containing a plasticized polyvinyl alcohol binder is shown in Fig. 20.19. In stage III of compaction, the parameter $K_{h/v}$ was approximately 0.4, indicating that some of the film of binder had been extruded from contacts between particles; the friction factor for plasticized polyvinyl alcohol binder was approximately 0.24 for a lubricated die and 0.30 for an unlubricated die.[*]

Two-directional pressing or pressing with a floating die effectively halves H_{max} in Eq. 20.3 and significantly reduces the pressure differential. Granules containing lubricants added during granulation or an external lubricant added on their surface after granulation rub lubricant onto the die wall during pressing. Empirical results indicate that the admixed lubricant (< 1%) should be added in proportion to the ratio of friction area/pressing area.[†] Lubricants must be admixed homogeneously and used sparingly, as they can increase springback. Pressure gradients during fast pressing are minimized by using a dwell time in proportion to the friction/pressing area.

[*] R. A. DiMilia and J. S. Reed, *J. Am. Ceram. Soc.* **66**(9), 667–672 (1983).
[†] F. Winterberg, *Powder Met. Int.* **1** 29–32 (1969).

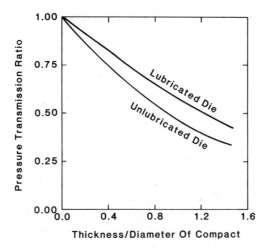

Fig. 20.19 Pressure transmission during the unidirection compaction of alumina powder containing a polyvinyl alcohol binder in a polished steel die with and without a stearic acid lubricant.

Studies of the movement of bulk powder during pressing indicate that the relative axial displacement is greater in the center than near the die wall, and the average movement decreases at a greater distance from the moving punch. Some radial translation is also observed, and the slippage plane is inclined relative to the axial direction. Pressure profiles determined for bulky powders using various microscale experimental techniques indicate general pressure gradients, as shown in Fig. 20.20 for unidirectional pressing. The diagonally directed pressure profile is sometimes called the Y thrust. The maximum pressure occurs near the top edge and diminishes in depth toward the central axis.

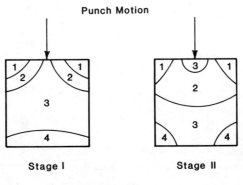

$$P_1 > P_2 > P_3 > P_4$$

Fig. 20.20 General pressure profiles for unidirectional pressing.

20.6 CONTROL OF COMPACT DEFECTS

The compact must survive ejection and handling without failure and should have a uniform microstructure. The most common defects in dry-pressed compacts are laminations and end capping (Fig. 20.21), which are caused by stresses produced by differential springback on ejection. Differential spring-back may have several causes:

1. Gradients in pressure transmitted within the compact, due to die wall friction.
2. Gradients in stored elastic energy due to nonuniform granules, nonuniform filling, or compressed air.
3. Frictional restraint of the die wall due to a high radial pressure in the compact, considerable elastic strain of the die, or poor die wall lubrication.
4. The ejected portion of the part possessing a greater springback than the die.
5. Undercutting of the die.

The tendency for laminations is decreased by reducing the average springback, changing the composition of the powder to increase the compact strength, reducing the pressing pressure or lubricating the die to decrease pressure gradients, and using a die of sufficient stiffness with a smooth wall and an entry bevel. Laminations near edges are often produced by the wedging of powder between the punch and die wall when the gap is too large relative to the minimum granule size; end ring capping is caused by a relatively high differential springback at the corner. A high average spring-back in the compact and a high die wall friction increase the tendency for an end cap or transverse laminations.

Cracks due to compressed air in pores occur when the compaction ratio

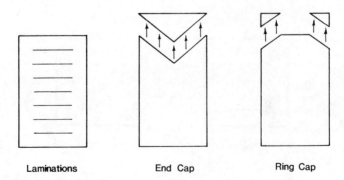

Laminations　　　　End Cap　　　　Ring Cap

Fig. 20.21 Common defects in pressed compacts.

and velocity are high, where the thickness of the compact is relatively large, and when the compact is of low gas permeability and low strength. Changes in the powder preparation such as granulation to increase the fill density and permeability and a longer pressing cycle can often eliminate this cause of defects. It is clear that the binder should be properly plasticized to facilitate granule deformation for compact strength without excessive springback. Less binder is required when it is properly plasticized.

Nonuniform shrinkage on sintering may be caused by density gradients in the pressed part. Top and bottom punch displacements are usually controlled to produce the minimum in the transmitted stress at the centroid of the part. Compositions and pressing procedures that improve stress transmission should reduce density gradients; the density varies less with pressure in stage III of compaction. A higher compaction rate may improve stage II stress transmission in parts with an unfavorable H/D ratio.

Surface smoothness is dependent on the smoothness of the die and punches, the absence of adhesion of particles on the surface of the punch, the size and deformation of the granules, and the pressing pressure. Higher pressing pressures reducc the size of the larger pores. Larger pores on the surface and within compacts are nested between relatively large granules and poorly plasticized granules; removal of coarser granules and control of T_g of the binder are required to eliminate large pores.

20.7 ISOSTATIC COMPACTION

Pressed products with one elongated dimension, a complex shape, or a large volume are not easily dry-pressed and are produced using isostatic pressing, often referred to as isopressing. Steps in isopressing are shown in Fig. 20.22.

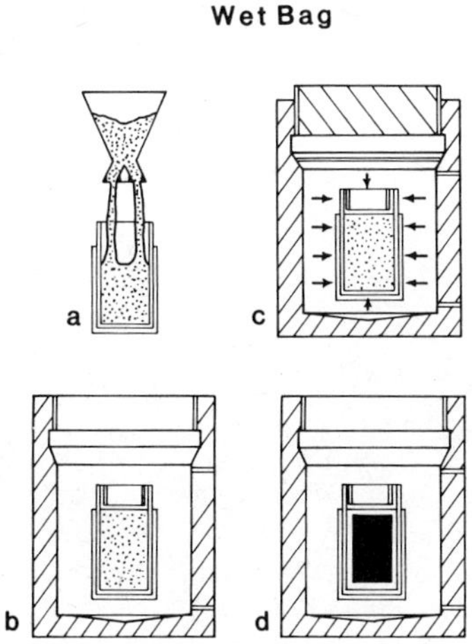

Wet Bag

Fig. 20.22 Stages in wet bag isopressing are (a) filling, (b) loading, (c) pressing, and (d) decompression prior to removal of the part. (Courtesy of Loomis Products Co., Levittown, PA.)

In the wet bag process, flexible molds are filled and sealed at a separate station. Molds may be filled while resting on a vibratory table to obtain a more uniform packing. Molds are often de-aired. The filled molds are then submerged in a liquid pressure chamber and pressed. After decompression, molds are removed, and the part is ejected. Dry bag isopressing (Fig. 20.23) imitates dry-pressing except that pressure is applied radially by means of the pressurized liquid medium between a flexible mold and a rigid shell. Wet bag isopressing is used for pressing complex shapes of various sizes and large refractories. Common products produced using dry bag isopressing are spark plug insulators, grinding media, and hollow tubes. Operating pressures up to 200 MPa are common, and wet bag presses operating to 500 MPa are used for pressing large blocks and billets.

Tooling for isopressing is formed from synthetic rubber, polyurethane, silicon rubber, and rigid steel end fixtures, mandrels, and perforated wet bag supports. Pressing powders are very similar to those used for dry pressing except that the granules are commonly softer. Shear stresses that deform granules are lower in isopressing owing to the different stress path. A relatively soft wax binder system is popular for isopressing.

Density gradients are minimized in isopressing thick parts and long parts, because the H/D ratio is inverted relative to that in dry pressing, and die wall friction is reduced. Sources of compact defects are similar to those described for dry-pressing except that the nonrigidity and greater elastic expansion (springback) of the tooling must be considered. Sudden decompression must be avoided to allow air to escape slowly and avoid a large differential air pressure. Slow decompression below 2 MPa is important when thick-walled tooling that has a relatively high springback is in contact with the part.

Dry Bag

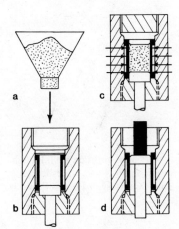

Fig. 20.23 In dry isopressing, (a) powder is loaded into (b) the opened die, (c) pressed, and (d) ejected after decompression. (Courtesy of Loomis Products Co., Levittown, PA.)

20.8 COMBINATION PRESSING

Large flatware items such as dinnerware having a thickness less than 0.5 cm and a diameter-to-thickness ratio ranging up to about 50 are pressed by a combination of dry-pressing and dry bag isopressing. The feed material for large flatware products is characteristically coarser (Fig. 20.3). In combination pressing, a relatively rigid steel or polyurethane plastic die component applies pressure to the simpler surface, and a flexible diaphragm pressurized by an internal liquid applies pressure to the surface with more complex relief (Fig. 20.24). Feed material in the die set is contoured using a rotating tool prior to pressing. Filling, contouring, pressing, and extraction occur simultaneously on a multistation, rotating table. Production rates may range up to several hundred pieces per hour. Advantages over plastic forming include a greater dimensional precision (tolerance in weight and shape $< \pm 1\%$), the elimination of drying and the drying shrinkage of the piece, and energy savings in spray-drying relative to the conventional drying.

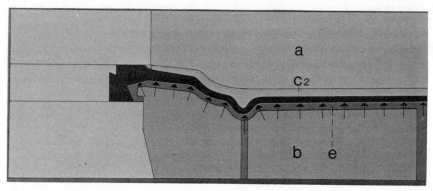

Fig. 20.24 In combination pressing, mechanical pressure produced by (a) the top die component and hydraulic pressure produced by the complex lower die component consisting of (b) a rigid support, and (e) pressurized hydraulic liquid pressing up on a flexible surface membrane are used to compact (c2) the flatware piece. (Courtesy of Dorst Maschinen und Anlagenbau, Kochel A. See, West Germany.)

20.9 ROLL PRESSING

Granulated powders may be compacted continuously between hard rolls into a thick sheet or segmented tablets such as substrates and chip carriers. Compaction pressures are high relative to dry-pressing, because the area of powder contacted by the rolls is very small. Production rates range up to several centimeters per second. Roll pressing is used to produce substrates up to 30 cm in width with a uniform thickness in the range of 0.5–1.5 mm and for the high-productivity forming of simple shapes in very large production lots.

Summary

Dry-pressing is a very important fabrication process used to produce products that are of relatively high density and dimensionally precise. Pressed products vary widely in size, shape, and composition. Isostatic pressing is used to press relatively large items and products having a more complex shape. Substrates and briquetted material are produced using roll compaction.

Feed material for pressing is commonly granulated. The system of processing additives and physical characteristics of the granules must be carefully controlled to produce pressing granules with the necessary flow and deformation properties. Pressing should eliminate large pores and produce a product having a uniform density and adequate strength to survive ejection and handling. Die wall friction causes a gradient in pressure through the thickness of the compact, which may produce a density gradient. The density gradient is reduced by using double-action pressing, by admixing a lubricant in the pressing powder, or by isostatic pressing. Elastic energy stored in the pressed part produces a volume expansion called springback when the pressed part is ejected. Springback must be controlled to prevent laminations and microcracks in pressed products but yet cause separation from the punch surface.

SUGGESTED READING

1. J. A. Mangels and G. L. Messing (eds.), *Advances in Ceramics Vol 9: Forming of Ceramics*, American Ceramic Society, Columbus, OH, 1984.

2. O. J. Whittemore Jr., Particle Compaction, in *Ceramic Processing Before Firing*, G. Y. Onoda Jr. and L. L. Hench (eds.), Wiley-Interscience, New York, 1978.

3. J. S. Reed and R. B. Runk, Dry Pressing, pp. 71–93 in *Treatise on Materials Science and Technology*, Vol. 9: Ceramic Fabrication Processes, Academic Press, New York, 1976.

4. Robert A. DiMilia and James S. Reed, Dependence of Compaction on the Glass Transition Temperature of The Binder Phase, *Am. Ceram. Soc. Bull.* **62**(4), 484–488 (1983).

5. Stress Transmission During the Compaction of a Spray-Dried Alumina Powder in a Steel Die, *J. Am. Ceram. Soc.* **66**(9), 667–672 (1983).

6. Robert A. Thompson, Mechanics of Powder Pressing. I. Model for Powder Densification, *Am. Ceram. Soc. Bull.* **60**(2), 237–243 (1981); II. Finite Element Analyses of End-Capping in Pressed Green Powders, *Ibid.* **60**(2) 243–247 (1981).

7. S. J. Lukasiewicz and J. S. Reed, Character and Compaction Response of Spray-Dried Agglomerates, *Am. Ceram. Soc. Bull.* **57**(9), 798–801, 805 (1978).

8. C. W. Nies and G. L. Messing, The Role of Binder Glass Transition Temperature During Powder Compaction, *J. Am. Ceram. Soc.* **67**(4), 301–304 (1984).

9. F. W. Dynys and J. W. Hallorn, Compaction of Aggregated Alumina Powder, *J. Am. Ceram. Soc.* **66**(9), 655–659 (1983).

10. Heinrich Niffka, Isostatic Dry Pressing of Flatware, *Am. Ceram. Soc. Bull.* **59**(12), 1220 (1980).

11. E. L. J. Papen, H. D. B. Raes, and G. Deplace, Development and Application of the Isostatic Process in the Refractories, Ceramic, and Electrical Porcelain Industries, *Interceram* **3**, 204–211 (1974).

12. P. Popper, K. Bloor, R. D. Brett, and D. E. Lloyd, Isostatic Pressing with Particular Reference to Tooling, pp. 101–107 in *Proceedings of the 3d CIMTEC International Meeting on Modern Ceramic Technologies, Rimini, Italy* (1976).

13. Felix Winterberg, The Pressing Technique in Powder Metallurgy, *Powder Met. Int.* **1**, 29–32 (1969).

14. Edward J. Motyl, Spray Drying, Pressing Lubricants Upgrade Ferrite Production, *Ceram. Age* **2**, 45–48 (1964).

PROBLEMS

20.1 What are the granule characteristics that produce a high fill density? Compare the size of interstices for monosize granules of 150 and 75 μm assuming orthorombic packing.

20.2 Calculate the compaction ratio for the alumina and the tile body granules in Fig. 20.6.

20.3 Calculate the packing factor on filling for the alumina and tile granules in Fig. 20.6. Assume that the granule density is 50% and 70% (inorganic basis) and that the inorganic phase density is 3.98 and 2.60 Mg/m^3 for the alumina and tile body, respectively.

20.4 Calculate the compaction constant m in eq. 20.1 from compaction results in Fig. 20.7.

20.5 Calculate the intergranular and intragranular porosity after filling and after pressing the alumina granules in Fig. 20.8.

20.6 Contrast the diametral expansion (mm) for compacts of 1.0 and 10 cm in diameter when the springback is 0.6 or 1.6%.

20.7 Estimate the percent reduction in ejection force for a lubricated die when the ejection rate is 1 cm/s in Fig. 20.14.

20.8 Contrast the friction area relative to the pressing area for an electronic substrate $30 \times 30 \times 2$ mm and the same substrate with 100 holes 0.5 mm in diameter.

20.9 When filling a large tile die, a tapered fill height is used to compensate for the differential fill density across the die. Sketch the tapered fill relative to the direction of motion of the fill shoe.

20.10 Calculate the diametral tensile strength of a compact if the diametral load causing failure is 5000 N and the compact is 50 mm thick and 30 mm in diameter.

20.11 Contrast the parameter $K_{h/v}$ for alumina (Poisson's modulus $\nu = 0.23$) and organic binder ($\nu = 0.4$) and a liquid ($\nu = 0.5$). For confined compression, $K_{h/v} = \nu/(1 - \nu)$.

20.12 Compare the pressure transmission ratio and differential pressure for the single action pressing of compacts with a H/Dia of 0.25 and 1.0 when the applied pressure is 100 MPa and the product $(K_{h/v}f)$ is 0.19.

20.13 The onset of stage III compaction is 40 MPa for a pressing powder. Calculate the necessary punch pressure to achieve stage III compaction at the midplane during double-action pressing of a compact with $H/Dia = 0.8$. Assume that $K_{h/v} = 0.39$ and $f = 0.26$.

20.14 Compare the pressure transmission for double-action dry pressing for cylinders with $H/Dia = 3/1$ and with $H/Dia = 1/3$. Assume that $K_{h/v}f = 0.15$.

20.15 If the applied pressure in problem 20.13 was 40 MPa, estimate the differential density between the pressed surface and midplane using Fig. 20.7 as a compaction diagram for the material.

20.16 Compare the elastic springback of hardened steel, an alumina compact (40% porous), and a rubber mold.

20.17 Calculate the location of the centroid for the part in Fig. 20.1 (bottom) if the height is 10 mm, the larger OD is 30 mm, the smaller OD is 20 mm, and the ID is 5 mm. The thickness of the wider portion is 5 mm.

20.18 Using the relationship in problem 14.15, does the contact stress in granules depend on the granule size for monosize granules in orthorhombic packing?

20.19 Using Fig. 20.7, explain why a uniform pore size cannot be obtained when a pressing powder contains a mixture of soft and hard granules.

20.20 A company is experiencing problems with laminations in parts pressed from an alumina powder containing 4 wt % of a polyvinyl alcohol binder. The parts pressed at 90 MPa are not of consistent strength, and weaker compacts are of a slightly lower density. What is your analysis of the problem and recommendations for solution?

20.21 Design a processing flow diagram for producing dry pressed, high performance structural alumina parts.

21

Plastic-Forming Processes

Plastic clay bodies have traditionally been formed by extrusion, plastic pressing, and jiggering. These processes are also used to a lesser extent for forming ceramic bodies plasticized with an organic binder. In plastic forming, feed material in a continuous, sectioned, or granular form that exhibits plastic behavior when compressed is consolidated and deformed into a particular shape.

"Wet-pressing" refers to the pressing of plastic granules in a rigid die; pressures are lower than for dry-pressing, and the die components are heated or spray-lubricated to facilitate ejection. Wet-pressing is used to fabricate a wide variety of refractory and porcelain parts of complex shape.

Extrusion is the shaping of the cross section of a cohesive plastic material by forcing it through a rigid die. Products formed by extrusion include structural refractory products such as hollow furnace tubes, honeycomb catalyst supports, transparent alumina tubes for lamps, very small tubular electronic and magnetic ceramics, electrical porcelain insulators and graphite electrodes ranging up to more than 1 ton in size, and flat substrates and tile products. Extrusion is also used to produce a de-aired plastic feed material of controlled volume for pressing and jiggering operations.

A section of extruded material may be reshaped by pressing between steel dies and deformed into a moderately thin product by pressing between permeable dies. The roughness and suction of the permeable dies cause the feed material to adhere during forming, and the pores permit the flow of water from the body during pressing and air flow for release of the product from the die. In jiggering, a section of de-aired, extruded body is compressed and sheared between the surfaces of a lubricated template or roller tool and a rotating permeable mold. Highly automated jiggering is used for making relatively thin-walled porcelain products such as institutional china and chemical porcelain. Thick-walled porcelain insulators are formed between a hot rotating metal alloy plunger and a permeable mold, a process that combines aspects of jiggering and pressing. Shrinkage on drying is commonly used to separate the piece from the permeable mold, but pneumatic separation is also used. Dimensional precision for plastic formed products varies with the shape and shrinkage but is generally lower than for dry-pressed products.

In injection molding, a thermoplastic polymer resin or wax system loaded heavily with a ceramic powder "filler" is formed into a complex shape by injecting heated material into a cooled metal die. Injection-molded parts may be formed with a high precision in size and shape, but this precision is not always maintained in parts with moderately thick cross sections because of problems in thermally removing the resin. Injection molding is used for fabricating complex shapes such as multivaned turbine rotors and small electronic components.

21.1 EQUIPMENT AND MATERIAL VARIABLES IN EXTRUSION

Plastic feed material is commonly prepared by directly batching and mixing the raw materials and additives in a high-shear mixer or, for a more homogeneous material, by mixing as a slurry and filter-pressing to form a plastic filter cake (Fig. 21.1). In some industrial operations where filter pressing is not feasible, plastic material is prepared by mixing feed slurry and spray-dried slurry in a high-shear mixer. Further processing commonly occurs in a pug mill and extrusion system where paddles or augers mix the feed material and force it through a shredder into a vacuum chamber (Fig. 21.2). Shredded material that is small in cross section is more uniformly de-aired without surface drying. De-aired material is consolidated, mixed, and extruded continuously using an auger and an extrusion die. Sectioned extrudate is used as a formed product, as a controlled feed for subsequent auger or piston extrusion, pressing, or jiggering, or is partially dried and surface-ground on a lathe.

Stages in extrusion are (1) material feeding, (2) consolidation and flow of the feed material in the barrel, (3) flow through a tapered die or orifice, (4) flow through the finishing tube of constant or nearly constant cross section, and (5) ejection. Piston and auger extruders are shown in Fig. 21.3. In piston extrusion, feed material may be a de-aired billet from the pug mill, which is sometimes enclosed in a sleeve, or segmented material. After insertion in the barrel and de-airing, the feed material is compressed and forced to flow down the barrel by the moving piston. For displacement of material in auger feeding, the material must not slip on the wall of the barrel, and the yield strength of the body must be less than the adhesive strength of the body on the surface. Axial ribs may increase the circumferential friction. The number of flights on the auger controls the number of feed columns displaced. A larger helix angle increases the potential delivery rate but reduces the compressive thrust on the material; a helix angle of

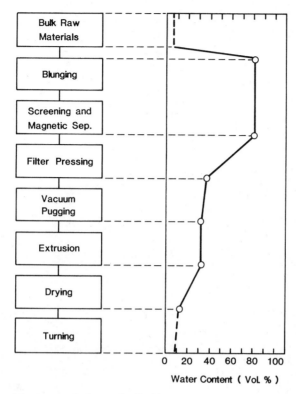

Fig. 21.1 Processing steps and change in liquid content during the processing of electrical porcelain.

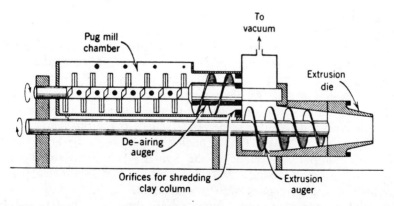

Fig. 21.2 Industrial pug mill with de-airing chamber and extrusion auger. (Courtesy of W. D. Kingery, *Introduction to Ceramics*, 1st Ed, Wiley, New York, 1960).

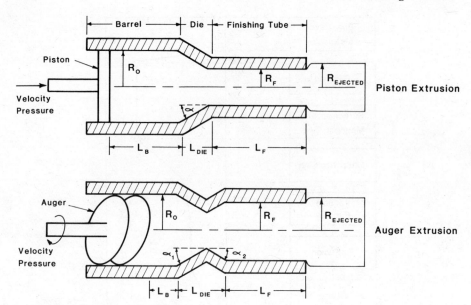

Fig. 21.3 Schematic diagram of (top) piston extruder and (bottom) auger extruder indicating design parameters.

approximately 20–25° is commonly used. The required ratio of the auger diameter to the product diameter increases as the yield strength of the material increases. Heating may change the adhesion of the body on metal surfaces and its flow properties.

The axial velocity of material discharged by the auger is a minimum at the hub and increases to a relative maximum near the surface of the flight of the auger. This differential velocity decreases as material moves down the barrel. The die alters the differential flow, reduces the cross section, and effects a particular cross-sectional geometry. Geometrical parameters of the die are the entrance angle α and the reduction ratio R_o/R_f (Fig. 21.3). Complex dies may contain small channels for injecting a die wall lubricant. For the conventional extrusion of hollow items, a system of arms supporting the central core rod called a spider is attached to the die, and the arms interrupt flow (Fig. 21.4). Flow through the finishing tube improves the knitting of material separated by the spider. The extrusion die must generate an internal pressure and flow pattern that eliminates defects. Measurable elastic springback occurs on ejection.

Industrial extrusion pressures range up to about 4 MPa for porcelain bodies and up to about 15 MPa for some organically plasticized materials. Capacities range widely, depending on the product size, but may approach 100 T/h. for large products. Industrial extrusion rates in terms of the velocity of ejected material also vary widely and are controlled in part by the rates of cutting and conveying the extruded material. An extrudate

Fig. 21.4 Spider in throat of die.

velocity of about 1 m/min is common for the extrusion of many porcelain bodies.

Ceramic substrates with a "honeycomb" cross section with up to 200 channels per square centimeter and a material thickness ranging down to 0.2 mm are now extruded using a special patented* square entry die. The die consists of a solid plate with entrance holes intersecting the center of perpendicular grooves which form the honeycomb shape. The body is of a relatively fine particle size, and the processing additives and rheology are very precisely controlled.

Compositions of plastic bodies are listed in Table 21.1. Plastic clay bodies usually approximate a close-packing size distribution with 25–40 vol % finer than 1 μm. Organic binders used as a body plasticizer are medium to high in viscosity grade; common binders used in aqueous extrusion are listed in Table 21.2. The liquid phase is commonly water, but liquid wax, petroleum oils, alcohol, etc. are sometimes used for nonaqueous extrusion systems.

Table 21.1 Examples of Compositions of Extrusion Bodies

Composition (vol %)					
Refractory Alumina		High Alumina		Electrical Porcelain	
Alumina ($<20\ \mu$m)	50	Alumina ($<20\ \mu$m)	46	Quartz ($<44\ \mu$m)	16
Hydroxyethyl		Ball clay	4	Feldspar (<44 mm)	16
cellulose	6	Methylcellulose	2	Kaolin	16
Water	44	Water	48	Ball clay	16
AlCl$_3$	<1	MgCl$_2$	<1	Water	36
(pH >8.5)				CaCl$_2$	<1

*Corning Glass Works, Corning, NY.

Table 21.2 Additives Used in Nonclay Aqueous Extrusion Bodies

Flocculant/Binder	Coagulant	Lubricant
Methylcellulose	$CaCl_2$	Various stearates
Hydroxyethyl cellulose	$MgCl_2$	Silicones
Polyvinyl alcohol	$MgSO_4$	Petroleum oil
Polyacrylimides	$AlCl_3$	Colloidal talc
Polysaccharides	$CaCO_3$	Colloidal graphite

Aqueous systems are typically coagulated using $< 1\%$ of a suitable additive. Internal lubricants are used when the die surface area/product cross section is high. The particle size distribution of non clay bodies may vary widely owng to constraints imposed by densification and grain sizes after sintering. The concentration of processing additives is somewhat dependent on the particle packing and the proportion of colloidal particles in the body. However, a minimum amount of colloidal particles and binder material of about 20 vol % appears to be requisite for plastic forming.

Extrusion bodies are elastic-plastic in nature. High-molecular-weight binders and flocculation increase the elastic compression and reduce the liquid migration. An adsorbed gel film of colloidal clay or binder molecules on the hard particles reduces the angle of friction between particles, and the colloid-liquid matrix restricts the degree of particle consolidation, which reduces the effective stress and the shear stress for flow. The binder phase provides flocculation for cohesion and strength after ejection.

21.2 EXTRUSION MECHANICS

Plastic behavior was discussed in Section 14.10. During extrusion, the driving force produced by the auger or piston must exceed the resistive force of the material and the die wall friction. The pressure motivating flow is highest in the barrel and decreases along the axis of the extruder, as is shown in Fig. 21.5. Flow may occur by the mechanisms of slippage at the wall and differential laminar flow in the extrudate (Fig. 21.6). Material flowing through the conical die is accelerated and translates toward the central axis. The particular flow behavior depends on the geometry of the die and the flow properties of the material. The permeability constant of an extrusion body must be very low, less than about 10^{-18} m^2, to prevent the migration of the pressured liquid through the material during its residence in the extruder. Internal defects, surface smoothness, and particle orientation in the extrudate are dependant on the flow behavior during extrusion.

In the barrel of the extruder, the summation of forces for the cylindrical element undergoing steady-state laminar flow (Fig. 21.7) provides a relation between the shear stress τ and the radial position r:

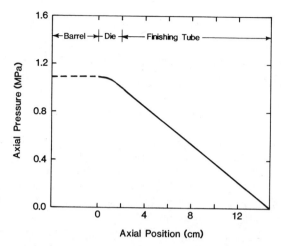

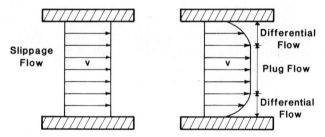

Fig. 21.5 Axial pressure gradient during the extrusion of an electrical porcelain body in a piston extruder ($R_o/R_f = 3$; $R_f = 0.6$ cm). (From D. Price, MS thesis, Alfred University, Alfred, NY.)

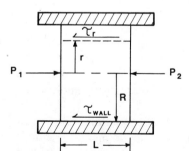

Fig. 21.6 Velocity profiles for (left) interfacial slippage and complete plug flow and (right) differential flow with central plug, for a Bingham plastic body.

Fig. 21.7 Axial stress during steady-state extrusion in a rigid tube.

$$\tau_r = \frac{r(P_1 - P_2)}{2L} = \frac{r \, \Delta P}{2L} \tag{21.1}$$

where $\Delta P/L$ is the differential axial pressure for the segment of length (L). The maximum shear stress occurs at the wall when $r = R$ and

$$\tau_{\text{wall}} = \frac{R \, \Delta P}{2L} \tag{21.2}$$

The flow of colored clay-water slurries and pastes in capillary tubes was studied by Bingham* and later by Scott Blair and Crowthers.† They observed that flow occurred first by slippage "en bloc" with a linear dependence of apparent velocity on pressure followed by a transition to differential laminar flow at a higher velocity. The tendency for slippage flow was observed to increase as the content of solids increased.

A plastic extrusion body may be considered to be a compressible, cohesive solid in which the shear strength for flow (τ_y) increases when the body is compressed (Eq. 14.17). When the imposed compressive stress exceeds that causing pore saturation and the properties are constant with position and time, the flow behavior may be approximated by the Bingham equation (Eq. 15.5). The steady-state flow of a Bingham material in a linear tube is described by the Buckingham-Reiner equation*,†

$$\bar{v} = \frac{R^2 \, \Delta P}{8L\eta_P} \left[1 - \frac{4}{3} \left[\frac{2L\tau_y}{R \, \Delta P} \right] + \frac{1}{3} \left[\frac{2L\tau_y}{R \, \Delta P} \right]^4 \right] + v_{\text{slippage}} \tag{21.3}$$

where $\bar{v}$ is the mean flow velocity and τ_y and η_P are the Bingham flow properties of the body.

Plug flow with slippage at the wall occurs when τ_y is greater than τ_{wall}; the equation for viscous slippage flow is†

$$v_{\text{slippage}} = \frac{\epsilon \, R \, \Delta P}{\eta_b 2L} = \frac{\epsilon}{\eta_b} \tau_{\text{wall}} \tag{21.4}$$

where η_b is the viscosity of a lubricating boundary film of thickness ϵ. For a particular slippage velocity, die wall lubrication increases ϵ/η_b and reduces τ_{wall}.

Differential laminar flow occurs when the body formulation and the flow velocity are such that $\tau_y < \tau_{\text{wall}}$. The shear stress developed increases radially and is equal to the yield stress at a position r_c (Fig. 21.8). Differential shear

*E. C. Bingham, *Bull. Bureau Standards*, **13**, 309–353 (1916–17).
†G. W. Scott Blair and E. M. Crowthers, *J. Phys. Chem.* **33**(3), 321–330 (1929).
*M. Reiner, *Z. Kolloid* **39**, 80–87 (1926).
†E. Buckingham, *Proc. Am. Soc. Test. Mat.* **21**, 1154–1161 (1921).

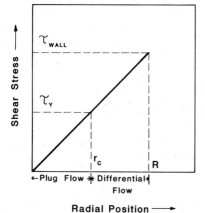

Fig. 21.8 Variation of shear stress with radial position for flow in a tube.

flow occurs in the region ($r_c \leq r \leq R$) where the shear stress exceeds the yield stress, and for a Bingham material the variation of the velocity is

$$v = \frac{1}{\eta_P} \left[\frac{\Delta P}{4L} (R^2 - r^2) - \tau_y (R - r) \right] + v \qquad (21.5)$$

Uniform flow without shear occurs in the center region of radius (r_c).

Material flowing through a tapered die is subjected to compressive and shearing stress, as shown in Fig. 21.9. The shear stress/compressive stress ratio at the wall decreases as the entrance angle (α) of the die increases. For a large die angle, the effective angle of shear lies in the material, and a region of static material occurs at the periphery of the die. The flow velocity is a function of both the radial position and axial location, and material nearer the central axis experiences more acceleration (Fig. 21.10). Experiments with striated feed of different color indicate that the differential flow velocity between the center and the wall increases as the entrance angle increases.

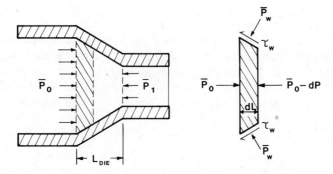

Fig. 21.9 Stress system for flow in the tapered die.

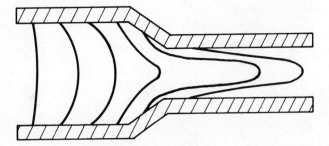

Fig. 21.10 Change of velocity profile on flow of an electrical porcelain body through the throat of a die.

The flow of a Bingham plastic material in a conical die involves both internal shear and slippage at the die wall and has been described by the general equation*

$$\Delta P_{die} = A\, f_1(\alpha, R_o, R_f)\, \eta_P Q + B\, f_2(\alpha)\tau_y + C f_3(\alpha)\tau_y \log(R_o/R_f)$$
$$(21.6)$$

where A, B, and C are constants for a die of initial radius R_o, R_o/R_f is the reduction ratio, $f_1(\alpha, R_o, R_f)$, $f_2(\alpha)$, and $f_3(\alpha)$ are complex functions of the die geometry, and Q is the volume flow rate.

When extruding through a conical die and finishing tube of constant cross section, the dependence of the pressure for extrusion on the flow velocity has the form shown in Fig. 21.11. The initial change of slope is a function of the increase in velocity and the parameters ϵ/η_b of the lubricating boundary film and η_P of the extrudate in the die. When extruding a clay porcelain body, a lubricating clay-water film may form on the surface of the extrudate (Fig. 21.12); for organically plasticized materials, a film of fine particles, liquids, and binder may provide lubrication. A lubricant admixed in the body or injected at the die wall may facilitate slippage flow and reduce the extrusion pressure significantly. At higher velocities, the extrusion pressure increases linearly with flow rate, as is expected when the properties of the extrudate and boundary layer are constant.

For moderate rates of extrusion, the apparent extrusion pressure varies directly with the ultimate shear strength (τ_y) of the extrudate determined in direct shear (Fig. 21.13) and the die geometry (Fig. 21.14), as described by Eq. 21.6.* The effect of a small addition of a liquid, lubricant, deflocculant and increasing the temperature of an electrical porcelain body is indicated in Table 21.3.

*A. Ovenston and J. J. Benbow, *Trans. Br. Ceram. Soc.* **67**, 543–567 (1968).
*G. Ackley and J. Reed, in *Advances in Ceramics Vol. 9: Forming of Ceramics*, J. A. Mangels and G. L. Messing (eds), American Ceramic Society, Columbus, OH, 1984.

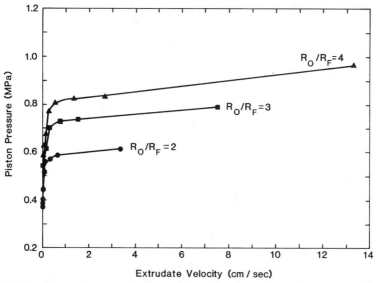

Fig. 21.11 Dependence of pressure for extrusion on the extrudate velocity and reduction ratio.

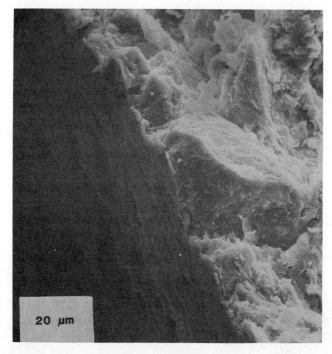

Fig. 21.12 Scanning electron micrograph of extruded surface and internal fracture surface of an electrical porcelain body. The smooth surface "skin" consists of oriented colloidal particles; large, angular particles are apparent below the surface film.

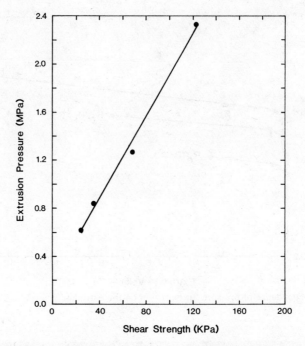

Fig. 21.13 Dependence of pressure for extrusion on the ultimate shear strength of the electrical porcelain body. (From G. Ackley, MS thesis, Alfred University, Alfred, NY, 1983.)

Table 21.3 Effect of Body Additives and Temperature on the Ultimate Shear Strength and Extrusion Pressure of an Electrical Porcelain Body

Body Variable	Shear Strength Decrease (%)	Pressure Decrease (%)
Water (0.5 wt %)	31	28
Na stearate (0.5 Wt %)	17	19
Na polyacrylate (0.5 wt %)	8	11
Temperature (+ 10°C)	4	10

Base values: shear strength, 48 kPa; extrusion pressure, 800 kPa.

In an extruded plastic clay body, the springback on ejection is very anisotropic owing to the orientation of clay particles tangential to the flow velocity profile. An axial springback ranging up to 10% and a radial springback of about 0% were measured for a compressed filter cake of the electrical porcelain body with a high degree of particle orientation. The anisotropy of springback is reduced by a moderate flow profile similar to

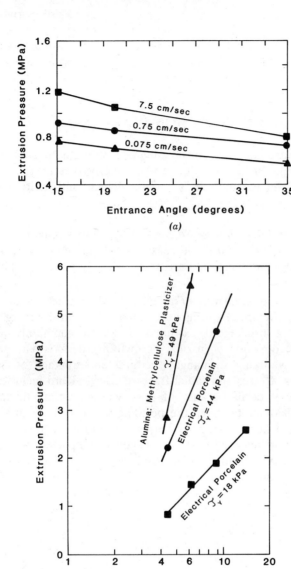

Fig. 21.14 Dependence of the pressure for extrusion on (a) the entrance angle for an electrical porcelain body and (b) the reduction ratio for alumina and porcelain bodies (ultimate shear strength (τ_y) determined using direct shear test). (After G. Ackley, MS thesis, Alfred University, Alfred, NY.)

that in Fig. 21.10. Springback of extruded alumina plasticized with 4 vol % methyl cellulose has been observed to be of the order of only 1% and approximately isotropic.†

21.3 CONTROL OF EXTRUSION DEFECTS

The ejected extrudate must have adequate strength to survive handling without slumping or deformation and a microstructure free from large pores, cracks, and laminations. Common defects and their causes are as follows:

1. *Insufficient Strength and Stiffness.* The yield strength of the plastic extrudate may be increased by decreasing the liquid content, increasing the content of binder and colloidal particles, coagulating the body, and using a binder that gels.

2. *Cracks and Laminations (Fig. 21.15).* Most cracks and laminations are caused by differential springback and/or differential drying shrinkage. The mean drying shrinkage of extruded products is commonly in the range of 1–5%. The common origin of a "crowfoot crack" is a dispersed hard inclusion of low drying shrinkage; the greater shrinkage of surrounding material creates stresses and radial fracture cracks. Hard inclusions may be agglomerates that occur because of poor wetting during mixing, foreign particles, or agglomerates from the periodic break-away of dry compacted material retained in the die. Dies with an entrance angle of less than 20° tend not to retain material that may become compacted and later break away. Another cause of differential shrinkage is liquid migration. The liquid permeability is reduced by increasing the colloidal content and the amount and molecular weight of the binder in the body. Increasing coagulation and the liquid content in the body may reduce the extrusion pressure, which is the driving force for liquid migration (Fig. 21.16).

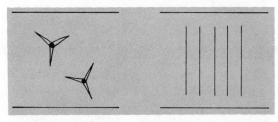

Crow Foot Cracks   Surface Laminations

Fig. 21.15 Surface "crowfoot" cracks and periodic surface laminations.

†R. Locker, PhD thesis, Alfred University, Alfred, NY, 1985.

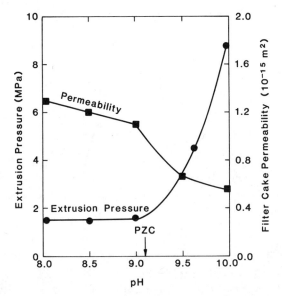

Fig. 21.16 Dependence of permeability of filter cake and extrusion pressure of the consolidated material on pH producing coagulation (pH reduced by adding 0.2 M AlCl$_3$ to the precursor suspension containing alumina and a hydroxyethyl cellulose binder).

Oriented platy particles cause anisotropy in the springback and drying shrinkage in proportion to the orientation and aspect ratio of the particles. The differential shrinkage between regions of different orientation may cause microcracks called checks. A sudden change in the particle orientation produced by differential flow can cause differential springback and differential drying shrinkage and characteristic laminations; a die design that produces a more gradual gradient in the final velocity profile (Fig. 21.10) aids in reducing the maximum stress. Adding clay particles of lower aspect ratio and increasing the concentration of fine isometric filler particles also aid in reducing the shrinkage anisotropy. A discontinuity in particle orientation is produced when extruding around a spider or core rod, as is shown in Fig. 21.17.

3. *Surface Craters and Blisters*. Air in the body dissolves in the liquid under pressure. On ejection, this air will rapidly migrate through small pore channels which join, and the air produces a crater or blister surface defect that is visible to the eye. Proper de-airing of the feed material is essential; the extruder must be well sealed, and feed material should be shredded to less than about 3 mm to reduce the air migration distance.

4. *Periodic Surface Laminations (Fig. 21.15)*. When the extrusion pressure is high and the die is poorly lubricated, slip-stick wall friction and a high elastic springback can cause periodic surface laminations or cracks perpendicular to the flow direction. This defect is eliminated by improving surface

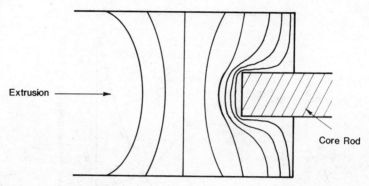

Extrusion

Core Rod

Fig. 21.17 Flow profile around a cylindrical core rod produces a discontinuity in the particle orientation near the corner of the rod.

lubrication, increasing the binder or liquid content, coagulating the body, and increasing the extrusion velocity.

5. *Nonuniform Flow Through the Die.* The die is normally centered on the auger system. A specific distance between the termination of the auger and the entrance to the die is needed to reduce the differential velocity of material coming from the auger. When extruding hollow items, the friction produced by the bridges and core rods must be balanced to prevent curling of the extrudate.

6. *Laminations Due to Divided Flow.* Material leaving an auger is separated by the hub; compression and radial twisting downstream creates a characteristic S-interface which should be eliminated during extrusion (Fig. 21.18). Divided flow is also produced by the spider used to form a hollow

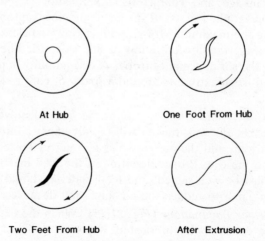

At Hub   One Foot From Hub

Two Feet From Hub   After Extrusion

Fig. 21.18 Development of S-crack during auger extrusion. (After R. Lester, p. 151 in *Technological Innovations in Whitewares*, Alfred University Press, Alfred, NY, 1982.)

product; a lamination may be produced where the flow lines join if the pressure downstream is insufficient. The persistence of a lamination is increased by air in the body and an insufficient pressure or shear flow to heal the lamination.

7. *Gradients in Extrudate Stiffness.* In piston extrusion, liquid migration down the column, called bleeding, will cause the material that is last out to be stiffer. Bleeding is reduced by decreasing the permeability of the body. Feed material should be uniform.

8. *Curling on Drying.* Dimensional distortion due to differential drying shrinkage is discussed in the chapter on drying.

21.4 PLASTIC TRANSFER PRESSING AND JIGGERING

In ancient times, bricks and pottery were formed by pressing a section of plastic material in a mold. Later, hollow ware was formed by pressing one's hands against material supported on a rotating wheel, which is called jiggering. These processes are now highly mechanized and are used for forming a wide variety of refractory and electrical porcelain pieces with surface relief, institutional porcelain, cooking ware, and fine china. Highly automated jiggering systems are used for the industrial fabrication of hollowware, as is shown in Fig. 21.19.

Fig. 21.19 Industrial automatic jiggering system. (Photo courtesy of R. Wahl, A. J. Wahl Inc., Brocton, NY.)

Feed material is commonly a section of de-aired extrudate. Plastic feed may be pressed in a metal die or in a permeable mold composed of relatively hard gypsum or another porous material. In jiggering, a porous mold or "bat" supports the body, and a roller tool or template is used to deform the plastic body and contour the outer surface. Small electrical porcelain insulators are commonly formed by forcing a rotating, heated metal tool into a section of material supported on a porous mold. Steam produced at the surface of the metal produces lubrication and facilitates release. Lubrication is produced by heating metal roller tools to produce steam or by spraying a lubricant onto a high-molecular-weight polymer tool, to minimize sticking. Separation of the piece from the mold is commonly effected by the differential drying shrinkage. Excess material is commonly eliminated by trimming during forming. Mechanized fettling is used to smooth surfaces.

Parameters in plastic pressing and jiggering include the thickness and diameter of the feed, the thickness reduction ratio, the material flow rate, surface features, the motion of the contouring tool and its lubrication, imposed vacuum, pressure, temperature, and the flow properties of the body. Fabrication begins as the feed material is cut and transferred to the cavity of the mold. On lowering the forming tool, the body makes contact with the mold and then flows against the mold and laterally in a plane between the tool and the mold (Fig. 21.20). In this stage, the shear strength of the body is only about 20–250 kPa. Material contacting the surface of the permeable die becomes static, but may slide across polished metal. Air may be forced into the permeable die.

Material flow is observed to be laminar except in protrusions such as the foot of a piece of dinnerwear. In plastic pressing, the closure rate is relatively rapid until the die cavity is filled. The velocity of laminar flow is higher near the center of the piece, and the velocity profile through the thickness is similar to that shown in Fig. 21.10. A roller tool produces a high degree of particle orientation because of the high shear flow near the surface. When pressing between permeable molds, consolidation and dewatering of the piece may occur if the permeability and dwell time under presure are sufficiently high; this increases the yield strength for handling and reduces the drying shrinkage (Fig. 21.21).

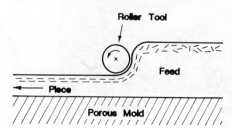

Fig. 21.20 Material flow and particle orientation near surface effected using a roller tool.

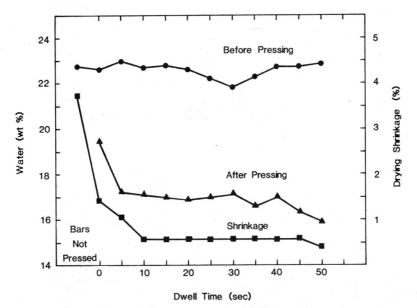

Fig. 21.21 Reduction of liquid content and drying shrinkage on pressing a hotel china body between permeable dies.

Defects in jiggered and pressed plastic pieces include those observed in extruded material. A differential liquid content between the surface and center of a piece pressed between permeable molds may sometimes cause "case hardening" and shrinkage cracks in the center. During jiggering, "winding" near the base of the piece occurs if the applied vacuum is insufficient or when the diameter of the "bat" of material is too large for the shape being formed. Splitting of a jiggered piece may occur when the speed of the mold or roller is too high or when the surface of the roller is insufficiently lubricated.

Plastic transfer pressing is commonly used for forming large shapes with detailed surface relief and relatively deep nonsymmetrical shapes. Jiggering is widely used for forming circular and elliptical shapes with a relatively thin wall (less than 5 mm) and deeper circular shapes.

21.5 INJECTION MOLDING

Injection molding is a major polymer fabrication process used especially for the high-productivity forming of parts with a thin wall and a complex shape. In forming a part in an injection molding machine, segmented feed material is consolidated by a plunger or screw, heated, and then forced through a nozzle and sprue into the cavity of a mold, as is shown in Fig. 21.22. Either a thermosetting or a thermoplastic resin may be used, but a thermoplastic

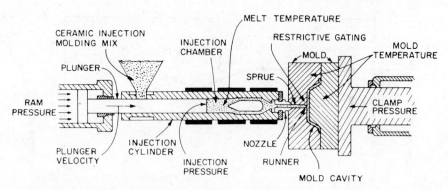

Fig. 21.22 Schematic diagram of a plunger type injection molding machine. (Reprinted from J. A. Mangels and W. Trela, in *Advances in Ceramics Vol.* 9: *Forming of Ceramics*, J. A. Mangels and G. L. Messing (eds), American Ceramic Society, Columbus, OH, 1984.)

system is generally preferred for reasons of storage stability, molding properties, and the recycling of scrap and material consumed in filling sprues and runners connecting multiple die cavities. Important variables that must be controlled and reproduced are the melt temperature, injection pressure, flow rates, and cooling rates. Temperatures must be precisely controlled to control the flow properties and deformation resistance of the formed part, and are monitored by thermocouples placed at critical points along the flow channels. Flow properties also vary with the flow rate, the injection pressure, and the design of the die. Flow fronts can join together into a monolithic part only if the cavity is filled rapidly and air is displaced into numerous small vents in the die.

The injection molding of ceramic parts is similar to the injection molding of plastics. For ceramics, special abrasion-resistant steels must be used in high-wear areas. The organic binder system must be formulated to facilitate both flow in forming and burnout during firing. The binder system usually consists of the major binder, a minor binder, a plasticizer/lubricant (or hardener), and a surfactant for particle wetting. Examples of components of binder systems are listed in Table 21.4. The minor binder should burn out at a lower temperature and provide pore channels for the escape of gas produced on decomposition of the major binder.* The major binder provides strength while gaseous products from the minor binder diffuse through the low-permeability structure. Ash content and carbon residue after burnout are also considerations in selecting a binder system. The plasticizer/lubricant is added to control the glass transition temperature and to improve flow and moldability.

*J. A. Mangels and W. Trela, in *Advances in Ceramics Vol.* 9: *Forming of Ceramics*, J. A. Mangels and G. L. Messing (eds), American Ceramic Society, Columbus, OH, 1984.

Table 21.4 Components of Binder Systems Used in Injection Molding

System	Major Binder	Minor Binder	Plasticizer/ Lubricant
Thermoplastic[a]	Polystyrene (high MW)	Resin (low MW)	Petroleum oil
	Polyethylene (high MW)	Paraffin wax	Stearate
Wax[b]	Paraffin	Beeswax	Stearic acid
		Liquid epoxy	Oleic acid

	Binder	Hardener	Plasticizer/ Lubricant
Thermosetting			
	Epoxy resin	Phenol formaldehyde	Wax

[a]Resin/oil/stearate in ratio of about 6/3/1.
[b]Wax/epoxy/stearic acid, 90/5/5 (K. W. French, J. T. Neil, and L. T. Turnbaugh, *Compositions for Injecting Molding*, U.S. patent No. 4,456,713 (1984)).

The binder system is proportioned to form a continuous matrix, and the volume fraction of particles in the mix is less than f_{cr}^v, as expressed by Eq. 15.27. The batch is similar to a casting slurry but contains a high-viscosity polymer that serves as a binder.

Both the yield strength and viscosity are very dependent on the proportions of resin and powder and the temperature. The ceramic powder and binder system are mixed hot under a vacuum in a thermally jacketed, high-shear mixer such as a sigma mixer or double-planetary mixer. Mixed material is cooled, crushed, and sized to serve as feed material for injection molding. Hot mixed material may be pressed into preforms for hot transfer molding. In the injection molding machine, feed is commonly heated to about 125–150°C, above the glass transition temperature of the binder system, and forced through a channel leading to the mold cavities using a plunger pressure in the range 30–100 MPa.* Compositions exhibit pseudoplastic behavior (Fig. 21.23), and the injection speed is a critical parameter. The size, shape, and location of runners should be designed to produce laminar flow and displace air into small vents. Shrinkage of the piece in the cool mold is dependent on the molding pressure and temperature. The yield strength of the cooled part is dependent on the coagulation of the resin and must be sufficiently high to resist deformation on handling.

A lower-pressure variation of thermoplastic injection molding involves the use of a wax binder system. In many respects it resembles injection casting. Waxes used soften in the range of 65–95°C. The relatively low viscosity of molten wax reduces the injection pressure to as low as 350 kPa, and wear rates are reduced rather significantly. Removal of the wax by

*Advances in Ceramics Vol. 9: Forming of Ceramics, J. A. Mangels and G. L. Messing (eds), American Chemical Society, Columbus, OH, 1984.

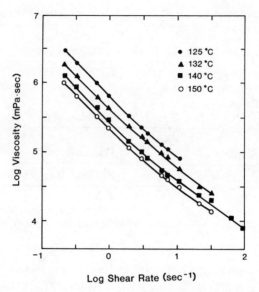

Fig. 21.23 Dependence of the apparent viscosity of an injection molding batch on shear rate and temperature. (After J. A. Mangels and W. Trela, in *Advances in Ceramics Vol. 9: Forming of Ceramics*, J. A. Mangels and G. L. Messing (eds), American Ceramic Society, Columbus, OH, 1984.)

heating without disruption or distortion of the part is a difficult and critical processing step in forming large cross sections.

The principal advantage of injection molding is the ability to form parts having a complex shape and surface relief that are dimensionally accurate and have smooth surfaces. Parts having a nominal thickness in the range of 0.5–5 mm, holes finer than 1.5 mm, and a tolerance of less than 0.1 mm are now produced industrially. Heated injection molding mixes are also used to form products by extrusion and transfer pressing.

Defects in parts include knit lines and air entrapment, as formed, and blisters, cracks, and shape distortion after binder burnout. Knit lines and entrapped air are eliminated by proper die design and control of material flow. Formulation of the binder system and control of the heating rate to regulate gas evolution and shrinkage on binder burnout aid in preventing blisters and cracks. Shape distortion can occur from anisotropy in the distribution of the binder or warping or sagging during binder burnout; the tendency for slumping is lower when the binder content is lower and decreases as binder is removed on heating.

Summary

Ceramic bodies exhibiting plastic mechanical behavior are fabricated by the methods of extrusion, plastic transfer pressing, jiggering, and injection molding. The plasticizing matrix may be either a low-viscosity liquid containing a medium- to high-molecular-weight polymer binder and/or colloidal particles or a thermoplastic or thermosetting resin system. Important variables in plastic forming are type and amount of plasticizer, particle size distribution, coagulation, lubrication, temperature, de-airing, and die design.

In extrusion, both differential laminar flow and slippage flow may occur, and the flow behavior is controlled by the die design and the yield strength of the body relative to the shear stress at the die wall. Extrusion is used to produce a wide variety of solid and hollow products with a uniform cross section and feed material for jiggering and transfer pressing. Plastically formed products are generally less precise in dimensions than dry-pressed products because of plastic deformation during handling and variable drying shrinkage. Defects in products may occur owing to differential drying shrinkage or springback, entrapped air, discontinuities in particle orientation, and the incomplete bonding of material separated and rejoined during flow.

In injection molding, the temperature and flow rate of the body and the design of the flow channels and molds must be carefully controlled to produce a satisfactory product. Molding pressures are reduced by using wax binders. Injection molding is currently used to form advanced ceramics of complex shape with thin sections.

SUGGESTED READING

1. *Advances in Ceramics Vol. 9: Forming of Ceramics*, J. A. Mangels and G. L. Messing (eds), American Ceramic Society, Columbus, OH, 1984.
2. G. C. Robinson, Extrusion Defects, in *Ceramic Processing Before Firing*, G. Y. Onoda and L. L. Hench (eds), Wiley-Interscience, New York, 1978.
3. F. H. Norton, *Fine Ceramics*, Robert E. Krieger, Malabar, FL, 1978.
4. Zehev Tadmar and Imrich Klein, *Engineering Principles of Plasticizing Extrusion*, Van Nostrand Rheinhold, New York, 1970.
5. David Price and James S. Reed, Boundary Conditions in Electrical Porcelain Extrusion, *Am. Ceram. Soc. Bull.* **62**(12), 1346–1350 (1983).
6. Angelo L. Salamone and James S. Reed, Extrusion Behavior and Microstructure Development in Alumina, *Am. Ceram. Soc. Bull.* **58**(12), 1175–1178 (1979).
7. A. Ovenston and J. J. Benbow, Effects of Die Geometry on the Extrusion of Clay-Like Material, *Trans. Br. Ceram. Soc.* **67,** 543–567 (1968).
8. William O. Williamson, Structure and Behavior of Extruded Clay. II. Piston and Auger Extrusion, *Ceram. Age* **31**, 27–29 (1966).

9. J. F. White and A. L. Clavel, Extrusion Properties of Non-Clay Oxides, *Am. Ceram. Soc. Bull.* **42**(11), 698–702 (1963).

10. F. Moore, The Physics of Extrusion, *Claycraft* **11**, 50–57 (1962).

11. E. Buckingham, On Plastic Flow Through Capillary Tubes, *Proc. Am. Soc. Test Mat'l.* **21**, 1154–1161 (1921).

12. Eugene C. Bingham, An Investigation of the Laws of Plastic Flow, *Bull. Bur. Stnrds.* **13**, 309–353 (1916–17).

PROBLEMS

21.1 Why may coagulation increase K_P of the precursor filter cake but reduce K_P of the body during extrusion?

21.2 Compare the effective stress $\bar{\sigma}$ for bodies for extrusion and injection molding.

21.3 For extrusion, the liquid content of the body must exceed some critical volume fraction. Explain (refer to Fig. 14.22).

21.4 For auger feeding, $\tau_{wall} > \tau_y$. But on extrusion through the die, the inequality may be $\tau_{wall} < \tau_y$. Explain.

21.5 Explain why the inequality $\tau_y > \tau_{wall}$ depends on both the body formulation and the flow velocity.

21.6 The ultimate shear strength of an extrusion body is 160 kPa. What is r_c for flow in the barrel ($R = 4.0$ cm, $\Delta P/L = 11$ MPa/m) and in the finishing tube ($R = 0.31$ cm, $\Delta P/L = 40$ MPa/m). Calculate τ_{wall} and the inequality between τ_y and τ_{wall} for each.

21.7 For the extrusion of the electrical porcelain body in the finishing tube of problem 21.6, the flow rate is 8 mm/s. Calculate Green's boundary ratio ϵ/η_b. If the thickness of the boundary layer is 0.5–1 μm, is the boundary layer simply water?

21.8 Does the ratio of $\bar{P}_{wall}/\tau_{wall}$ depend on the entrance angle (α) of the die? Use your answer to explain the dependence of the velocity profile on α.

21.9 Estimate the extrusion pressure for a body with $\tau_y = 30$ kPa when the reduction ratio is 10 for the extruder used to obtain the results in Fig. 21.14b.

21.10 Illustrate the state of stress in the extrudate surrounding the inclusion in Fig. 21.15.

21.11 Illustrate the orientation of platelike particles in Figs. 21.10 and 21.17.

21.12 A two-stage deformation rate program is commonly used in plastic transfer pressing and jiggering. Should the initial rate be higher or lower than the final rate? Explain.

21.13 What is the dependence of the dwell time to produce a constant water content on the permeability K_p and thickness of the body for plastic pressing between permeable dies? In Fig. 21.21, why does the minimum shrinkage remain about 0.5%?

21.14 Calculate the relative change in viscosity on heating from 125 to 150°C and on increasing the shear rate from 1 to $10\,s^{-1}$ for the injection molding material in Fig. 21.23.

21.15 Assuming that the body in Fig. 21.23 is Bingham in behavior, estimate τ_y and η_p at 150°C.

21.16 Illustrate the change in velocity profile as material flows from the auger through the barrel and die in Fig. 21.3.

21.17 Design a processing flow diagram for producing extruded high performance zirconia tubes.

22

Casting Processes

Casting processes are used to produce a self-supporting shape called a cast from a specially formulated slurry. The yield strength of the cast may be increased by the partial removal of the liquid, which concentrates the solids, or by the gelation, polymerization, or crystallization of the matrix phase. An aqueous slurry containing fine clay has been traditionally called a slip, and conventional casting is commonly called slip casting. In slip casting, the slurry is poured or pumped into a permeable mold having a particular shape; capillary suction and filtration concentrate the solids into a cast adjacent to the wall of the mold (Fig. 22.1). The motivating force for the separation of the liquid may also be pressure applied to the slurry (pressure casting), a vacuum applied to the mold (vacuum-assisted casting), or centrifugal pressure. Slip is drained from the mold after the desired wall thickness has been achieved using a drain casting mode, as is shown in Fig. 22.1. In solid casting, filtration continues until a nearly solid cast with a center pipe. having the shape of the cavity of the mold, has been developed. Drain casting and solid casting are the traditional processes used to produce a variety of porcelain products with a complex shape, such as bathroom fixtures, chemical porcelain components, dense refractories that are of a complex shape or massive in cross section, highly porous thermal insulation, and traditional fine china and dinnerware. Some high- performance ceramics with a complex shape are now produced by casting, because additives and agglomerates in the submicron powder are conveniently controlled in the slurry, and the capital investment for casting is relatively low.

In tape casting, a controlled film of slurry forms on a substrate. The evaporation of liquid from the film during controlled drying transforms the film of slurry into a flexible, rubbery tape or sheet. Tape casting is now widely used for the high-volume, continuous production of substrates 0.03–0.06 mm in thickness for modern multilayer electronic packaging and for multilayer capacitors (Fig. 1.1).

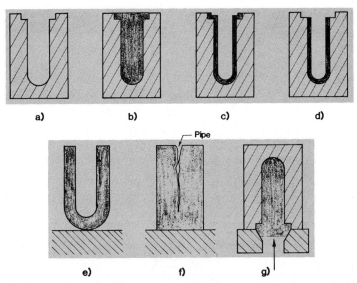

Fig. 22.1 In drain casting, (a) a porous mold is (b) filled with slurry; after a period of casting, (c) excess slurry is drained from the mold leaving a cast which (d) may be trimmed in place if plastic. A longer casting time produces (e) a thicker wall which can produce (f) a nearly solid cast. Injection of mold (g) with pressurized slurry increases the casting rate.

Slurries containing a bonding additive such as a hydraulic cement, a reactive bond, or a binder that gels may be cast in molds or "in place," as in a furnace lining or a dental cavity. Mechanical vibration is often used to reduce the liquid requirement for material flow and mold filling. In a process called gunning, castables containing a sticky binder are mixed with water and blown through a pipe to apply the material as a coating or to patch a refractory lining.

22.1 SLIP CASTING IN A PERMEABLE MOLD

Control parameters in slip casting are the slurry rheology during mold filling, the casting rate, the cast density and yield strength, flow rheology on draining, the shrinkage and release of the cast from the mold, and the strength and plasticity of the piece during trimming and surface finishing.

Casting slurries are typically formulated to be pseudoplastic with a small yield point, because the viscosity must be low enough for convenient pumping, mold filling, and the escape of bubbles; in conventional slip casting, the viscosity is less than 2000 mPa s at a shear rate of $1-10 \, s^{-1}$. The density and yield point of the slurry should be high enough during casting to minimize particle settling; a higher solids loading also reduces the difference in liquid concentration between the slurry and the cast. Slurries containing liquids, solids, and processing additives may be prepared in a variety of mixers and in ball mills and vibratory mills. Batch sizes may range from a fraction of a liter for a special ceramic to several thousand gallons when casting porcelain products. Intensive mixing and milling disperse agglomerates more completely and shorten the mixing time; however, gentler mixing for a longer time is used when solid materials can be comminuted in mixing, and the proportion of submicron sizes must be relatively low. Additives that can be degraded during mixing, such as binders of medium to high molecular weight, are often added near the end of the mixing cycle, and their potential degradation may influence the selection of the mixing operation. Casting slips containing up to 50% clay are often blunged for 8–48 h. The long blunging produces an "aged" time stable slip with a more controlled colloid content and a more reproducible casting behavior. Postmixing processes include classification, separation of magnetic material, temperature control, and de-airing. Mixed slip may be conveyed in the mixing container or pumped into portable holding tanks for distribution and mold filling.

Permeable molds for casting are commonly gypsum with 40–50% porosity. The casting time may range from a few minutes for a porcelain product with a thin wall to about 1 h for a thin but dense cast from a well-defloccuated slurry of fine powder or a relatively thick (2 cm), more porous cast from a partially flocculated whiteware casting slip. The casting time for a dense solid cast refractory of fine particle size may range up to several weeks for thicknesses of 30 cm. Casting times are reduced significantly when using vacuum, pressure, and centrifugal casting.

In drain casting, the excess slurry must drain readily from the surface of the cast, but the cast must be relatively rigid. The marked change in flow behavior is effected by a differential liquid content between the slip and cast, but in some systems, such as clay products, gelation of the cast is more important.

Separation of the piece from the mold normally occurs from the shrinkage on drying; air may be blown back through the mold for pneumatic release. When casting porcelain slips, molds are stripped 20–60 min after draining. The partially dried pieces must be strong enough to resist fracture and plastic deformation during handling. Bodies that are plastic may be trimmed and surface-finished while damp, and then loaded onto conveyors or racks for the final drying. Brittle casts are surface-finished after drying, called fettling, or after partial or final sintering.

Examples of compositions for porcelain and refractory alumina casting slurries are listed in Table 22.1. In the porcelain slip, the clay content of 15–20% is needed to obtain the proper fired microstructure. The clays used should be sufficiently coarse and relatively free of bentonite (see Table 5.5) so that the total submicron content of the body is less than about 30% when completely dispersed. Generally, clay slips are only partially deflocculated, and the working viscosity is higher than the minimum that is possible for the

Table 22.1 Examples of Compositions of Casting Slurries

Whiteware Slip			Refractory Slurry	
Material	Concentration (vol%)		Material	Concentration (vol%)
Nonplastics	25–30		Alumina ($<45\ \mu$m)	40–50
Clay	15–25		Ball clay	0–10
Water	45–60		Water	50–60
		Additives[a] (wt%)		
Na_2SiO_3, Na polyacrylate, Na lignosulfonate	<0.5	Deflocculant	NH_4 polyacrylate	0.5–2
$CaCO_3$	<0.1	Coagulant	$MgSO_4$	0–0.1
$BaCO_3$	<0.1			
Clay $<1\ \mu$m	Variable amount	Binder	NH_4 alginate, carboxymethyl cellulose, methyl cellulose, hydroxyethyl cellulose	0–0.5

[a] Percentage of weight of solids in slurry.

solids loading (Fig. 22.2). The agglomeration of particles on adding a coagulant to the deflocculated slip reduces the fraction of submicron material, as indicated in Fig. 22.3 for a casting slip just before pouring. The type and amount of additives used to control the state of partial deflocculation depend on the type and amount of colloids present, soluble impurities in the raw materials, and impurities in the water.

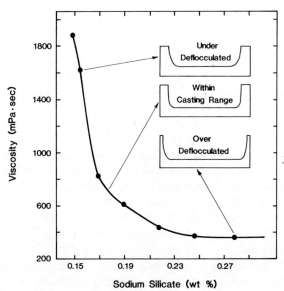

Fig. 22.2 Effect of deflocculation on the casting behavior of a whiteware slip (viscosity at $\dot{\gamma} = 3\ \text{s}^{-1}$ just before filling).

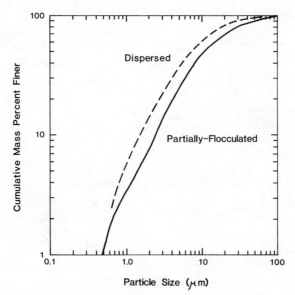

Fig. 22.3 Particle size distribution in a partially deflocculated industrial porcelain slip immediately after mixing and for a completely deflocculated sample.

Slurries for casting oxides, carbides, nitrides, etc. range more widely in composition. A fraction of a percent of an organic binder increases the pseudoplasticity of the slurry and the strength of the cast but reduces the permeability of the cast and the casting rates (Fig. 18.10); a slurry that is only partially deflocculated will cast more rapidly, but the cast will be more porous. Binder migration, which reduces the liquid permeability of the cast, is reduced by using a binder of medium molecular weight. A deflocculated slurry with adequate colloids and a minimal binder content will produce a cast with a narrow range of pore size and an adequate green strength for careful handling; however, the casts are brittle and relatively intolerant of surface finishing prior to sintering. Compositions containing a small amount of a fine clay and an organic binder are more plastic and more tolerant to stresses in finishing and handling.

Ceramic materials such as magnesium oxide and calcium oxide, which quickly hydrate in water, are slip-cast using a nonaqueous liquid such as an alcohol, trichlorethylene, or methyl ethyl ketone. Nonoxide ceramics such as silicon carbide and silicon nitride are also cast as a nonaqueous slurry. However, carbide and nitride powders handled in air commonly have an oxide surface layer, and stable aqueous slurries can often be deflocculated by particle charging.

The most commonly used permeable mold material is gypsum ($CaSO_4 \cdot 2H_2O$) formed from the reaction between plaster of Paris ($CaSO_4 \cdot 0.5H_2O$) and water;

$$CaSO_4 \cdot 0.5H_2O + 1.5H_2O \rightarrow CaSO_4 \cdot 2H_2O \qquad (22.1)$$

Gypsum is used for reasons of its relatively low cost, ability to cast molds with good surface smoothness and detail, high ultimate porosity and small pore size, short setting time, and small dimensional expansion (about 0.17%) on setting, which aids release from models.

According to Eq. 22.1, the weight ratio of water to plaster required for hydration forming gypsum is 18.6/100. A range of 60/100 to 80/100 water/plaster is used in slurries for production molds. During setting, an interconnected network of needle and platelike arrangements of gypsum crystals is formed and gives the mold strength. Molds with a maximum pore size of about 5 μm and an apparent porosity of 40–50% are formed after drying, as is seen in Fig. 22.4. The high water/plaster ratio used for an industrial casting mold produces the highest porosity and a slightly larger pore size, which increases the water absorption but lowers the strength (Fig. 22.5). Stronger molds for pressing are prepared at a water/plaster ratio of about 40/100; air must be blown through the mold after the initial set to maintain continuous pore channels. The setting time and crystal structure in the gypsum mold depend on the temperature and electrolytes in the water and the mixing time, and these variables must be controlled to obtain molds with reproducible behavior.

Disadvantages of gypsum molds are their low compressive strength when partially saturated with water, erosion in use due to their low abrasion resistance and the significant solubility of gypsum in water, their relatively

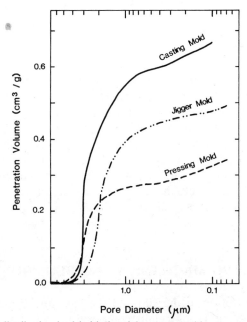

Fig. 22.4 Pore size distribution in dried industrial gypsum molds, prepared using water/plaster ratios of 80, 65, and 40/100. (From L. Hollern, M.S. thesis, Alfred University, Alfred, NY, 1969.)

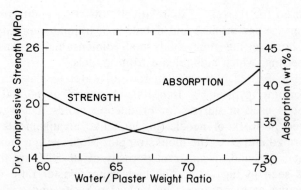

Fig. 22.5 Dependence of the absorption, and compressive strength on the water/plaster ratio used in fabricating gypsum molds.

Table 22.2 Properties of Conventional Gypsum Molds

Desiccation	Dehydrates in dry air
Thermal expansion (°C^{-1})	155×10^{-7}
Compressive strength (MPa)	
Dry	14
Wet	7
Tensile strength (dry, MPa)	3
Solubility[a] in water at 25°C (g/L)	2.6

Source: C. M. Lambe, in *Ceramic Fabrication Processes*, W. D. Kingery (ed), MIT Press, Cambridge, MA.

[a] Increases with temperature and sodium sulfate in water.

low thermal shock resistance, and the potential desiccation of the gypsum when heated above 40°C during drying, as is indicated in Table 22.2. Improved mold materials such as a porous fired ceramic, a porous polymer resin, and paper products have been developed but are not yet widely used because of difficulties in fabrication and in controlling dimensional tolerances and surface detail. The life of gypsum molds is lower when using acidic aqueous slurries or an alcohol medium. Mold coatings such as alginate, talc, and paper pulp are sometimes applied on the mold surface to facilitate separation of a cast part during drying.

22.2 SLIP-CASTING MECHANICS AND BEHAVIOR

The liquid concentration in the slip and the cast for an industrial alumina refractory and a sanitary porcelain composition are shown in Fig. 22.6. When casting porcelain ware, the casting rate of the partially deflocculated slip is relatively high, but the difference in liquid content between the cast

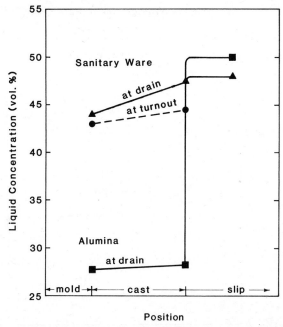

Fig. 22.6 Liquid profile for slip casting an industrial porcelain slip and an alumina refractory slurry. (After W. Brodie, *Technological Innovation in Whitewares*, Alfred University Press, Alfred, NY, 1982.)

and the slip is quite small; a liquid gradient in the cast is measurable. In contrast, the casting rate of the fine alumina cast from a well-deflocculated slurry is relatively low, the liquid differential between the slurry and cast is much larger, and the liquid concentration is more uniform across the section.

Whiteware casting slips are generally only partially deflocculated and are yield-pseudoplastic and thixotropic, as indicated in Fig. 22.7. During casting, the slight removal of water and gelation increase the apparent viscosity and apparent yield strength of the cast. The partially deflocculated slip contains agglomerates (flocs) which must be of a controlled size and solids content. Gel formation in the cast is critical, and the gel strength varies with the proportions of raw materials used, colloid content, flocculation (colloid coagulation) caused by chemical additives, and casting time (Fig. 22.8). A cast with a balance between firmness, density, and plasticity is requisite to prevent fracture or shape distortion during trimming, handling, drying, and surface finishing. A more uniform distribution of both the liquid and particle sizes throughout the cast reduces stress and distortion on drying and firing. Gelation minimizes mold staining and improves mold release by reducing the mobility of colloids through the cast. Very poor deflocculation causes even larger, more porous agglomerates in the slip, and the cast is more porous with larger pores.

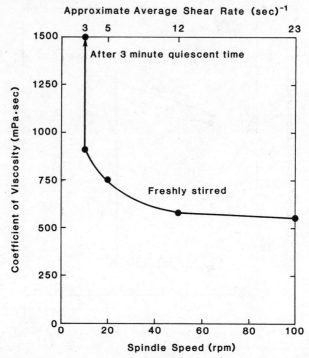

Fig. 22.7 Pseudoplastic behavior of a porcelain slip; gelation causes the viscosity to increase when at rest after pouring (53 vol% solids in slip).

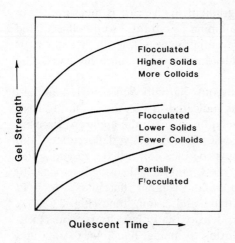

Fig. 22.8 Trends in gel strength with time for slips of different solids content, colloid content, and degree of flocculation.

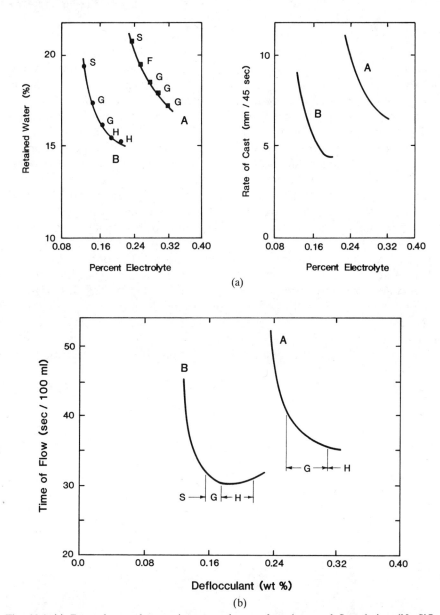

Fig. 22.9 (a) Dependence of water in cast and rate of casting on deflocculation (Na_2SiO_3 deflocculent; S, F, G, and H are symbols for a soft, fair, good, and hard cast, respectively) and (b) dependence of flow time of slurry through an orifice on deflocculation for two whiteware casting slips. (After G. W. Phelps and J. van Wunnik, United Clay Mines Inc. Manual Cyprus Industrial Minerals, Sandersville, GA.)

When the viscosity of the slip is used as a control parameter, it is commonly observed that slips within a narrow range of viscosity produce casts with the best compromise of casting rate, cast density, and handling properties (Fig. 22.9a). A dense, brittle cast is produced from a slip that is nearly completely deflocculated, and a relatively porous, soft cast is produced from a very poorly deflocculated slip. In industrial casting, a satisfactory slip must be tolerant to small fluctuations in the particle size distribution and soluble impurities in materials and chemical additives; the behavior of relatively tolerant (wide casting range) and intolerant slips is shown in Fig. 22.9b. The control of the microstructure of the slip, gelation, and the microstructure of the cast when using industrial minerals and unrefined water is not an easy task. Water used for molds and casting slurries is filtered and deionized in some industrial operations. Some clays are received as an aged slurry. Ball clays contain both complex organic (lignite) and inorganic colloids, which are altered during mixing. The proper balance of the type and amount of colloid and coagulating electrolyte is critical to the gelation and microstructure development in the cast (Fig. 22.10). Coagulating ions such as Ca^{2+} and Mg^{2+} are precipitated as silicates on the addition of Na_2SiO_3 and sulfate ions on the addition of $BaCO_3$. The concentration of divalent cations and anions and amount of added electrolytes used depend on the purity of the water and solubility of impurities and fine particles in raw materials.

The casting rate (L^2/t) is constant with time when the parameters in Eq.

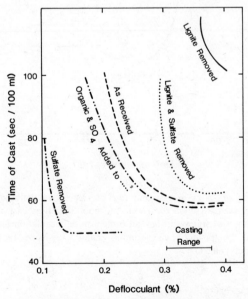

Fig. 22.10 Effect of organic matter and soluble sulfate on the casting time of an industrial whiteware casting slip deflocculated using Na_2SiO_3 deflocculent. (After G. W. Phelps, *Am. Ceram. Soc. Bull.* **38**(5), 246–250 (1959).)

18.10 and the mold suction and resistance are invariant with time. The effective suction of gypsum casting molds is in the range of 120–180 kPa. In use, the absorption rate of a gypsum mold decreases owing to the presence of water in finer pores and the gradual enlargement of pores as gypsum is dissolved, as is indicated in Fig. 22.11. The "staining" of the mold may increase its resistance to liquid migration (Fig. 22.12). The time to achieve a particular wall thickness is proportional to $(\eta_L/JK_P\Delta P)$. The permeability K_P is dependent on the porosity and pore structure of the cast and, therefore, the particle size distribution and agglomerate size in the slurry; the J parameter is very dependent on the relative difference in the solids content of the slurry and cast (Fig. 18.9).

Slurries for casting refractories and fine-grained technical ceramics are partially to completely deflocculated. Systems containing colloidal clay may be handled much like standard whiteware compositions. Binders such as carboxymethyl cellulose and sodium or ammonium alginate of moderate to high molecular weight may be added as a fraction of a percent of the weight of inorganic particles. The adsorbed ionic binder that coats the particles and attracts water retards settling in the slurry and causes a gelling mechanism in the cast. However, its presence retards liquid migration during casting and drying, and the amount must be very limited to obtain a compromise of slurry properties, casting rate, cast properties, and drying behavior. The quantity of colloidal inorganic particles and their interaction with the organic binder are very important in the casting process.

Heating an aqueous slip can increase the deflocculation and widen the deflocculation range, which would reduce L^2/t, but the reduction of the η_L is greater, and L^2/t may be increased by as much as 25%. Control of the mold temperature and slip temperature is required to control the firmness of the

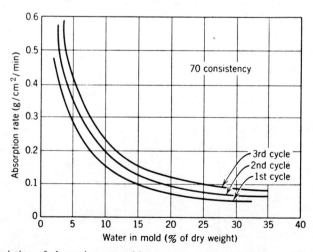

Fig. 22.11 Variation of absorption rate with water content in mold for several casting cycles. (From C. M. Lambe, p. 38 in *Ceramic Fabrication Processes*, MIT Press, Cambridge, MA, 1958.)

Fig. 22.12 Scanning electron micrograph of clay film on the surface of a stained gypsum mold. (Figure courtesy of Ward Votava, Alfred University, Alfred, NY.)

cast and the draining behavior. Adjustments in slip temperature and deflocculation are commonly based on the mold temperature.

During casting, the effective stress in the particulate structure is nearly zero in the layer of particles just deposited, but it increases in the depth of the cast owing to the drop in liquid pressure along the pore channels leading to the mold. A consequence is the density gradient in the cast, which is larger when the cast is more compressible and when the liquid pressure gradient is greater (Fig. 22.13).

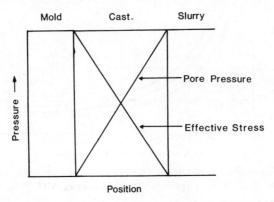

Fig. 22.13 Nominal profile of pore pressure and effective stress during casting.

22.3 NONTRADITIONAL CASTING

Equation 18.10 indicates that the casting rate can be increased by increasing the pressure inducing liquid migration. The capillary suction of a gypsum casting mold is less than about 0.2 MPa, and the motivating pressure for casting can easily be increased fivefold by introducing an external pressure or suction. This principle is employed in vacuum casting and pressure casting.

Vacuum Casting

Vacuum casting is widely used for forming very porous refractory insulation having a complex shape. The slurry typically contains partially deflocculated ceramic powder and chopped refractory fiber. The fiber increases the viscosity and liquid requirement and increases the porosity of the cast. An expendable, permeable preform is coupled to a vacuum line and then submerged in the slurry. After the desired wall thickness has been cast, the preform is withdrawn and removed for drying and a chemical set. The external surface of a gypsum mold may also be subjected to a vacuum to effectively increase the parameter ΔP_T and the casting rate in slip casting.

Pressure Casting

Pressure casting has been investigated as a forming technique for porcelain and complex refractory shapes, and mechanized, relatively automatic systems are now used in some industries. In pressure casting, the mold serves as a filter, and the casting time is controlled by regulation of the external pressure (Fig. 22.1); the molds need not be dried, and the humidity of the workplace is lower.

   In studies of the pressure casting of porcelain ware,[*] it was found that a slurry pressure of up to 1.5 MPa increased the cast thickness by a factor of 2–3. Pressure casting reduced the water content of the cast and the drying shrinkage from about 3–3.5% to 1–1.5% and increased the shrinkage anisotropy. The casting time for a wall approximately 1 cm in thickness cast from a heated slip containing very little deflocculent was reduced from 1 to 2 h to less than 20 min. Compressed air was also used for drying and to release the piece from the mold. Results for other compositions are listed in Table 22.3. Disadvantages of pressure casting include problems with conventional mold materials and the greater capital cost. When pressure-casting deflocculated slurries of submicron powders, density gradients in the cast occur because of the high average drying shrinkage and the greater effective stress across the cast.

---

[*] T. M. Gainer and J. A. Carter, *Am. Ceram. Soc. Bull.* **43**(1), 9–12 (1964).

Table 22.3 Casting Rate (L^2/t) for Different Casting Pressures

Casting Pressure (MPa)	Casting Rate (mm^2/min)			
	Porcelain[a]	Alumina[b]	Alumina[c]	Zirconia[d]
0.14	2.0	2.4	0.3	—
0.28	3.1	4.9		
0.56	5.2	10		0.1
1.40		25		0.3
2.80		50		0.8

[a] Semivitreous procelain.
[b] Calcined alumina (no binder).
[c] Calcined alumina (with 0.5% PVA binder).
[d] 39 M^2/g; deflocculated and ultimate particle size < 50 nm.

A variation of pressure casting is used to form hard ferrite magnets of a simple shape that have a highly oriented microstructure. The ferrite particles are submicron tabular crystallites that have a preferential magnetization in the c-direction. A die set with pistons that are magnetically permeable and also drilled to permit the escape of liquid is charged with a well-dispersed slurry. After a magnetic field has been applied to orient the particles, the pistons are forced into the die, and a dense cast forms as liquid is separated.

Thixotropic Casting

Casting slurries containing a gelling, reactive, or hydraulic binder are used to produce a variety of refractory shapes and molds for metal casting, to repair refractory structures, to obtain dental impressions and fill cavities, and to produce concrete structural materials. For dense materials, the mold or cavity is usually filled with a very concentrated slurry of minimal liquid content, and mechanical vibration is often used to induce flow and eliminate air pockets. The yield strength and viscosity of the cast are increased by the initial "set" and the ultimate strength when the binder system hardens.

Bond contents relative to the amount of ceramic materials depend on the strength required. For castable refractories, 20–30 vol% phosphate bond or 25–35 vol% hydraulic cement is commonly used. Foamed concretes are produced to reduce the bulk density, increase the resistance to disintegration during freezing and thawing conditions, and reduce the bulk thermal conductivity. Foaming is produced by adding a foaming agent that stabilizes air bubbles during mixing or by the introduction of an additive that reacts with one of the components just before the material "sets" to produce gas bubbles. Bond contents are higher in dental cement applications where finer powders and maximum possible strength are desired.

Electrophoretic Casting

The electrophoretic deposition of particles onto a conducting substrate has been investigated as a means for forming thin tubular shapes and coatings on metals.* Cross sections must be limited in thickness, because the casting rate is relatively slow.

22.4 CONTROL OF CAST DEFECTS

Common defects in drain cast products are pinholes, wreathing, warped shapes, uneven thickness, and cracks on drying.

Pinholes are formed from small air bubbles trapped at the slip-mold interface. Air bubbles and froth are eliminated by the proper wetting of materials in mixing, de-airing, reducing the viscosity of the slurry, filling the mold without turbulence, and rotating the filled mold to displace bubbles at the slip-mold interface.

Wreathing is a wavelike pattern on the drained surface and is caused by variations in the casting rate in different regions of the cast. This problem is minimized by using a mold having a uniform pore structure and adsorption, and by careful mold filling.

Warping is caused by differential shrinkage on drying due to a gradient in the liquid content (density) or average particle size. Warping is minimized by reducing the average shrinkage, the liquid gradient, and the segregation of particles by size or shape, which produces a gradient in the average spacing between particles. A linear drying shrinkage of 2–4% is rather typical for cast products; pressure casting may reduce the average shrinkage to <1%. In products containing only colloidal particles, the average drying shrinkage may exceed 5% and warpage is more of a problem.

Other defects in cast ware are an uneven thickness of the cast wall due to gravitational settling, the uneven absorption rate of a mold, and uneven draining. Proper formulation of the slip and design of the mold is required to minimize these defects. Cracks may appear in dried casts for the reasons presented in Chapter 21. An anisotropic drying shrinkage may occur where two surfaces of the cast meet at a sharp angle when using a slurry containing dispersed, anisometric particles; the juncture is a potential site for a crack on drying. Casts that are brittle, lack strength, or are difficult to fettle may also be considered defective for some types of products.

22.5 TAPE CASTING

Tape casting is the process of forming a film of controlled thickness when a slurry flows down an inclined substrate or under a blade onto a supporting

---

* Robert W. Powers, Ceramic Aspects of Forming Beta Alumina by Electrophoretic Deposition, *Am. Ceram. Soc. Bull.* **65**(9), 1270–1277 (1986).

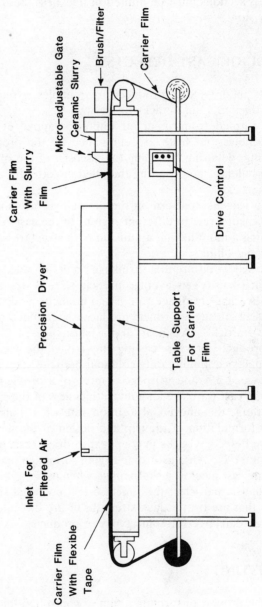

Fig. 22.14 Schematic of continuous tape casting apparatus.

surface. The process is generally referred to as continuous tape casting when the blade is stationary and the supporting surface moves, as is shown in Fig. 22.14, and batch doctor blade casting when the blade moves across the stationary, supporting surface covered with slurry. The cast film is dried to an elastic, leathery state, slit for proper handling width, and shaped by punching. Tape casting may be used to produce ceramic substrates with relatively smooth surfaces that are thin (0.01–1 mm), flat, uniform, and somewhat compressible. The surface of substrates produced by extrusion or pressing a granulated powder is less smooth; below 2 mm in thickness, breakage, nonuniform density, and nonparallelism of surfaces become a problem when pressing.

Tape casting is a continuous, high-productivity process and is now used to produce substrates for multilayer ceramic electronic packaging, multilayer titanate capacitors, piezoelectric devices, thick and thin film insulators, ferrite memories, and catalyst supports. In producing electronic packaging, metal circuit patterns and film resistors are commonly printed on the green sheet and cofired. Multilayer packaging is produced by stacking printed sheets, laminating these under pressure, punching again for size and shape, inserting metal via interconnections, and sintering (see Fig. 1.1).

Tape Casting Process

The first step in the tape-casting process is the preparation of the casting slurry consisting of a liquid (solvent), binder, plasticizers, dispersant, and the ceramic powder. Examples of batch formulations used industrially are listed in Table 22.4. A mixture of liquids is used to balance the dissolving and adsorption of additives and to extend the volatilization range on drying. The concentration of binder and plasticizers is higher than in dry-pressing. The high-molecular-weight binder must have satisfactory handling characteristics, produce a film that resists migration and is tough and flexible after

Table 22.4 Examples of Compositions of Tape-Casting Slurries

Component	Alumina Tape	Composition (vol%)	Titanate Tape	Composition (vol%)
Ceramic powder	Alumina powder*	27	Titanate powder*	28
Liquid system	Trichlorethylene	42	Methylethyl ketone	33
	Ethanol	16	Ethanol	16
Deflocculant dispersant	Menhaden fish oil	1.8	Menhaden fish oil	1.7
Binder	Polyvinyl butyral	4.4	Acrylic	6.7
Plasticizer	Polyethylene glycol	4.8	Polyethylene glycol	6.7
	Octyl phthalate	4.0	Butyl benzyl phthalate	6.7
Surfactant			Cyclohexanone	1.2

Source: Advances in Ceramics, J. A. Mangels and G. L. Messing (eds), American Ceramic Society, Columbus, OH.

* <5 μm, includes sintering aids, grain growth inhibitor.

drying, and burn out cleanly over a sufficient temperature range. The plasticizer reduces the glass transition temperature of the binder and the temperature required for laminating. A wetting agent promotes the dispersion of agglomerates and the elimination of pinholes.

Mixing and dispersion are usually accomplished using two-stage milling. The powder, liquids, dispersant, and wetting agent are mixed for 12–24 h, and then the binder system is introduced, and milling is continued for another 2–24 h. Two-stage milling is used to maximize dispersion but minimize the degradation of the binder.

After milling, the slurry is heated, filtered, and de-aired to remove agglomerates and bubbles. Slurry is cast on a clean, smooth, impervious, insoluble surface such as Mylar®, Teflon®, or cellulose acetate. The thickness of the tape is a function of the height of the blade, viscosity of the slurry, speed of the carrier film, and drying shrinkage. The casting speed is dependent on the thickness of the tape, evaporation rate, and length of the machine, all of which control the drying time. For a casting speed of 1–2 cm/s, the shear rate is about 15–$80\,s^{-1}$. A pseudoplastic slurry with a viscosity of about 2500–5000 mPa s in the range of 15–$80\,s^{-1}$ is used for casting, as is shown in Fig. 22.15.

The viscoelastic tape is formed as liquids evaporate. The slurry dries slowly as the counterflow of filtered air passes above the tape; the drying temperature is controlled below the flash point of the solvent system. The dried tape must not bond to the carrier film but must adhere sufficiently to control shrinkage and to eliminate curling and peeling. Tape formed in contact with the carrier film is smoother. The drying shrinkage occurs essentially through the thickness, and the dried thickness is commonly about

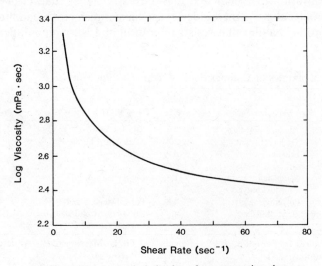

Fig. 22.15 Pseudoplastic behavior of a tape-casting slurry.

one-half the blade height. Oriented textures are produced when the powder particles are anisometric in shape. Dried tape may be used directly or stored on a take-up reel. After elimination of solvents, the tape is about 50 vol% powder, 35 vol% organics, and 15 vol% porosity. The concentration of binder, plasticizer, and pores must be controlled to produce a viscoelastic tape having adequate strength and elasticity for handling yet sufficient permanent compressibility for deformation around metallization and interfacial bonding during the lamination process. Also, the amount of binder and porosity affect the permeability and ease of binder burnout during firing.

Tape defects include cracks, camber, local regions of low density, and surface defects consisting of unacceptable average roughness and large surface pores. Cracks are caused by a differential shrinkage. Low-density regions are caused by retarded sintering in regions containing powder agglomerates. Surface roughness is reduced as the sintered grain size and the size of surface pores are reduced. The average roughness as measured by profilometry may be less than 2 μm (CLA). Surface pores larger than 5 μm are undesirable, as they may interrupt metallization lines; pores are caused by bubbles in the slurry, segregated organic, and pullouts on separating the tape from the carrier.

22.6 CASTING OF MONOSIZE, SPHERICAL COLLOIDS

The recent availability of spherical colloidal particles has stimulated research interest in the as-formed and fired microstructures that may be produced. Handling in slurry form provides a means for controlling surface forces and agglomeration; packed structures have been formed by pressure casting, centrifugal casting, and using coating techniques (see Chapter 25). Deposition of the spheres in casting produces domains consisting of ordered particle arrangements separated by interdomain regions of higher porosity (Fig. 22.16), similar to the packing of atoms in polycrystalline materials.

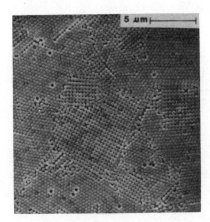

Fig. 22.16 Surface of consolidated monodisperse silica spheres. (From E. Barringer et al., p. 321 in *Ultrastructure Processing: Ceramics, Glasses, and Composites*, Wiley-Interscience, New York, 1984.)

SUMMARY

Casting processes are used to produce both traditional and more advanced ceramics. Casting modes include drain casting and solid casting in molds and tape casting onto a substrate. In conventional slip casting in a porous mold, the cast is formed by filtration, and the driving force for the removal of the liquid is the capillary suction of the mold. The slip must be properly formulated to produce a balance between the gelation rate and the casting rate to produce a plastic cast with adequate density, strength, and casting rate. Pressure casting and vacuum casting reduce the casting time and drying shrinkage of the cast. Organic binders used in casting nonclay ceramics reduce the casting rate but may improve the mechanical properties of the cast. Casting slurries containing a binder system that develops a chemical set are used to produce porous and dense refractory structures, dental materials, and other structural materials.

Thin substrates for modern electronic packaging are produced by tape casting. Slurries for tape casting are formulated with a relatively high organic content to produce a tape having requisite mechanical properties for handling, surface characteristics enabling the printing of electrical circuit paths, and compressibility and deformation for interlaminar bonding during lamination.

SUGGESTED READING

1. *Advances in Ceramics Vol. 9: Forming of Ceramics*, John A. Mangels and Gary L. Messing (eds), American Ceramic Society, Columbus, OH, 1984.

2. Edward Adams, Slip Cast Ceramics, in *High Temperature Oxides Vol. IV*, A. Alper (ed), Academic Press, New York, 1971.

3. C. F. Cooper and S. F. A. Miskin, The Slip Casting Kinetics of Alumina by Gamma-Ray Absorption, in *Proceedings of the British Ceramic Society Vol. 3: Fabrication Science*, British Ceramic Society, Stoke-on-Trent, England, 1965.

4. R. Mistler, D. Shanefield, and R. Runk, Tape Casting of Ceramics, in *Ceramic Processing Before Firing*. G. Onoda and L. Hench (eds), Wiley-Interscience, New York, 1978.

5. J. C. Williams, Doctor Blade Process, in *Treatise on Materials Science and Technology Vol 9: Ceramic Fabrication Processes*, F. Y. Wang (ed), Academic Press, New York, 1976.

6. Edmond P. Hyatt, Making Thin, Flat Ceramics—A Review, *Am. Ceram. Soc. Bull.* **65**(4), 637–638 (1986).

7. M. D. Sacks, Properties of Silicon Suspensions and Cast Bodies, *Am. Ceram. Soc. Bull.* **63**(12), 1510–1515 (1984).

8. J. Williams, Evolution of Ceramic Substrates for Thick and Thin Film Components and Circuits in the United States, *Am. Ceram. Soc. Bull.* **56**(6), 580–584 (1977).

9. R. A. Gardner and R. W. Nufer, Properties of Multilayer Ceramic Green Sheets, *Solid State Technol.* **5**, 38–43 (1974).

10. T. J. Fennelly and James S. Reed, Mechanics of Pressure Slip Casting, *J. Am. Ceram. Soc.* **55**(5), 264–268 (1972).

11. R. P. Heilich, G. Maczura, and F. J. Rohr, Precision Cast 92–97% Alumina Ceramics Bonded with Calcium Aluminate Cement, *Am. Ceram. Soc. Bull.* **50**(6), 548–554 (1971).

12. R. R. Rowlands, A Review of The Slip Casting Process, *Am. Ceram. Soc. Bull.* **45**(1), 16–19 (1966).

13. T. M. Gainer and J. A. Carter, Pressure Casting of Sanitary Ware, *Am. Ceram. Soc. Bull.* **43**(1), 9–12 (1964).

14. D. S. Adcock and I. C. McDowall, The Mechanism of Filter Pressing and Slip Casting, *J. Am. Ceram. Soc.* **40**(10), 355–360 (1957).

15. P. E. Rempes, B. C. Weber, and M. A. Schwartz, Slip Casting of Metals, Ceramics, and Cermets, *Am. Ceram. Soc. Bull.* **37**(7), 334–338 (1958).

16. Von A. Dietzel and H. Mostetzky, Vorgange biem Wasserentzug aus einem keramischen Schlicker durch die Gipsform, *Ber. Deut. Keram. Ges.* **33**, 7–18 (1956).

PROBLEMS

22.1 A whiteware casting slip has a bulk density of 1.82 Mg/m^3. Estimate the volume fraction of solids if the mean particle density is 2.60 Mg/m^3.

22.2 What is the effect of a relatively high or low viscosity or thixotropy produced by differences in deflocculation on the casting time, cast density, draining, drying time, and fettling.

22.3 Derive the equation $\tau_y \geq \frac{2}{3}a(D_p - D_s)g$ indicating the τ_y of the slurry of density D_s required to suspend particles of size a and density D_p.

22.4 For the slurry in problem 22.1, what is the minimum yield stress that will suspend -325 mesh particles?

22.5 What would be the effects on slip casting behavior when using (1) a cold mold and a warm slip, (2) a warm mold and a cold slip, and (3) a warm mold and a warm slip?

22.6 Do humis and colloids in lignite in a ball clay aid or hinder deflocculation of the slip?

22.7 Estimate the porosity and the capillary suction of the mold materials in Fig. 22.4. The density of gypsum is 2.32 Mg/m^3.

22.8 Estimate the thermal shock resistance ΔT of gypsum if its tensile strength is 3.5 MPa.

22.9 Why does the thickness of a whiteware cast, after some standard casting time, depend on the water content in the cast?

22.10 Explain the behavior illustrated in Fig. 22.2 in terms of the J parameter and settling behavior.

22.11 Explain why the adsoprtion rate of a plaster mold is higher when the water/plaster ratio used in making the mold is higher. (Consider the driving force and the migration rate in the mold.) Why does the adsorption rate decrease with the water content of the mold?

22.12 Calculate $J = $ vol cast/vol filtrate from the results in Fig. 22.6.

22.13 What properties are important for wet trimming? How would you determine quantitative values?

22.14 The casting time for a large solid cast refractory block is 2 weeks when cast conventionally. Estimate the casting time if a pressure of 1.6 MPa is applied to the slurry and other factors are constant.

22.15 Compare the yield stress values of a freshly mixed porcelain casting slurry with that of the plastic body after partial drying.

22.16 Draw a diagram showing the qualitative variation of the apparent consistency and the as-formed density with liquid content for a castable containing a reactive bond. Use this diagram to explain why the cured strength achieves a relative maximum at a particular liquid content.

22.17 Estimate the relative difference in effective pore size for the flocculated filter cake and deflocculated casting body for the porcelain bodies in Table 18.1.

22.18 Estimate the shear rate during tape casting when the substrate velocity is 1.5 cm/s and the blade height is 0.05 mm.

22.19 Explain why the amount of deflocculant for proper deflocculation depends on the solids loading of the slurry as well as the adsorption propensity.

22.20 Alumina tape after drying, containing 12 wt% organics, has a mean bulk dentisy of 2.55 Mg/m^3 and an apparent mean density of 2.95 Mg/m^3. Estimate the open porosity.

22.21 Design a connected-block, processing flow diagram for producing pressure cast silicon nitride.

22.22 Design a processing flow diagram for producing slip cast and pressure cast porcelain.

23

Molecular Polymerization Forming

The recent availability of a wide spectrum of organometallic compounds and the interest in obtaining higher-performance ceramics has stimulated much research in molecular polymerization methods for forming novel ceramic materials. These techniques, in principle, provide a means for designing and controlling the composition and structure at the molecular level. The two processes we will consider are the sol-gel process and chemical vapor deposition.

The advantages of molecular polymerization processing are the potential improvements in purity, homogeneity, lower firing temperatures, much finer grain size, and new crystalline phases.* Disadvantages include the much higher raw-material cost, the large volume shrinkage during processing, longer processing times, and residual impurities. Because of the relatively high product cost per mass, research has been directed toward films, coatings, porous products, and fibers where the specific surface area of the product is high. Product possibilities include thinner functional films in electronic components, molecular filters, catalysts, and coatings on fibers used in composites.

23.1 SOL-GEL PROCESSING

Sol-gel processing was introduced in Chapter 4 as a means for producing novel powders. Precursors that have received the most interest are the alkoxides ($-OC_nH_{2n+1}$) commonly represented as ($-OR$); when $n = 2$, 3, and 4, the specific compounds are ethoxide, isopropoxide, and butoxide,

---

*J. D. MacKenzie, in *Ultrastructure Processing of Ceramics, Glasses and Composites*, Wiley-Interscience, New York, 1984.

respectively. Partial hydrolysis of an alkoxide as $Ti(OR)_4$ on mixing in water (solvent) occurs by the reaction

$$Ti(OR)_4 + 4H_2O \rightarrow (RO)_3 Ti-OH + ROH \tag{23.1}$$

Condensation produces a polymerized oxide aquagel of much higher viscosity, and having a yield point

$$(RO)_3 Ti-OH + HO-Ti(OR)_3 \rightarrow RO_3-Ti-O-Ti(OR)_3 + H_2O \tag{23.2}$$

The hydrolysis of alkoxides has been used to prepare gels of several metals, including zirconium and aluminum.

If a multicomponent alkoide solution or single alkoxide and a soluble inorganic salt are mixed in solution, a multicomponent alkoxide or complex alkoxide may form. The reaction of a partially hydrolyzed alkoxide ROM–OH with an alkoxide specie $M'(OR')_4$ may form the double alkoxide*

$$(RO)_3M-OH + R'O-M'(OR')_3 \rightarrow (RO)_3M-O-M'(OR')_3 + R'-OH \tag{23.3}$$

Equations (23.1) and (23.2) may be summarized by the net reaction

$$Ti(OR)_4 + 2H_2O \rightarrow TiO_2 + 4ROH \tag{23.4}$$

which indicates that water is consumed. The reaction indicated by Eq. 23.1 is rate-limited by the availability of water, and the reaction indicated by Eq. 23.2 is rate-limited by the availability of products from the first reaction. Sakka† has shown that in the sol-gel preparation of SiO_2, in the presence of a low water content and an acid catalyst which speeds hydrolysis, long polymer chains form to produce the gel; but in a base-catalyzed sol, three-dimensional polymerization forms spherical particles, and these gel by agglomeration. In an acid-catalyzed sol, there was a progressive decrease in linearity of the polymerized gel with increasing content of water. Fibers can be drawn from the former type.

Another system that has been used successfully for forming films is a nonaqueous solution of an organic acid, such as acrylic acid R–COOH, and a dissolved metal salt. Polymerization (see Eq. 11.12) on drying may produce a tenacious film that is several 100 nm in thickness. Other potential sol-gel systems are described in Chapter 4.

*S. R. Gurkovich and J. B. Blum, in *Ultrastructure Processing of Ceramics, Glasses and Composites*, Wiley-Interscience, New York, 1984.
†S. Sakka, *Am. Ceram. Soc. Bull.* **64**(11), 1463–1466 (1985).

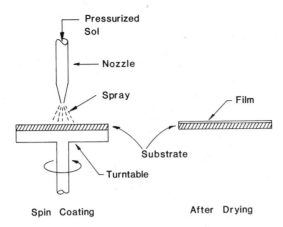

Fig. 23.1 The spin coating technique.

The three techniques used for applying gel films on product surfaces are dipping, spinning, (Fig. 23.1), and spraying. Dipping has been used to produce uniform fired coatings about 100 nm in thickness on nonporous surfaces. The thickness varied directly with the viscosity and the pulling rate. Properly formulated gels with a flexible, cross-linked polymer structure were observed to shrink only in thickness when dried and fired and did not exhibit cracks.

Sol-gel processing may also be important as a means to coat fibers in ceramic-ceramic composites; coatings on the fibers are requisite to control the bonding of the fiber to the ceramic matrix during subsequent firing. Sol-gel processing is also being investigated as a means of forming thick monolithic ceramics and matrices for composites. Because of the very fine pore size, drying of the gel into a xerogel is difficult unless special precautions or hypercritical drying is used. The xerogels commonly have extremely fine pores, in the range of 1–50 nm, a specific surface area often exceeding $100 \, \text{m}^2/\text{g}$, and a porosity in the range of 50–70%.

23.2 CHEMICAL VAPOR DEPOSITION

Vapor deposition is the condensation of an element or a compound from the vapor state, forming a solid deposit. In physical vapor deposition, the deposit is of the same composition as the vapor, such as the vacuum deposition of metal films onto substrates (thin film technique) or sputtering in a rarefied gas. In chemical vapor deposition (CVD), a compound of the source material is vaporized and thermally decomposed or reacted with other gases or vapors to produce a nonvolatile reaction product which deposits on a substrate as a coating. Chemical vapor deposition is a molecular process in that the deposit is built up molecule by molecule. It has the capability of forming high-density deposits of uniform thickness on surfaces of complex shape.

Chemical vapor deposition is used to form coatings of borides, carbides, nitrides, oxides, and silicides on the surfaces of metals and ceramics. A common CVD process for the deposition of oxides is the hydrolysis of a metal chloride by water formed at a high temperature by the reaction of hydrogen and carbon dioxide. An example for the formation of alumina at about 1000°C is*

$$3H_{2(g)} + 3CO_{2(g)} \rightleftharpoons 3H_2O_{(g)} + 3CO_{(g)} \tag{23.5}$$

$$2AlCl_{3(g)} + 3H_2O_{(g)} \rightleftharpoons Al_2O_{3(s)} + 6HCl_{(g)} \tag{23.6}$$

Silicon carbide may be deposited below 1000°C from an organosilane compound such as CH_3SiCl_3 in the presence of hydrogen; the overall reaction is†

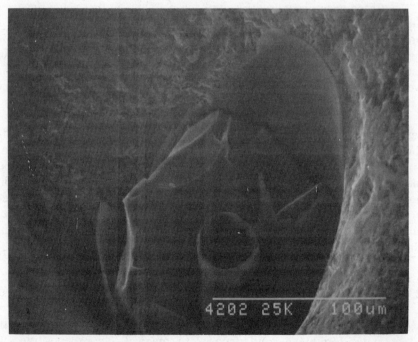

Fig. 23.2 Microstructure of a ceramic fiber composite (boron nitride coating on carbon surface of silicon carbide monofilament, silicon matrix). (Courtesy of H. T. Godard and K. T. Faber, Ohio State University.)

*R. Colmet et al., p. 18 in *Chemical Vapor Deposition*, 1981, Electrochemical Society, Pennington, NJ, 1981.
†F. Christin et al., p. 502 in *Chemical Vapor Deposition*, 1979, Electrochemical Society, Pennington, NJ, 1981.

$$CH_3SiCl_{3(g)} \xrightarrow{H_2} SiC_{(s)} + 3HCl_{(g)} \qquad (23.7)$$

but decomposition of the organosilane in the hydrogen atmosphere precedes the formation of the silicon carbide. The purity of the deposited phase and decomposition rate are functions of the initial compositions and the pressure and temperature of the reaction. Deposition rates are of the order of a few microns per minute, and CVD is used as a means of forming surface coatings with electrical, corrosion, or wear-resistant functions and for coating fibers for composites (Fig. 23.2).

SUMMARY

In sol-gel processing, molecular polymerization and drying produce a very porous but elastic xerogel that is densified subsequently on firing. Chemical vapor deposition forms dense ceramic material by condensing the product of a gas phase reaction. The process of chemical mixing and molecular addition provides special opportunities for obtaining a very high purity and for controlling the structure at the molecular level. These processes can be used to form films and coatings and shaped products.

SUGGESTED READING

1. *Ultrastructure Processing of Ceramics, Glasses, and Composites*, Larry L. Hench and Donald R. Ulrich (eds), Wiley-Interscience, New York, 1984.
2. *Better Ceramics Through Chemistry*, C. Jeffrey Brinker et al. (eds), North-Holland, New York, 1984.
3. S. Sakka, Sol-Gel Synthesis of Glasses: Present and Future, *Am. Ceram. Soc. Bull.* **64**(11), 1463–1466 (1985).
4. W. LaCourse, et al., Factors Controlling the Sol-Gel Conversion in TEOS, *J. Can. Ceram. Soc.* **52**, 18–23 (1983).

PROBLEMS

23.1 Describe a procedure for preparing a mixed $TiO_2 + SiO_2$ gel. Illustrate the structure.

23.2 Illustrate the sol and gel structure produced when (1) linear polymers form in sol and form a network in the gel and (2) spheres form in the sol and agglomerate in the gel.

23.3 Explain why the sol-gel dip coating is thicker when the withdrawal velocity is faster.

23.4 Describe a process for extruding a sol to produce ceramic fibers.

23.5 Write a chemical reaction for the chemical vapor deposition of ZrO_2.

23.6 What is a chemical reaction for the chemical vapor deposition of Si_3N_4?

PART VIII

DRYING, SURFACE PROCESSING, AND FIRING

Products formed by casting, paste processing, and plastic forming must be dried in a controlled manner to remove the interstitial liquid phase prior to firing in a furnace. The processes of green machining and surface grinding are used for trimming and to contour the surface of a partially dried product with adequate strength or a partially sintered product. Damp sponging and dry fettling are used to smooth surfaces. Green ceramic tape for electronic substrates is shaped by cutting and punching. Thick films, which ultimately provide important electrical functions or decorations, are applied to the surface using printing techniques or decals. Spraying, pouring, dipping, and banding are also used to apply controlled decorations and glazes on the surface of an unfired or fired product. These important postforming processes are discussed in Chapters 24 and 25.

The transformation from a relatively fragile green product containing particles, decomposable phases, and pores into a finished product that we recognize as a ceramic material is accomplished by firing in a kiln or furnace. In Chapter 26 the many chemical and physical changes in the microstructure produced during firing are examined in terms of the mechanisms involved in different systems and the dependence of the final microstructure on the characteristics of the unfired materials and the heating schedule and atmosphere.

24

Drying

Drying is the removal of liquid from a porous material by means of its transport and evaporation into a surrounding unsaturated gas or, in some cases, a desiccating liquid. It is an important operation prior to firing in processing bulk raw materials, products shaped by plastic forming and casting, and decorations and coatings on surfaces.

The evaporation of processing liquids is relatively energy-intensive, and drying efficiency is always an important consideration. Drying must be carefully controlled, because stresses produced by differential shrinkage or gas pressure may cause defects in the product. In this chapter we will consider drying practices, mechanisms of drying, and causes of defects.

24.1 DRYING SYSTEMS

Drying costs are a significant factor in the selling price of industrial minerals, and the natural drying action of the sun and wind is considered in mining and storing these materials. Wet-processed raw materials are commonly dried in large rotary dryers in which the material is tumbled, on belts in continuous tunnel dryers, by spray-drying, or supported on trays in a chamber dryer. The flow of a warm dry gas through a permeable material, called fluidized bed drying, is used when a granular material is very temperature-sensitive. Material that has been partially dewatered mechanically and is in the form of a ribbon, pellets, granules, etc., is usualy dried more uniformly.

Freshly cast plaster molds and very large wet-processed shapes are commonly dried slowly by "open-air drying." Shaped ceramic products requiring drying and working molds are usually dried in a controlled manner in fabricated metal dryers. Chamber dryers are used for drying large, free-standing shapes and smaller products supported on shelves or suspended from the structural framework on dryer cars. Continuous dryers may convey the ware up and down through a baffled chamber by means of shelves supported on continuous chains in a mangle dryer, or on rack dryer cars in a tunnel dryer (Fig. 24.1). Air circulation is maintained and controlled by means of fans. Heat sources include direct fired air heaters, steam coils, waste warm air from kilns and furnaces, and infrared or microwave radiation.

Damp material may be heated by the mechanism of convection, conduction, and radiation (Fig. 3.10). When using convective heating, which is most common, the product temperature during drying is lower than the temperature of the hot circulating gas. Conductive heating in which heat passes to the product from a heated surface supporting the product is used in drying some thin substrates and slurries in belt and rotary drum dryers. Infrared radiation that is of a long wavelength does not penetrate deeply into wet ceramics but may be absorbed by the liquid and transported into the interior by conduction. It is more often used for drying thin substrates, films, and coatings. Heating is produced by the coupling between the infrared radiation of 4–8 μm and the molecular vibration of O–H groups. Radiation of very long wavelength in the microwave range penetrates deeply into most ceramics; energy dissipation from the polarization of the water molecules absorbs the radiation and heats the liquid. Dielectric and microwave drying are used for the drying of liquid saturated products where

Fig. 24.1 Products supported on gypsum molds entering a continuous dryer. (Photo courtesy of A. J. Wahl, Brocton, NY.)

the drying must be relatively rapid, the maximum temperature of solids must be relatively low, or the product integrity is sensitive to liquid concentration gradients and capillary stresses. Large cast and extruded products containing water have also been partially dried by inserting electrodes and passing a high *dc* current at low voltage through the piece.

The design and capacities of industrial dryers vary considerably and are a function of the size, shape, and drying behavior of the product; the dryer loading; setting configuration; the production rate; the temperature, humidity, and velocity of the drying air; and the mode of heating. The temperature and humidity of inlet air and the circulation of air within the dryer must be monitored to control the performance of an industrial dryer.

24.2 MECHANISMS IN DRYING

Drying involves the transport of energy into the product; liquid is transported through pores to the meniscus, where evaporation occurs, and by vapor transport through pores. In a drying system, heat energy must be brought to the surface of the product, and vapors must be carried away. The mechanisms of evaporation and mass and thermal transport must be considered before discussing the drying process.

Liquid placed in a closed container will evaporate to establish a pressure of vapor above the liquid. This pressure will increase with the temperature of the liquid and is dependent on the radius of curvature of the meniscus (Eq. 2.7). The constant vapor pressure at a particular temperature is referred to as the saturation vapor pressure. Because evaporation involves the loss of molecules having the highest kinetic energy, evaporation is a cooling process, and heat must be supplied to maintain an isothermal condition. The difference between the temperature of thermometers with dry and wet bulbs in flowing air indicates the cooling due to evaporation. Heats of vaporization for several processing liquids are listed in Table 24.1. The boiling point of a liquid is the temperature at which its vapor pressure becomes equal to the external pressure acting on the surface of the liquid,

Table 24.1 Latent Heat of Vaporization of Several Processing Liquids

Temperature (°C)	Heat of Vaporization (kJ/kg)		
	H_2O	CH_3OH	C_2H_5OH
20	2.45	1.17	0.91
40	2.40	1.14	0.90
60	2.36	1.11	0.88
80	2.31	1.06	0.85
100	2.26	1.01	0.81

which for water is 100°C at 760 mm Hg, or 1 atm. The critical temperature and pressure of a liquid are the conditions for which the physical properties of the liquid and vapor become identical (no meniscus); for water these are 374°C and 220 atm.

The evaporation rate of water into air below its boiling point varies directly with the temperature and surface area of the liquid and inversely with the concentration of that liquid in the air. Above the boiling point, the evaporation rate is dependent on the rate at which heat is supplied and independent of the concentration of liquid vapor in the air. The humidity is defined as the mass of liquid/mass of air, and the relative humidity as the humidity/maximum possible humidity at that temperature. The dew point is the temperature at which air of a certain humidity becomes saturated and precipitates as liquid droplets; it is lower when the content of liquid in the air is lower. During evaporation into moving air, a static boundary layer of air and vapor exist between the mobile air and the stationary product. Diffusion through this boundary layer controls the liquid transport between the air and the product. The apparent evaporation rate at a particular temperature varies indirectly with the salt content of the liquid and the forces bonding liquid on the surfaces of solids.

Thermal transport to the product may occur by convection, conduction, and radiation. Convective heating of the product is limited by the heat transfer coefficient (h_b) for the static boundary layer between the moving air and the static liquid on the surface. Mobile air of higher velocity produced by forced convection reduces the thickness of the boundary layer and increases thermal transport and the apparent evaporation rate. If this heating is the only source of energy for evaporation, the rate of evaporation (R_E) is

$$R_E = \frac{h_b(T_A - T_L)}{L_E} \tag{24.1}$$

where $T_A - T_L$ is the difference in temperature across the boundary layer and L_E is the latent heat of evaporation. In contrast, radiation transport to the product is relatively independent of the static boundary layer and the airflow conditions.

Within the product, thermal transport occurs by conduction and radiation. Conduction through the solid particles occurs by phonon propagation, and in the liquid between particles and within pores by molecular vibrations. The thermal conductivity of the ceramic particles is significantly higher than the conductivity of water. Infrared radiation does not commonly penetrate deeply into the product, because the waves are scattered by the particle system; microwaves of much longer wavelengths may penetrate deeply into an electrically insulating ceramic material but are absorbed by water.

Liquid in a porous solid may be chemically or physically adsorbed on solid surfaces, exist as bulk liquid in pores of microscopic size, and be distributed as an external surface film. The migration of liquid to the surface may occur by means of capillary flow, chemical diffusion, and thermal diffusion. Capillary migration is motivated by capillary forces and, as discussed in Chapter 2, is very dependent on the radius of the pores and the surface tension and viscosity of the liquid. Liquid and vapor may diffuse along a gradient in concentration. Thermal diffusion is the migration of liquid or vapor along a thermal gradient and is important when using conduction heating and in dielectric and microwave drying.

24.3 THE DRYING PROCESS

Drying is often regarded as occurring in three stages corresponding to the ranges of liquid content for which the drying rate is increasing, constant, and decreasing, as is shown in Fig. 24.2.

In a saturated material, liquid is initially removed by evaporation from the external surface (Fig. 24.3A). The drying rate, expressed as a weight loss per unit time, increases on heating when the relative humidity is less than 100%. The drying rate is strictly constant when the evaporation rate and evaporation surface area are constant. In this stage, the product temperature is normally equal to the wet-bulb temperature of the environment. The mass of water evaporating per unit (area·time) (R_E) is

$$R_E = K_E (P_w - P_o) \tag{24.2}$$

where K_E is the evaporation constant that is dependent on air flow conditions, P_w is the vapor pressure of the liquid at the evaporation temperature,

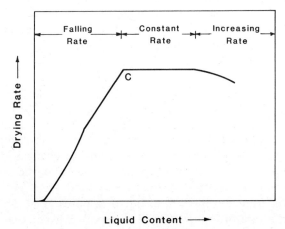

Fig. 24.2 Changes in apparent drying rate when drying a product under moderate conditions. The critical point C corresponds to the termination of shrinkage in all regions only when the liquid is distributed uniformly throughout the volume.

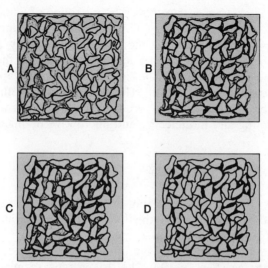

Fig. 24.3 Apparent volume and distribution of liquid among particles during slow drying: (A) as cast with interparticle contacts, (B) just preceding the end of the constant rate period, (C) on entering the falling rate, and (D) near the end of the falling rate period (liquid in capillaries with a sharp meniscus and liquid physically adsorbed on surfaces remains).

and P_o is the partial pressure of liquid in the surrounding atmosphere. Evaporated liquid may be replenished by interparticle liquid transported to the external surface by fast diffusion and capillary flow. The loss of interparticle liquid coupled with external capillary stress may cause dimensional shrinkage (Fig. 24.3B) and an increase in the plastic shear strength. Adsorbed binders, binders of higher molecular weight, and gelled binder will resist migration with the liquid, but dissolved salts and dispersed colloidal particles or molecules may migrate to the surface. Shrinkage and chemical migration may cause a perceptible lowering of the drying rate. For the constant rate (CR) period, the practical drying rate (R_{CR}) is given by the equation

$$R_{CR} = \frac{W_1 - W_2}{A\,t} \tag{24.3}$$

where W_1 and W_2 are the liquid contents before and after drying for a time t, and A is the surface area for evaporation. Dynamic changes in the temperature of the product or absorptive capacity of the atmosphere can reduce or preclude the appearance of a constant rate period.

Evaporation from the menisci of liquid in pores (Fig. 24.3C) begins to occur when the rate of internal liquid transport is lower than the evaporation rate and when the pores in the body become unsaturated; the concomitant decrease in the drying rate is called the falling rate period (Fig. 24.2). The temperature of the surface of the product may rise rapidly to the dry-bulb temperature in those areas where the evaporation surface recedes into the pores. Thermal diffusion into the body is retarded by heat consumed in supplying the latent heat of vaporization. Transport by vapor diffusion through pores increases when the liquid menisci recede into the body. The termination of capillary migration may cause a noticeable change in slope in the falling rate period. Large pores and large accessible interstices are emptied first (Fig. 24.3D) in an isothermal body; an increasing amount of energy is needed to remove water from smaller interstices and interparticle contacts where the radius of curvature of the liquid meniscus is relatively small. Dissolved salts in residual liquid may also reduce the vapor pressure slightly.

A larger capillary stress is produced as the menisci recede. The shape of the drying curve during the falling rate period is strongly influenced by the driving force for liquid evaporation and the pore structure of the product which moderates liquid and vapor transport. Spatially nonuniform drying may occur in bodies with a nonuniform microstructure.

Adsorbed liquid remains after the bulk liquid has been removed (Fig. 24.3D). Heating above the boiling point of the liquid or long exposure to a desiccating environment can remove physically adsorbed liquid and may sometimes remove chemically combined liquids such as the water of crystallization in some inorganic salts and gypsum molds.

During drying, heating increases both the vapor pressure of the liquid and the adsorptive capacity of the drying air. The forced convection of hot air of lower relative humidity maintains the product temperature and flushes away air of higher humidity. Impinging air thins the boundary layer and increases the evaporation rate, particularly during the constant rate period. Setting patterns and the circulation of air into cavities in products must be considered to improve the uniformity of drying as well as the efficiency. Heating becomes more important than air flow during the falling rate period.

24.4 DRYING SHRINKAGE AND DEFECTS

Shrinkage occurs during drying as the liquid between the particles is removed and the interparticle separation decreases. When the shrinkage is isotropic, the volume shrinkage $(\Delta V/V_o)$ is related to the linear shrinkage $(\Delta L/L_o)$ by the equation

$$\frac{\Delta V}{V_o} = 1 - \left(1 - \frac{\Delta L}{L_o}\right)^3 \qquad (24.4)$$

The linear shrinkage is proportional to the mean reduction in interparticle spacing $\Delta \bar{l}$ and the mean number of interparticle liquid films per unit length N_l when particle sliding and rearrangement do not contribute significantly to the shrinkage:

$$\frac{\Delta L}{L_o} = \bar{N}_1 \Delta \bar{l} \qquad (24.5)$$

A variation of N_1 or Δl with direction due to particle orientation or liquid gradients may cause the linear shrinkage to be anisotropic; variations with position produce differential shrinkage.

Shrinkage during drying can be reduced by forming the product at a lower liquid content to reduce $\Delta \bar{l}$ and increasing the mean particle size to decrease N_1. Bonds that develop a chemical set may reduce or eliminate Δl and the shrinkage on drying.

For common extruded and cast products, the linear drying shrinkage is usually in the range of 1.5–4%. The dependence of the volume of the piece on the liquid content for homogeneous drying where the liquid content is spatially uniform is shown in Fig. 24.4; shrinkage ceases at a particular liquid content, which is called the leatherhard liquid content.

During drying, a liquid concentration gradient at the beginning of the falling rate period may cause differential shrinkage and a tensile stress in the surface. The transition from a plastic body to an elastic body occurs with a loss of liquid content of about 5–15 vol % (Fig. 24.5). Unfired ceramics are normally quite weak and can sustain stresses of only a few 100 kPa without deformation or local fracture. Surface cracks may form when material near the surface becomes brittle and the differential shrinkage produces a stress that exceeds the tensile strength. Regions of low strength such as laminations are especially susceptible. Stresses produced during drying may be increased by the pressure of vapor in pores. Small cracks called checks are

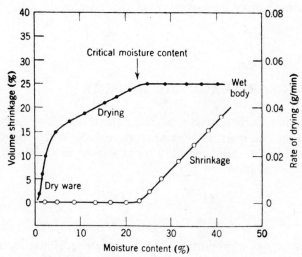

Fig. 24.4 Change in bulk volume on drying a ceramic body. (From W. D. Kingery, *Introduction to Ceramics*, Wiley-Interscience, 1960).

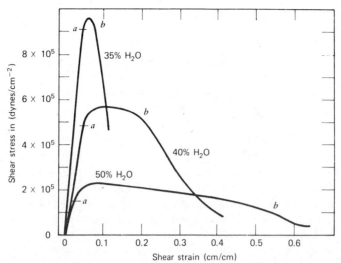

Fig. 24.5 Stress-strain diagram for an uncompressed plastic clay with different liquid contents (liquid concentration in wt.%). (After F. H. Norton, *Elements of Ceramics*, Addison-Wesley Press, Cambridge, MA, 1952.)

produced by differential shrinkage in a small region. Parameters that increase the dry strength, discussed in Chapter 14, may increase the stress required to form cracks. Case hardening occurs when shrinkage of surface material has ended, but internal material with a higher liquid content continues to shrink; tensile stresses may produce internal cracks. Differential shrinkage may be produced by gradients in particle size and at junctures between differentially oriented particles of high aspect ratio, as discussed in Chapter 22.

Differential shrinkage due to a differential liquid content when the product was formed or a differential drying rate across the surface of the product may also produce warping (Fig. 24.6). Warping is caused by stresses

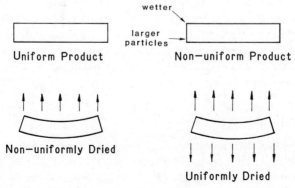

Fig. 24.6 Distortion produced by nonuniform drying shrinkage.

accompanying nonsymmetrical shrinkage, which produces plastic elongation in regions with a lower shrinkage rate. Warping is reduced by increasing the uniformity of drying and reducing the average drying shrinkage of the body. The tendency for warping may be increased by nonuniform external films or coatings, particle orientation, or binder migration, which produce a nonuniform surface permeability, and by nonuniformities in the circulation or temperature of drying air due to setting patterns and supporting hardware. Ideally, drying should occur symmetrically in an isotropic material, and all shrinkage should end before entering the falling rate period. An insufficient rate of transport to the surface relative to the evaporation rate will reduce the constant rate period and increase the transition range during which the deleterious differential shrinkage can occur.

The mechanical restraint of shrinkage may also produce stress and cracks. Contact friction between a rigid support and the shrinking product may cause a crack, especially when the product is heavy and the surface is rough. Restrained shrinkage is also produced between sections drying at different rates, as in a nonuniform cross section, and when material adheres to a porous mold. When drying shapes with a deep cavity, the forced convection of drying air into the cavity may be required to prevent the formation of a crack due to differential drying rates or the restraint of material adhering to the mold.

Other sources of defects are the migration of colloids to the surface producing a skin having different properties and internal gas pressure produced by too rapid volatilization of liquids and insufficient permeability of the pores.

24.5 MODES OF DRYING

Conduction and convection drying are commonly used for drying ceramic products, as discussed in Section 24.1. Floors or shelves supporting the product may be heated by waste heat or steam. Convection drying is used to heat the product and remove vapors, and the air circulation may be designed to facilitate the drying of a particular size, setting, or shape. Infrared drying can be used to reduce the drying time and for the drying of coatings such as decorations and glazes. Vacuum-assisted drying reduces the partial pressure of vapors. Other modes of drying deserving special comment are discussed below.

Controlled Humidity Drying

For every product with a finite drying shrinkage, there is some critical drying rate that will cause defects in the product. When drying thick products or a product of very low permeability (K_P), drying must be conducted in such a way that the liquid concentration gradients are not large. The isothermal capillary flow rate to the drying surface of a slab is proportional to K_P/η_L. The safe drying rate may be increased by reformulating the body to increase K_P or by using controlled humidity drying. Heating the product in humid air arrests the surface evaporation and reduces the viscosity of the liquid (η_L) prior to drying. On reducing the humidity of the drying air, drying can occur at a greater rate without an increase in the liquid concentration gradient.

Microwave Drying

Microwaves are generally reflected by electrical conductors, transmitted by electrical insulators, and adsorbed by dielectrics. Liquid water behaves like a dielectric, because the molecule is polar and the direction of polarization cycles when subjected to a microwave field. Microwave adsorption causes heating in proportion to the field strength and the product of the frequency and dielectric loss factor. At a particular field strength, penetration of the dielectric varies inversely with power adsorption. Microwave energy may be used to heat and evaporate liquid in large cross sections relatively rapidly, independently of the thermal conductivity of the solid. When drying ceramic insulators, the microwaves are preferentially adsorbed by the water, and the product temperature during drying may never exceed 50°C; i.e., the high surface temperatures in conventional drying are avoided. The apparent penetration of microwaves increases as water is vaporized and diffuses as a gas to the surface. Potential uses for microwave drying are in the processing of temperature-sensitive products, more rapid drying, drying products of large cross sections and large gypsum molds, and drying products containing colloidal materials such as gels, pigments, and clay of extremely low liquid permeability.

Slurry Drying

Slurried raw materials supported on a metal belt and tape-cast films are dried continuously and relatively rapidly. In drying a mineral slurry on a metal belt, heat is supplied by conduction through the belt and by the hot drying air, which may exceed 400°C. Turbulence produced by boiling may retard the formation of a low-permeability surface layer, as is shown in Fig. 24.7; the drying rate is dependent on the solid's concentration of the slurry and the viscosity and permeability of the partially dried material.

In tape casting, the thin slurry film cannot be dried too rapidly, because the drying is one-directional and the liquid permeability of the tape is very low. The concentration of liquid in the air at the beginning of drying is relatively high, as in controlled humidity drying. The maximum liquid temperature must be well below the boiling point of the liquid during the constant rate period and is commonly less than 50°C for nonaqueous systems. Filtered air enters the exit end and is solvent-loaded on exiting above the entering tape. The air flow and temperature are carefully control-

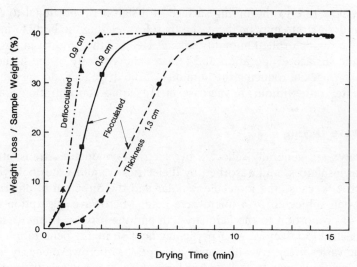

Fig. 24.7 Drying behavior on rapidly heating to 500°C for a slurry that develops a low-permeability skin.

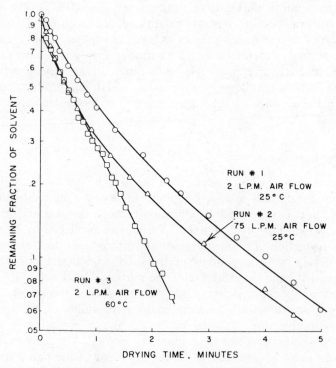

Fig. 24.8 Drying behavior for a tape-cast slurry. (After R. E. Mistler et al., p. 432 in *Ceramic Processing before Firing*, G. Onoda and L. Hench (eds), Wiley-Interscience, New York, 1978.)

led to maximize the constant rate period and minimize liquid concentration gradients until shrinkage has ceased (Fig. 24.8). Non aqueous solvent vapors are collected and reclaimed.

Supercritical Drying and Freeze-Drying

Supercritical drying may be used to minimize effects of the surface tension of the liquid during drying. In supercritical drying, the product is heated in an autoclave until the liquid becomes a supercritical fluid. Supercritical drying occurs during isothermal depressurization. Freeze-drying (discussed in Chapter 4) is used for drying products where a liquid formation and product heating must be avoided.

Spray-Drying

Spray-drying is relatively efficient compared to the convection drying of formed products, because the material is well dispersed in the drying medium, the diffusion path is shorter, and the high specific surface area contributes to a higher rate of evaporation per unit mass of product.

SUMMARY

Drying is an important process in producing ceramic raw materials and shaped products ready for firing. During drying, heat is transported to the liquid in the body, and evaporated liquid is transported into the surrounding atmosphere. The drying rate depends on the temperature of the liquid in the body and the temperature, humidity, and flow rate of the drying air. Radiation may be used to augment conduction and convective heating or as the primary heating source. After initial heating, the product dries at a constant rate during which shrinkage commonly occurs. Transition to a decreasing drying rate occurs when the external surface of the product is incompletely covered with liquid. When the drying rate is very fast, or nonuniform, the constant rate period is relatively short, and the differential shrinkage can cause cracks. Warping is produced by nonuniform drying when the body is shrinking and can deform plastically. Dried products are commonly hygroscopic and may readsorb moisture in proportion to the relative humidity of the atmosphere.

SUGGESTED READING

1. R. B. Keey, *Introduction to Industrial Drying Operations*, Pergamon Press, Elmsford, NY, 1978.
2. F. H. Norton, *Fine Ceramics*, Robert E. Krieger, Malabar, FL, 1978.

3. A. R. Cooper, Quantitative Theory of Cracking and Warping During the Drying of Clay Bodies, in *Ceramic Processing Before Firing*, George Y. Onoda and Larry L. Hench (eds), Wiley-Interscience, New York, 1978.

4. W. E. Brownell, *Structural Clay Products*, Springer-Verlag, New York, 1976.

5. Rex W. Grimshaw, *The Chemistry and Physics of Clays*, Wiley-Interscience, New York, 1971.

6. R. W. Ford, *Institute of Ceramics Textbook Series Vol. 3: Drying*. Maclaren and Sons, London, 1964.

7. Pavel Hrma, Effect of Surface Saturation on Transfer of Water in a Saturated Body, *J. Am. Ceram. Soc.* **66**(5), (1983).

8. J. E. Funk, Simultaneous Weight Loss and Shrinkage of Clays, *Am. Ceram. Soc. Bull.* **53**, 450–452 (1974).

9. W. O. Williamson, Dimensional Changes and Microstructures of Unfired Clay Products, *Interceram* **31**, 199–204 (1968).

PROBLEMS

24.1 Construct a graph indicating specimen weight as a function of liquid concentration for drying corresponding to Fig. 24.2.

24.2 The diametral shrinkage of a product is 3.6%. Calculate the volume shrinkage (1) assuming the shrinkage is isotropic and (2) assuming the shrinkage in thickness is twice the diametral shrinkage.

24.3 Derive a general relationship for the dependence of the volume shrinkage on linear shrinkage assuming the shrinkage is anisotropic.

24.4 Contrast the specific energy required to heat water from 20 to 100°C and the energy to evaporate it at 100°C. Compare the energy for the evaporation of ethyl alcohol at 50°C and water at 100°C on an equal weight basis.

24.5 If the energy for evaporation of water in problem 24.4 were used to heat an equal mass of alumina, what temperature would the alumina achieve on heating from 20°C?

24.6 At what point in drying does the sheen disappear from the surface?

24.7 A product is dried such that the drying air impinges on the top surface causing very rapid drying, but very little air flow occurs along the sides. Contrast the drying rate curves for the different surfaces.

24.8 How would the aqueous drying behavior of the alumina body containing particles coated with a methyl cellulose binder differ from the drying of the alumina particles coated with wax?

24.9 When initially heating an unsaturated product in the dryer, liquid may be driven into the center of the product. What are the drying conditions that will minimize this?

24.10 Calculate the thickness of the liquid films between particles 3 μm in diameter when the linear shrinkage is 2.5%. If particles with a square platelike shape 0.5×2.0 μm have this film thickness, estimate the drying shrinkage for an oriented microstructure.

24.11 An extruded product has an interstitial porosity of 36% which is 95% saturated when it enters the dryer. What is the relative volume of liquid removed between entering the dryer and the termination of shrinkage, when the linear shrinkage is 1.4% and isotropic?

24.12 Sketch Fig. 24.3D for the case of rapid microwave drying, and comment on the difference relative to slow drying. (Assume liquid still remains in the body in each case.)

24.13 A cylindrical capillary (pore) nearly filled with a wetting liquid extends from the surface to the interior. In which direction would capillary flow occur when (1) the product is initially heated and the interior is cooler and (2) the product has been uniformly heated at 100% relative humidity and the humidity is quickly reduced?

24.14 Does the stress causing drying cracks depend on pore size and wetting angle of the liquid?

24.15 Compare the tensile stress produced when the differential shrinkage on drying is 4% and 1.5%. Assume Young's modulus of the dry surface material is 1.4 GPa. Would the stress cause cracks for the body in Fig. 14.19?

24.16 Does the drying rate during the falling rate period depend on the thickness of the product (for the same liquid content)? Explain.

24.17 A flatware shape cast in a gypsum mold is dried using forced convection while still in contact with the mold. If the center dries in advance of the rim, how is the shape distorted on complete drying?

24.18 Calculate the radii of liquid emptied by evaporation (in the falling rate period) for a relative humidities of 0.3, 0.6, and 0.9 and temperatures of 45 and 90°C.

24.19 The differential concentration $C(w, t) - \bar{C}(t)$, where w is one/half the thickness of a slab, t is time, and $\bar{C}$ is the mean liquid concentration, for the constant rate period has been given (Ref. 3) as $C(w, t) - \bar{C}(t) = 1/3\, J_w/D$ where J is the evaporated flux and D is the apparent liquid diffusivity. Calculate the maximum flux J as a function of w when $D_w \approx 6 \times 10^{-4}$ cm^2/s and $J \leq D/w$ for safe drying.

25

Shaping, Surface Finishing, Film Printing, and Glazing Processes

Many ceramic products are not completely formed until the surface has been modified and functional or decorative coatings have been applied. We may smooth the surface of a product using the processes of sponging or fettling. Surface grinding and turning are used to produce undercut recesses, contoured surfaces, and threaded surfaces. Green ceramic tape for electronic substrates is sized and shaped by blanking and punching.

Screen printing processes are used widely for printing thick film conductors, resistors, and dielectrics on electronic substrates and decorative films on tile, institutional ware, and a variety of household ceramic products. Pad transfer printing and transfer decals are used for printing on flat or contoured surfaces. Other processes for decorating surfaces and applying glaze coatings are banding, brushing, pouring, dipping, and spraying.

25.1 TRIMMING, SMOOTHING, AND GRINDING

The machinability of an unfired or lightly sintered ceramic material depends on its mechanical strength and toughness. When the liquid content of the product exceeds that corresponding to the leatherhard state, as in a slip cast object just after draining, the friction of the trimming tool causes distortion and tearing, because the product does not have sufficient strength. In the leatherhard state, a material with sufficient binder will exhibit limited plastic behavior but significant toughness and can be trimmed using a knife or a contoured cutting tool much like the machining of a metal. After a reduction in the liquid content, a reduction of the temperature (relative to the glass transition temperature of the binder), or heating to the initial stage of sintering, the stiffness of the material increases, but the material loses toughness (Fig. 24.5). Materials with a low toughness may be machined by surface grinding or turning when the stress produced at the surface causes particle attrition but not bulk fracture.

426

Trimming is commonly performed when the material is leatherhard and is a common operation in producing slip-cast whiteware products. Damp sponging is used to remove high spots, and the paste fills in low spots and smooths the surface; this may be done automatically as the product passes under rotating sponges. Burnishing is smoothing when a polished tool is pressed against the surface and platy particles are forced into alignment.

Contour grinding is widely used in producing small spark plug insulators; large, high-tension electrical porcelain insulators; and other parts with a complex shape such as bioceramic implants, wear-resistant inserts, and threaded structural components (Fig. 1.1). In rotary machining operations, the blank is fixed on a spindle and surface-ground using a contoured grinding wheel or hard cutting tool. Spark plug insulators are commonly contoured using a porous grinding wheel. The contact force during grinding should remove particles from the surface layer without causing fracture of the piece. The failure rate due to fracture is high when the green strength of the part is less than about 2 MPa (i.e., when the part is fragile) or when the green strength is high and the tool force for grinding is high.

Parameters controlling the intrinsic green strength were discussed in Chapter 14. Microstructural defects such as microcracks, laminations, extremely large pores, inhomogeneous binder, etc. can reduce the strength and impair grinding behavior. For each product and grinding system, there is an intermediate range of green strength for which the scrap rate is a minimum. Coarser particles in the material increase the grinding rate but also accelerate wheel wear and reduce surface smoothness; submicron particles enter pores in the wheel and reduce the grinding rate.[*] Both the fraction coarser than about 10 μm and the submicron fraction must be controlled for high-productivity machining using grinding wheels. Grinding also depends on the density of the part and the amount and hardness of the binder. When using wax binders, both the melting point and shear strength of the binder are important.[†]

The grinding rate also increases when the grit size, angular velocity, and mechanical load on the wheel increases. These parameters and the porosity of the wheel must be controlled to achieve the optimum balance of surface smoothness, grinding rate, and scrap rate.

Parts containing a submicron average particle size are contoured using multiple-stage grinding or by turning on a lathe. Large, cylindrical, electric porcelain insulators are contoured on an automatic lathe, using a carbide or a ceramic cutting tool for shaping. In these bodies containing approximately 50% clay, the green strength is dependent on the content of clay collids and the moisture content (Fig. 14.19). Control of the microstructural

---

[*] D. B. Quinn et al., pp. 4–15 in *Advances in Ceramics Vol. 9: Forming of Ceramics*, American Ceramic Society, Columbus, OH, 1984.
[†] *Ibid.*

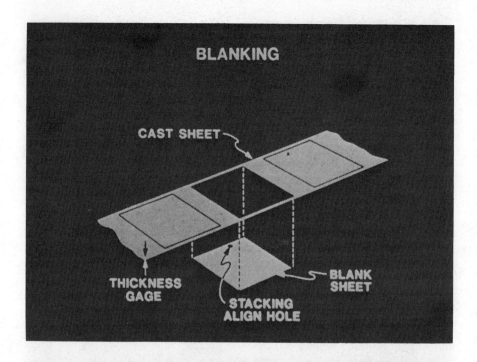

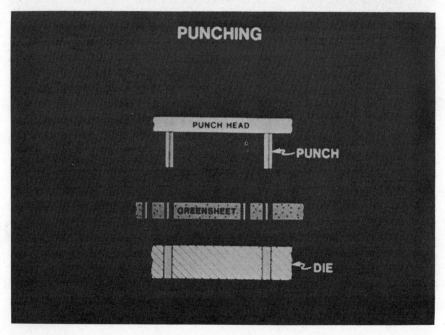

Fig. 25.1 The blanking and punching of ceramic tape. (From R. A. Rinne and D. R. Barbour, *Electrocomponent Sci. Technol.* **10**, 40 (1982).)

homogeneity and green strength is absolutely essential for high-productivity contouring and a good surface finish.

Flashing and irregularities on the edge of a part may be removed by dry fettling using an abrasive pad or belt or a carbide tool. The surface of fetted clay products is commonly polished in an automated, damp-sponging operation.

25.2 BLANKING, PUNCHING, AND LAMINATING

In producing electronic packaging from tape-cast substrates, blanks are cut to the desired size and shape in a blanking press or using longitudinal and transverse wheel knives. Holes larger than about 0.5 mm are often punched during blanking (Fig. 25.1). Smaller holes are commonly punched in a subsequent operation in a single stroke for each substrate layer. Holes smaller than 0.3 mm are often produced by laser drilling after firing; a series of these holes in a line is called laser scribing. Deviations in the location of holes due to punch alignment or wear and nonuniform shrinkage must be carefully controlled, especially for via holes between layers which are filled automatically by injection of a metal paste.

Finished sheets of tape heated to 50–80°C are stacked and laminated at a pressure in the range of 3–30 MPa. The interlaminar bond strength is important for maintaining integrity in later handling and during binder burnout when gas pressure is produced internally. The bond strength (σ_b) is approximated by the equation*

$$\sigma_b = K_1 + K_2 \ln(PTt) \tag{25.1}$$

where P and t are the pressure and time of lamination at a temperature (T) and K_1 and K_2 are constants for a particular organic system. Laminated tape is commonly 2–8% higher in density. The stack is then sawed or punched into a finished module for firing.

25.3 PRINTING PROCESSES

In screen printing, an open pattern in a stencil screen defines the printed pattern in the printing process; a thick paste is forced through a stencil screen onto the surface below, using a squeegee or a traversing paste reservoir and nozzle assembly. Process variables include mechanical indexing, screen variables, squeegee variables, composition and rheology of the paste, and surface roughness.

* R. A. Gardner and R. W. Nufer, *Solid State Technol.* **5**, 38–43 (1974).

Equipment and paste compositions for screen printing are produced commercially. Characteristics of common screen printing are print thickness in the range of 2–25 μm, print patterns ranging up to about 15×15 cm, line widths as small as 0.25 mm, squeegee speeds up to about 25 cm/s, and a cycle time of about 2 s. Special techniques are required for filling vias and achieving line widths finer than 0.25 mm. Paste compositions include the primary metal or ceramic powder as indicated in Table 25.1, a powder sintering aid which typically vitrifies such as a lead borosilicate frit, an organic liquid of low volatility, a high-molecular-weight binder, and other additives such as a lubricant. Specific compositions are highly proprietary. Powders are typically finer than 10 μm; mixtures must be well milled and mixed to produce a homogeneous dispersion. Soluble impurities must be carefully monitored and controlled.

To print a specific image, printing paste must be allowed to pass only through specific areas of the screen. A popular means for producing the stencil pattern is to utilize a photosensitive emulsion. The liquid emulsion is applied in sequential layers to the stretched screen until a smooth, uniform coat has been achieved. After drying, the emulsion is exposed to the photographic positive of the printing image. Areas of the emulsion exposed to light become hardened owing to cross-linking of the polymer, but unexposed areas remain soft and are washed out of the screen. The exposure intensity and time must be carefully controlled to produce an accurate stencil.

Screen frames must be dimensionally stable. The fabric of the screen must be of a regular weave, elastic, and abrasion-resistant to register a shape image with dimensional reproducibility. Popular fabrics are monofilament nylon and polyester, stainless steel, and metalized polyester filament. For precise registration, polyester is used because it is less elastic and absorbs very little moisture. Stainless steel is used for printing very abrasive or thermoplastic pastes on flat surfaces. Metalized monofilament polyester is used to improve the abrasion resistance.

The mesh and thread diameter of the screen define the width, height, and spacing of the rectangles of paste contacting the substrate. As these rectangles flow and merge into a continuous pattern, the thickness becomes one-fourth to one-third of the filament diameter (Fig. 25.2). The choice of the filament thickness and mesh size depends on the detail of the image, rheology of the paste, shape and roughness of the surface, and print

Table 25.1 General Compositions of Powders Used for Printing

Conductor	Resistor	Dielectric	Colorants
Au, Au/Pt	Pd/Ag	$BaTiO_3$	Transition metal oxides
Ag, Ag/Pd	RuO_2	Glass	ZrO_2: doped with V, Pd
Ni, Cu, Mo	$Ri_2Ru_2O_7$	Glass-ceramic	$ZrSiO_4$: doped

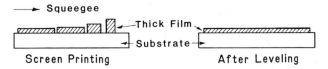

Fig. 25.2 Leveling and merging of screen printed paste.

thickness. The mesh opening must be several times larger than the particle size of the paste composition but small enough to prevent bleeding of the printed images. Printed paste that is 60–80 μm thick on contact is 20–30 μm thick after merging and drying.

The action of the squeegee is shown in Fig. 25.3. For a uniform print, the squeegee pressure must be uniform. This is achieved using a squeegee that does not excessively overlap the stencil opening and an adequate clearance between the end of squeegee and the screen frame. A soft squeegee will conform more readily to an irregular surface or a previously printed surface and produce a more uniform print. The print thickness is more uniform when the surface is flat and the stroke of the squeegee is parallel.

Important properties of the paste for printing are the coefficient of viscosity, yield strength, and rate of drying. The shear rate during flow through the screen is in the range of 100–1000 s^{-1}. The coefficient of viscosity of the paste for this shear rate is usually in the range of 70–150 Pas and ideally independent of temperature between 20 and 30°C. The paste must be pseudoplastic and thixotropic so that under pressure of the squeegee it will flow readily through the apertures of the screen and the print blocks will merge into a monolithic film. Ethyl cellulose or an acrylic resin of high molecular weight dispersed in terpineal or butyl carbitol is commonly used as a binder-liquid system. The viscosity must increase several hundred percent after leveling to prevent bleeding. For a particular screen, the viscosity of the binder and the solids loading can be coordinated to provide some flexibility in the fired thickness. The solids loading and binder viscosity can be adjusted to produce a particular paste viscosity;

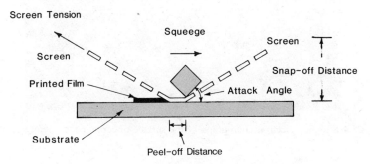

Fig. 25.3 Parameters in the screen printing process using a squeegee.

however, the use of a lower-viscosity binder and a higher powder loading will produce a thicker film after firing. Multiple printings are used to produce the desired combination of electrical components.

Defects include voids and pinholes in prints, improper print thickness, variation in print thickness, and poor line resolution. Voids and pinholes are caused by entrapped air, clogged screen apertures, and poor peel. Clogging is minimized by good mixing, control of the vapor pressure of the liquid solvent and air flow effecting drying, and the use of a larger mesh opening. Peeling behavior is dependent on the shear resistance between the paste and filament, screen tension, and the snapoff distance which controls the force normal to the print surface. The screen tension must be high enough for peeling but not too high to cause nonuniform squeegee pressure. The snapoff distance should be about 5% of the screen size.* The peeling speed should match the squeegee speed. Control of the screen mesh and filament diameter, emulsion thickness, and squeegee parameters is requisite for control of the print thickness. The uniformity of the print thickness is higher when the stroke of the squeegee is parallel to the surface and the hardness and pressure on the squeegee is carefully controlled. Carefully controlled

Fig. 25.4 Scanning electron micrograph of the surface of ceramic tape before printing.

* R. J. Bacher, *Insulations Circuits* **28**(8), 21–25 (1982).

peel and the use of a screen mesh having a finer filament diameter and concomitantly larger openings are required when printing fine lines. A surface with some roughness aids in paste transfer and adhesion, but the coating thickness becomes unacceptable above a particular roughness (Fig. 25.4).

Pad transfer printing is used for decorating surfaces that may vary widely in shape and surface texture and has the ability to print multicolors without intermediate drying. In this process, thin paste called ink in the recesses of an engraved plate is transferred to a surface by means of a rubber pad that first contacts the plate and then the surface. Photoengraved plates are commonly made by coating a lapped tool steel plate with a thin layer of photosensitive etching resist, exposure using a film positive, and then etching to a depth of 30–60 μm. Most pads are of silicone rubber. The inks used in pad printing are less viscous and more slowly drying than paste used in screen printing.

Pad transfer printing depends on the ink's having some affinity for the rubber necessary for peel up but a greater affinity for the surface necessary for transfer. The plate is first flooded with ink using a squeegee. The descending pad contacts the plate for a specific time and then rises. During the transition of the pad to the printing surface, the plate is again flooded with ink. The pad descends and the contacts the indexed part for a specific time, during which most of all of the ink is transferred. The pad then ascends, and the process is repeated. Cycle times may be as short as about 2 s. Image sizes may range up to about 15×15 cm.

Because the pad is flexible, the surface of the part need not be flat, and printing on surfaces with moderate relief and imprecise shapes may sometimes be done routinely. Multiple printing without drying is accomplished when the ink has a much greater affinity for the dry surface than for the pad, and the pad does not deform the previously printed image.

An indirect process for surface printing is the decal process. Decalcomanias or printed transfers, commonly called decals, are designs printed on specially prepared paper from which the design may be transferred (Fig. 25.5). Ceramic decals are printed using a prepared formulation similar to that used for screen printing. A simplex decal consists of the design printed on a coating of binder adhesive supported on a porous paper. After the simplex decal has been saturated with warm water or suitable solvent, the design is slid off onto the surface and smoothed with a sponge or cloth. Most ceramic decals are of the duplex type and consist of a tissuelike paper supported on a heavy paper backing, coated with the binderadhesive; the design is printed on the adhesive and then dried. Using a solvent-mount decal, the tissue and design are stripped off of the paper backing and soaked for a specific time in a suitable solvent; the softened decal is squeezed onto the surface, and the tissue is removed using a damp sponge. In the varnish mount process, the surface of the product is coated with a varnish or similar adhesive or sizing. The decal on tissue paper is stripped from the backing

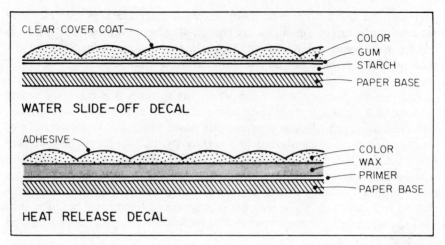

Fig. 25.5 Cross-sectional views of (top) solvent release and (bottom) heat release decals. (From J. J. Svec, *Ceramic Industry Magazine* **12**, 49 (1980).)

and applied printed side down on the varnished surface, using a damp sponge, roller, or brush, and the tissue is removed. The varnish mount process has been adapted for application on surfaces of unfired whiteware products. Heat release decals consist of a paper base, a special primer coat, a layer of wax on which the ceramic color is printed, and a heat-sensitive top coat. The application of the decals can be highly automated; heat from the product activates the adhesive and softens the wax, enabling transfer of the design and removal of the paper backing.

25.4 COATING PROCESSES

Spraying is the controlled atomization of a slurry and the directed flow of the droplets onto a surface. On impact, the droplets deform and coalesce into a thick film. The cross section of the spray is controlled by the design of the nozzle and the surrounding spreader, the air pressure and rheology of the slurry, and the working distance. Using a needle-type spray nozzle, the air flow and nozzle opening are varied simultaneously. The spreader determines the shape of the spray pattern, which is usually elliptical and of a particular aspect ratio. Substrates commonly translate normal to the direction of the spray sweep. Hollow-ware shapes are commonly rotated during the spray application. The thickness of the sprayed film is dependent on the spray geometry, solids content of the slurry, working distance, spraying time or sequence, rebound loss, and film flow. The reproducibility of the thickness of sprayed coatings is lower than in screen or pad printing. Film patterns may be controlled by using templates or masks or by coating the surface with a film such as wax that prevents wetting.

The orifice of the nozzle must be much larger than the largest particles in the slurry, and slurries are commonly screened before being pumped to the nozzle. A smaller nozzle tends to produce a narrower distribution of droplet sizes and a more controlled spray. The slurry should be pseudoplastic (Fig. 25.6) to permit flow through the nozzle but resist running due to gravity, air currents, or mechanical vibrations of the film on the product. The yield strength and viscosity of the film are increased by the adsorption of liquid from the film into the product, gelation, and drying. When using a slurry containing a thermally gelling binder as methyl cellulose, the slurry may be sprayed onto a heated product. Liquids used in nonaqueous slurries must not be too volatile. An example of an industrial glaze formation is presented in Table 25.2. Spraying machines automatically spray, continuously moving and rotating items. Disk atomization eliminates problems associated with

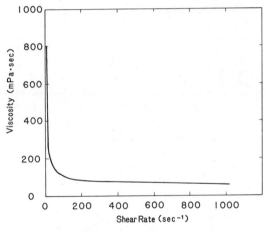

Fig. 25.6 Pseudoplastic rheological behavior of a glaze for spraying.

Table 25.2 Example of a Spray Glaze formulation (−325 Mesh)

Component	Weight (%)
Frit	46
Quartz	20
Feldspar	15
Whiting	14
Clay	5
Bentonite	0.25
Methylcellulose	0.40
Ethylene glycol	0.22
Deflocculant	0.03

compressed air and can produce coatings equal to those produced by spraying. A typical film thickness for a dried glaze is in the range of 0.1–0.2 mm.

Other methods for applying coatings and decorations are listed in Table 25.3. Porous ceramics can adsorb a thin, relatively uniform film of slurry, and translation of a product under a waterfall (Fig. 25.7) or above a fountain may be used to glaze one surface; advantages over spraying include a greater thickness uniformity and smoothness and significantly lower collector losses.*

Table 25.3 Glaze Application Processes

Wet	Dry
Spraying	Screen
Waterfall/fountain	Press
Dipping	
Slinger	
Splatter	
Flash/mist	
Dipping	

GLAZE WATERFALL CURTAIN

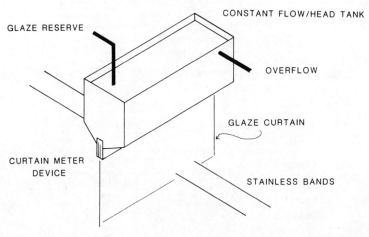

Fig. 25.7 Glaze waterfall curtain. (From R. F. Jaeger, in *Technical Innovations in Whitewares*, Alfred University Press, Alfred, NY, 1982.)

* R. P. Jaeger, Wall and Floor Tile Glaze Preparation, *Innovations in Whitewares*, J. S. Reed et al. (eds), Alfred University Press, Alfred, NY, 1982.

Dipping submerges the product, and a continuous coating is formed on the product. The thickness (L) of the viscous glaze film is[†]

$$L = K_1 \left(\frac{\eta_S v}{D_S} \right)^{1/2} + K_2 t^{1/2} \tag{25.2}$$

where η_S and D_S are the viscosity and density of the slurry, respectively, v is the withdrawal speed, t is the submersion time, K_1 is a constant that depends on the surface tension and angle of inclination, and K_2 is a constant that varies directly with the rate of adsorption of liquid into the product and the solids content of the slurry.

Slinger, splatter, and flash/mist applications are used to produce coarse, fine, and irregular surface texture or color effects. Dry applications are relatively unrefined but may be used to produce particular aesthetic effects. Lines and bands are often applied as the piece translates against a brush, sponge, or other porous material filled with slurry.

Common defects in the applied coating are voids and pinholes, thickness variations, and a wavy surface. Voids and pinholes are caused by air trapped in the film and an insufficient flow time during leveling or a spot contaminant or dust particle which prevents adhesion. Thickness variations are caused by improper application and leveling, a nonuniform adsorption into the porous surface, and gravity flow on inclined surfaces. Waviness is produced by cyclic air movement or interrupted flow on inclined surfaces while the glaze is thickening.

Conductive substrates may be coated using electrophoretic or highly automated electrostatic techniques. Advantages include a high degree of thickness uniformity and smoothness and a better coverage of sharp edges and surfaces of drilled holes or small cavities. In electrophoretic deposition, a slurry containing about 25 vol% solid, is well deflocculated to provide maximum particle mobility; the dispersed powder must be extremely fine to minimize settling. For an aqueous slurry, the substrate is normally the anode; the conductivity of the slurry must be very low to suppress the electrolysis of water producing O_2. Electrostatic application is now widely used for applying paints and enamels on metal. In the dry process, powder suspended in air is pumped through a power gun where ion bombardment charges the particles; charged particles accelerate to the surface of the grounded substrate forming a deposited coating. When a potential of about 100 kV is applied to the tip of a spray nozzle, charged droplets are formed, and these will be attracted to a grounded substrate. The solids loading of the slurry used in electrostatic spraying is lower than in conventional spraying, which reduces the air pressure for atomization and air flow in the spray booth.

In flame spraying and plasma spraying, powder feed melted on passing

[†] L. D. Landau and V. G. Levich, *Acta Phys.-Chem. URSS* **17**, 42 (1942).

through a special nozzle forms a thin film when deposited onto the surface of the product. This technique is used to form conductive or refractory coatings that may be complex in design when a mask is used.

SUMMARY

Products formed by plastic forming and casting that have mechanical toughness and adequate strength are trimmed in the leatherhard state. More brittle material produced on drying or heating to the initial stage of sintering is surface-ground using a porous grinding wheel or a hard cuting tool; grinding occurs by attrition, and the strength of the bond post between particles in the product is critical. Green ceramic tape is shaped and laminated after drying. Control of the temperature of each operation relative to the glass transition temperature of the binder is important to minimize distortion in shaping and to produce a strong bond between layers. Surface films serving as conductors, resistors, dielectrics, or decorations are commonly applied using a printing process or a printed decal. The paste used for printing must be time-stable and carefully formulated to develop the pseudoplastic-thixotropic flow behavior for printing without bleeding and adhesion sufficient for attachment. A glaze slurry or slip may be applied in a variety of ways. As for a printing paste, the processing additives are critical to developing the requisite flow for application, adhesion strength, and dry impact strength.

SUGGESTED READING

1. *Advances in Ceramics Vol. 9: Forming of Ceramics*, Jonh A. Mangels and Gary L. Messing (eds), American Ceramic Society, Columbus, OH, 1984.

2. Richard F. Jaeger, Wall and Floor Tile Glaze Preparation, Application and Control: Recent Evolution, in *Technical Innovations in Whitewares*, J. S. Reed et al. (eds), Alfred University Press, Alfred, NY, 1982.

3. F. H. Norton, *Fine Ceramics*, R. E. Krieger, Malabar, FL, 1978.

4. Albert Kosloff, *Ceramic Screen Printing*, Signs of the Times Publishing, Cincinnati, 1977.

5. *Decorating in Glass Industey*, Alexis G. Pincus and S. H. Chang (eds), Magazines for Industry, New York, 1977.

6. R. W. Vest, Materials Science of Thick Film Technology, *Am. Ceram. Soc. Bull.* **65**(4), 631–636 (1986).

7. Discovery in Decoration—New Riches for Glass, Ceramics, *Ceramic Industry Magazine* **9**, 24–31 (1984).

8. Rudolph J. Bacher, Thick Film Processing for Higher Yield, 1: Screen Printing, *Insulations/Circuits* **28**(8) 21–25 (1982).

9. David A. Karlyn, Pad Transfer Decorating, *Am. Ceram. Soc. Bull.* **59**(2), 247 (1980).

10. John R. Larry et al., Thick Film Technology: An Introduction to the Materials, *IEEE Transactions CHMT-3* **2**(6), 211–225 (1980).

11. U. C. Pack and V. J. Zaleckas, Scribing of Alumina Material by YAG and CO_2 Lasers, *Am. Ceram. Soc. Bull.* **54**(6) 585–588 (1975).

PROBLEMS

25.1 Draw a qualitative graph showing the change in toughness with liquid content on drying a slip-cast whiteware body and a cast alumina body containing no binder.

25.2 Draw a qualitative graph showing the dependence of surface-grinding failure rate on fracture strength of the product.

25.3 How would an increase in the glass transition temperature of the binder affect the lamination pressure or time?

25.4 Illustrate several stacked layers of green ceramic tape with printed metallization (1) just before laminating, (2) after laminating but incomplete bonding, and (3) laminated and completely bonded.

25.5 Lamination involves deformation and consolidation. Are the stresses during lamination uniform?

25.6 Explain why the ink in pad printing must be of a lower viscosity than the ink used in screen printing. Why dosen't the lower viscosity ink bleed on the surface?

25.7 What is the relation between snapoff distance and the force for peeling in screen printing?

25.8 When spraying a glaze onto an inclined surface, where should the initial coating begin?

25.9 For waterfall coating, what is the dependence of thickness of the coating on the translation velocity, viscosity, and slurry density when adsorption is insignificant?

25.10 How does gelation alter the yield strength, viscosity, and pseudoplastic behavior of a paste?

25.11 Design a processing flow diagram for preparing a cured ceramic resistor thick film on an alumina substrate.

26

Firing Processes

Products that have been dried and surface-finished, traditionally called "green products," are heat-treated in a kiln or furnace to develop the desired microstructure and properties. This process, called firing, may be considered to proceed in three stages: (1) reactions preliminary to sintering, which include binder burnout and the elimination of gaseous products of decomposition and oxidation; (2) sintering; and (3) cooling, which may include thermal and chemical annealing.

"Sintering" is the term used to describe the consolidation of the product during firing. Consolidation implies that within the product, particles have joined together into an aggregate that has strength. The term sintering is often interpreted to imply that shrinkage and densification have occurred; although this commonly happens, densification does not always occur. Highly porous, refractory insulation products may actually be less dense after they have been sintered.

In this chapter we will consider general aspects of firing and principles involved in sintering simple powders and relatively complex, multiphase materials.

26.1 FIRING SYSTEMS

Ceramic materials and products are fired in a variety of kilns and furnaces* that are designed to operate either intermittently or continuously. Intermittent kilns for industrial firing, commonly called periodic kilns, are usually of

---

* "Furnace" is a general term for an enclosed chamber in which heat is produced. The more specific term "kiln" is commonly used in the field of ceramics. A product is heated in contact with the hot combustion gases in a conventional kiln but is insulated from the combustion gases in a muffle kiln. The firing atmosphere may be varied more widely in a muffle kiln. Electrical heating is used in an electric kiln.

the shuttle or elevator type. When loading a shuttle kiln, individual product items, product items supported on refractory setters and supporting members (kiln furniture), or product contained in refractory saggers are set on refractory shelves that are supported on a thermally insulated kiln car. The car mounted on a rail is pushed into the kiln for firing and withdrawn for unloading (Fig. 26.1a). An elevator kiln lined with relatively low mass thermal insulation is raised for setting the product and then lowered for the firing cycle; a variation is a car with a hearth that may be elevated into or lowered from the kiln.

Continuous tunnel kilns of the car, sled, roller hearth, and continuous belt type are commonly used for firing high-volume production items. When firing in a car or sled kiln, the product is set in a manner similar to that for a

Fig. 26.1 Photographs of (a) a shuttle kiln and (b) a three-tier roller hearth kiln. (Figure a, courtesy of Temtek Allied Div., Eisenman Corp.; Fig. b, courtesy of Siti Kilns.)

shuttle kiln, but the cars or sleds move through the thermal gradient in the kiln. In a roller hearth kiln (Fig. 26.1b), large flat products such as ceramic tile or smaller products supported on refractory setter tile are conveyed on refractory rollers through the heated tunnel. A continuous refractory fiber carpet or mesh belt is also used to support and convey low-mass items through a kiln. A rotary kiln is an inclined, rotating or oscillating refractory cylinder used to continuously fire granular materials.

When firing products with a large mass or setting surface, the product is supported on mobile refractory supports, a granular bed, or partially fired "shrink plate" that can accommodate the differential shrinkage between the product and the setter. Products that have a tendency to deform during firing (slump) are supported on contoured refractory setters. In some setting configurations, the product is arranged so that shrinkage gradients and dimensional distortion during firing can compensate for density gradients or oversize dimensions in the green product. Elongated products are sometimes hung so that their weight acts to maintain linearity on firing.

Periodic kilns are used when a wide variety in product firing schedules or a relatively small or intermittent production volume requires flexibility in firing. Products from a continuous kiln that have had a minor body or glaze defect repaired are also often refired in a periodic kiln.

The car-type tunnel kiln (Fig. 26.2), introduced in 1916, is used widely for firing large products and heavy loads when the firing cycle exceeds about 6 h. Sled and roller hearth kilns provide very good temperature uniformity for single tier settings and cycle flexibility. A sled kiln can be used for firing to a higher temperature than when using a roller kiln but requires a greater investment in refractories and a sled transfer system. Sled, roller hearth, and small periodic kilns are commonly used when the firing cycle is as short as 0.5 h.

Product heating is commonly produced by the combustion of natural gas or fuel oil or by electric heating. Radiant heating improves the temperature uniformity throughout the setting. When firing in a gas-fired muffle kiln or an electric kiln, the setting pattern should facilitate radiation transport to the center of the configuration, as shown in Fig. 26.2. Direct-fire combustion kilns also utilize convection heat transfer. In a car-type kiln, the setting pattern and hollow refractory supports are configured to facilitate the convection of hot gas throughout the car. High-velocity burners provide a higher velocity of convection and a better heat transfer and temperature uniformity.

Combustion heating is commonly produced from the ignition of a mixture of natural gas or fuel oil and air. The combustion of methane, the primary constituent in natural gas, in air is

$$CH_4 + 2O_2 \rightarrow 2H_2O + CO_2 \qquad (26.1)$$

Excess air is needed for an oxidizing atmosphere. Oxygen may be intro-

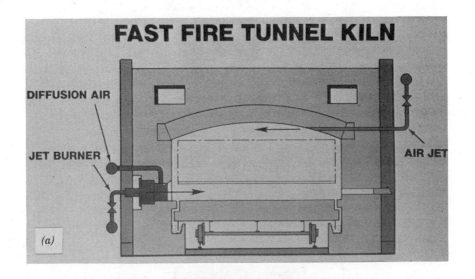

FAST FIRE TUNNEL KILN

DIFFUSION AIR

JET BURNER

AIR JET

(a)

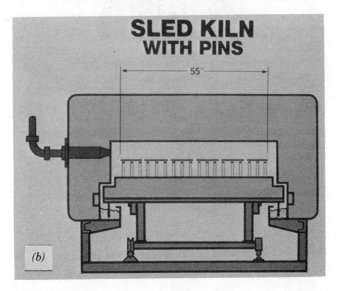

SLED KILN
WITH PINS

55"

(b)

Fig. 26.2 Cross section of (a) car-type and (b) sled-type tunnel kilns, and plan views of product settings for

443

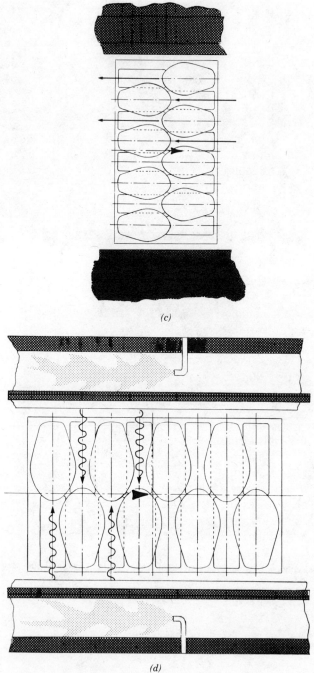

(c)

(d)

Fig. 26.2 (c) axial convection heat transfer and (d) a gas-fired kiln with a muffle wall and a setting for radiant heat transfer. (Figures a and b, courtesy of Bickley Furnaces Inc.; Figs. c and d, courtesy of Swindell Dressler International, Subsidiary of Rust International Co.)

duced at the burners to reduce the volume of air, to achieve a higher temperature, and to increase the oxygen partial pressure. A furnace temperature exceeding 1700°C may be achieved using oxygen and natural gas or another gas such as propane of higher caloric value.

Refractory metal alloy heating elements are used for heating to about 1150°C, silicon carbide to about 1550°C, and molybdenum silicide rods for heating to 1650°C in an oxidizing atmosphere. Molybdenum, tungsten, and graphite may be used for heating above 1700°C in an inert or reducing atmosphere. Induction, microwave, and plasma heating have been investigated. Significant improvements in the temperature uniformity and thermal efficiency have been achieved by using a greater amount of high-performance refractory insulation of much lower thermal mass and thermal conductivity, and higher-strength kiln furniture of thinner construction; these innovations enable the volume of product/furniture to increase in a setting and a faster firing when the kiln furniture is rate limiting.

26.2 PRESINTERING PROCESSES

Sintering does not commonly begin until the temperature in the product exceeds one-half to two-thirds the melting temperature, which is sufficient to cause significant atomic diffusion for solid-state sintering or significant diffusion and viscous flow when a liquid phase is present or produced by a chemical reaction. Material changes on heating prior to sintering may include drying, the decomposition of organic binders, the vaporization of chemically combined water from the surfaces of particles and from within inorganic phases containing water of crystallization, the pyrolysis of particulate organic materials introduced with the raw materials or as contamination during processing, changes in the oxidation states of some transition metal and rare-earth ions, and the decomposition of carbonates, sulfates, etc. introduced as additives or as a constituent of the raw materials.

In this stage, stresses from the pressure of gas evolved or from differential thermal expansion of phases must not cause cracks or fracture of the fragile product. When firing a laminated structure or a product with a surface coating, evolved gas must be eliminated while the surface permeability is high. These presintering reactions are commonly investigated using the thermal analysis techniques discussed in Chapter 6.

Initial heating may remove any liquid remaining after forming and drying the product and any moisture adsorbed from the atmosphere during transporting and setting. The adsorption of moisture can be quite significant when a warm product with a high specific surface area is removed from the dryer but cools in moist air before firing. The adsorbed moisture may persist in the product up to a temperature exceeding 200°C.

Binder burnout is very dependent on the composition and the structure of the binder material, the composition of the gas surrounding and in the pores

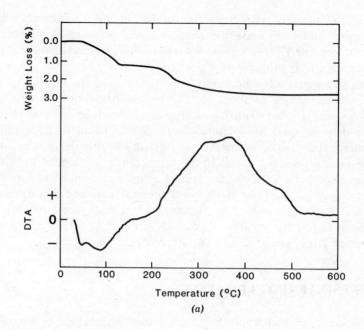

(a)

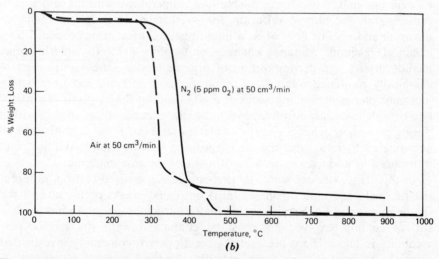

(b)

Fig. 26.3 Thermal analysis of (a) powder compact containing 1.5 wt% polyvinyl alcohol with 1.25 wt% adsorbed water in air and (b) methyl cellulose in air and nitrogen (From N. Sarkar and G. Greminger, *Am. Ceram. Soc. Bull* **62**(11), 1284 (1983).)

of the product, and the rates of diffusion of the decomposition product gases and furnace gas through the product, which is dependent on the permeability K_p. The burnout of polyvinyl alcohol from a compact of barium titanate powder in air is shown in Fig. 26.3. The initial endothermic reaction up to about 150°C corresponds to the evaporation of adsorbed water. From about 150 to 400°C, side groups of the binder are gradually eliminated as indicated by the reaction

$$
\begin{array}{c}
\underset{\displaystyle \overset{\displaystyle H}{|}}{\overset{\displaystyle H}{|}} \quad \underset{}{} \quad \underset{}{} \quad
\end{array}
\quad \xrightarrow{150\text{–}250°\text{C}} \quad
\mathrm{C-C{=}C-C} + \mathrm{H_2O}_{(gas)}
\qquad (26.2)
$$

Depolymerization and oxidation of the vinyl chain continue the exothermic behavior, and a weight loss is observed up to about 500°C. The volume of the gaseous products may be several hundred times larger than the volume of the compact. Similar results are observed on heating vinyl binders with other common side groups and cellulose binders except that the burnout range may be narrower. Waxes and polyethylene glycols melt at a relatively low temperature and vaporize over a narrow temperature range. Binder systems with a narrow decomposition range coupled with oxidation heating can cause a rapid increase in temperature, especially when the binder loading is high. The ash content is higher for natural binders and binders containing metal ions. Refined binders such as polyvinyl alcohol and cellulose binders leave a few percent ash. The residue after burnout of synthetic polymerized glycols and acrylic binders is a fraction of 1%.

The burnout process may cause a slight volume expansion when the ceramic particles are in contact, as in a dry-pressed compact (Fig. 26.4), but a volume contraction when the particles are separated by the organic phase as in a tape cast or injection molded material; for the latter, the contraction varies directly with the binder loading. The apparent burnout temperature on heating in air is higher for a thicker product and when K_p is lower (Fig. 18.7). Burnout is suppressed when the oxygen pressure in the furnace is reduced.* When firing nitride and carbide ceramics in an inert atmosphere, the carbon residue from the binder aids in removing oxygen impurity from the surface of the particles prior to sintering. For products containing less than about 10 vol% binder, gas permeation to the surface is important, and the safe rate of burnout depends on the permeability (Fig. 18.7). For much higher binder contents, burnout proceeds from a reaction front that recedes into the product. A two-phase binder system containing a minor binder that

---

* J. P. Polinger and G. L. Messing, Thermal Analyses of Organic Binders for Ceramic Processing, pp. 359–370 in *Advances in Materials Characterization II*, R. L. Snyder et al. (eds), Plenum, New York, 1985.

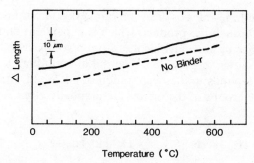

Fig. 26.4 Thermal dilatometric analysis of a compact of a 1-μm oxide powder containing 5 vol% polyvinyl alcohol plasticized with polyethylene glycol.

burns out relatively early opening pore channels and increasing K_P has been reported to facilitate binder burnout in injection molded products.

In bodies containing kaolin, the water of crystallization is eliminated between 450 and 700°C by the reaction producing metakaolin $Al_2Si_2O_6$

$$Al_2Si_2O_5(OH)_4 \rightarrow Al_2Si_2O_6 + 2H_2O_{(gas)} \qquad (26.3)$$

The gas pressure may produce a volume expansion, and the heating rate must be carefully controlled in bodies containing a high clay content. Other dehydration reactions producing gas are the dehydration of aluminum hydrates in the range 320–560°C and talc at 900–1000°C. The decomposition of magnesium carbonate at 700°C and dolomite at 830–920°C produce CO_2. Magnesium sulfate decomposes as low as 970°C, but the decomposition of calcium sulfate occurs at a much higher temperature, in excess of 1050°C. The termination of decomposition depends on the particle size, the gas permeability of the product, and reactions between the salt and other phases present. Anion impurities in compacts of chemically prepared powders may persist into the sintering stage. Fine organic matter in ball clay is commonly oxidized between 200 and 700°C, but coarse particulate carbon in the material may not be completely oxidized at 1000°C even when firing with excess air. Persistent carbon discolors the microstructure and causes the effect called "black coring" that may be observed on a fracture surface. Persistent carbon and a deficiency of oxygen produces carbon monoxide (CO) in the pores of the product. A chemical dopant such as a transition metal oxide in its higher oxidation state, such as Mn_2O_3, is added in some products as an internal oxidizing agent;

$$Mn_2O_3 + CO_{(gas)} \rightarrow 2MnO + CO_{2(gas)} \qquad (26.4)$$

Carbon monoxide in the product is not inert and may change the stoichiometry and properties of a transition metal oxide pigment or a magnetic ferrite;

$$CO_{(gas)} + Fe_2O_3 \rightarrow 2FeO + CO_{2(gas)} \qquad (26.5)$$

26.3 SOLID-STATE SINTERING

Sintered ceramic products represent a wide range of material systems that may vary widely in the number of components, particle characteristics, complexity of chemical reactions, and densification mechanisms during sintering. In this section we will examine sintering within the categories of solid-state sintering, sintering of glass particles, sintering with reactions and dissolving, and the sintering of a glassy coating (glaze).

Examples of solid-state sintering include the sintering of a crystalline single phase such as α-Al_2O_3 and the sintering of a single phase containing a refractory dopant such as Al_2O_3:0.5% MgO, ZrO_2:3% Y_2O_3, and SiC:2% B_4C. An example of the densification behavior of a dense compact of a well-dispersed, nominally 1 μm magnesia-doped alumina powder during a constant rate of heating is shown in Fig. 26.5. The density increase of about 2% corresponding to the upturn of the curve is called the initial stage; the range of major densification, the intermediate stage; and the range of rapidly decreasing rate and cessation of densification, the final stage.

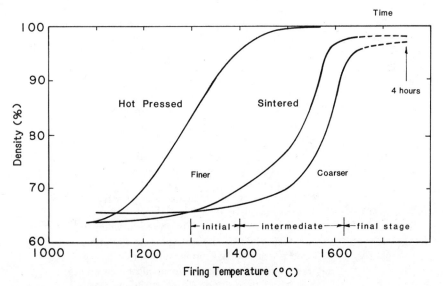

Fig. 26.5 Densification behavior of compacts of two log-normal reactive alumina powders ($\bar{a}_{g_M}$ = 1.3 and 0.8 μm) and the initial, intermediate, and final stages of sintering for the coarser powder.

Microstructural characteristics or changes observed in these stages in rather uniform packings of particles are listed in Table 26.1. In a very heterogeneous compact containing particles, agglomerates, and pores of widely different size and shape or chemical microinhomogeneities, the characteristic transitions between stages may proceed at different rates in different microscopic regions. Microstructural characteristics are determined using the techniques described in Chapter 6.

The driving force for sintering is the reduction in the total free energy ΔG_T of the system

$$\Delta G_T = \Delta G_v + \Delta G_b + \Delta G_s \qquad (26.6)$$

where ΔG_v, ΔG_b, and ΔG_s represent the change in free energy associated with the volume, boundaries, and surfaces of the grains, respectively. The major driving force in conventional sintering is ΔG_s but the other terms may be significant in some stages for some material systems.

Mechanisms for mass transport during sintering are listed in Table 26.2. Surface diffusion is a general transport mechanism that can produce surface smoothing, particle joining, and pore rounding, but it does not produce volume shrinkage. In materials where the vapor pressure is relatively high, sublimation and vapor transport to surfaces of lower vapor pressure also produce these effects. Diffusion along the grain boundaries and diffusion through the lattice of the grains produce both neck growth and volume

Table 26.1 Microstructural Changes Observed in the Initial, Intermediate, and Final Stages of Solid-State Sintering (Powder Compact)

Stage	Observations
Initial	Surface smoothing of particles
	Grain boundaries form, neck growth
	Rounding of interconnected, open pores
	Diffusion of active, segregated dopants
	Porosity decreases <12%
Intermediate	Shrinkage of open pores intersecting grain boundaries
	Mean porosity decreases significantly
	Slow grain growth
	(Differential pore shrinkage, grain growth in heterogeneous material)
Final (1)	Closed pores containing kiln gas form when density is $\approx$92% (>85% in heterogeneous material)
	Closed pores intersect grain boundaries
	Pores shrink to a limited size or disappear
	Pores larger than grains shrink relatively slowly
Final (2)	Grains of much larger size appear rapidly
	Pores within larger grains shrink relatively slowly

Table 26.2 Mass Transport Mechanisms in Sintering

Mechanism	Densification
Surface diffusion	No
Evaporation-condensation	No
Boundary diffusion	Yes
Lattice diffusion	Yes
Viscous flow	Yes
Plastic flow	Yes

shrinkage. The mechanisms of bulk viscous flow and plastic deformation may be effective when a wetting liquid is present and when a mechanical pressure is applied, respectively.

The shrinkage of an interstice in a uniform packing of uniform crystalline spheres (Fig. 26.6) has been considered by many investigators. Because of a difference in chemical potential, the concentration of vacancies beneath a concave surface is higher than beneath a flat or convex surface. The transport of vacancies from a concave surface can occur by the mechanisms of lattice and boundary diffusion, with a concomitant flow of atoms in the opposite direction. The effect is pore rounding and a decrease in the total surface free energy. Burke and Rosolowski* have pointed out that in the initial stage, pore rounding causes a decrease in ΔG_s in proportion to the reduction of the surface area, but the formation of a grain boundary causes an increase in ΔG_b. The angle of intersection of the pore at the pore-grain boundary juncture is

$$\cos\left(\frac{\phi}{2}\right) = \frac{\gamma_b}{2\gamma_s} \qquad (26.7)$$

where γ_b is the interfacial tension of the boundary and γ_s is the surface tension. Neck growth should occur when $\gamma_b < \sqrt{3}\gamma_s$. For most materials, $\gamma_b < \gamma_s$, and the angle ϕ is large; however, in nonoxide materials, which sinter with difficulty, a dopant that concentrates in the boundary and

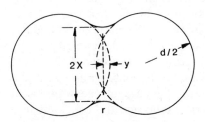

Fig. 26.6 Two-sphere initial-stage sintering model.

* J. E. Burke and J. H. Rosolowski, Chapter 10 in *Treatise on Solid State Chemistry Vol. 4: Reactivity of Solids*, N. B. Hannay (ed), Plenum, New York, 1976.

reduces γ_b may be required for neck growth. Pore shrinkage occurs by the diffusion of vacancies to the grain boundaries and their annihilation there. The pore is a source of vacancies, and the grain boundary is a vacancy sink. Microcreep at the grain boundary eliminates vacancies and enables the centers of particles to approach, producing pore shrinkage.

The shrinkage $\Delta L/L_0$ with time in the initial stage for uniformly packed spheres has been modeled by numerous investigators[*] and is described by the equation

$$\frac{\Delta L}{L_0} = \left[\frac{KD_v \gamma_s V_v t}{k_B T d^n} \right]^m \tag{26.8}$$

where D_v is the apparent diffusivity of the vacancies of volume (V_v), d is the grain diameter, K is a constant dependent on the geometry, and m and n are constants that depend on the mechanisms of mass transport. The value of n is commonly observed to be approximately 3, which implies that surface diffusion is the predominant mechanism, and m is commonly in the range 0.3–0.5. The shrinkage is very temperature-dependent in that the diffusivity (D_v) varies exponentially with temperature [$\exp(-Q/k_B T)$]. Equation 26.8 also indicates that smaller particles packed identically will produce an equivalent shrinkage in a shorter time.

In the intermediate stage, modeling of shrinkage is complicated by grain growth and a change in the pore geometry. More than one mass transport mechanism may be contributing significantly to the changes in microstructure. Sintering in this stage depends on the parameters in Eq. 26.8 but is also very dependent on the size, shape, and packing of the particles and chemical dopants that increase the vacancy flux of the slower-diffusing specie. Particle aggregates are a common source of packing heterogeneity and inhomogeneous sintering (Fig. 26.7). As is seen in Fig. 26.8, the dispersion of particle aggregates by milling the powder before fabrication increased the compact bulk density only slightly; however, the particle packing and the intermediate sintering behavior were improved, and the transition to the final stage of sintering occurred at a significantly higher density in the more homogeneous compact.

In the intermediate and final stages, the densification behavior is very dependent on the association of pores with grain boundaries and the rate and mode of grain growth. Diffusion of atoms across the grain boundary, the disordered region between grains, causes the grain boundary to be displaced. Heating causes some grains to grow at the expense of others which shrink, and the net effect is an increase in the mean grain size and a reduction in the total grain boundary area. Because of the topological similarities between the grain boundaries in a solid and the cell walls in a

[*] R. L. Coble, *J. Am. Ceram. Soc.* **56**(9), 461–466 (1973); R. W. Lay and R.E. Carter, *J. Am. Ceram. Soc.* **52**(4), 189–191 (1969).

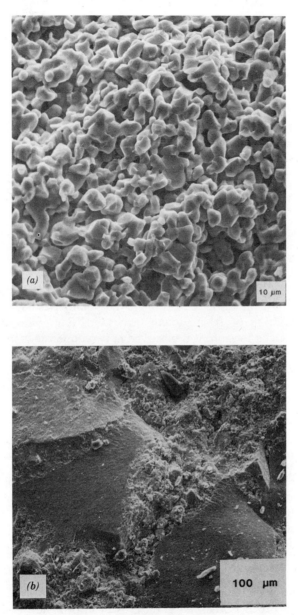

Fig. 26.7 Scanning electron micrograph showing (a) homogeneous sintering behavior in a compact of spinel powder and (b) inhomogeneous sintering in a compact of zirconia powder.

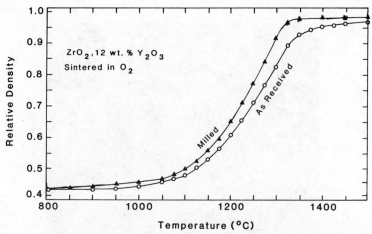

Fig. 26.8 Densification of compacts of submicron yttria-stabilized zirconia powder containing aggregates and with the aggregate size reduced by milling. (From C. Scott, Ph.D. thesis, Alfred University, Alfred, NY, 1977.)

foam, a model for the displacement of curved boundaries has been used as a classical description of grain growth in a solid. For grains with a nearly isotropic interfacial tension, it is observed in two dimensions that grains with fewer than six sides have convex boundaries and shrink and that those with more than six sides have concave boundaries and grow, as is indicated in Fig. 26.9. Diffusion across the boundary will cause the boundary to be displaced toward its center of curvature at a velocity (v_b) given by the expression

$$v_b = M_b \gamma_b (1/r_1 - 1/r_2) \tag{26.9}$$

where M_b is the mobility of the boundary, which varies as $[\exp(-Q_b/k_bT)]/T$, and r_1 and r_2 are the principal radii of curvature. When the mean grain size is directly proportional to the radius of curvature and M_b is a constant, the dependence of the mean grain size on time t is[*]

$$d_t^n - d_0^n = 2AM_b\gamma_b t \tag{26.10}$$

where d_0 is the original mean grain size and A is a constant that depends on the geometry. Much isothermal grain growth information is approximated by Eq. 26.10, and n is observed to have a value in the range of 2–3. A log-normal grain size distribution is commonly observed during regular grain

[*] J. E. Burke and J. H. Rosolowski, Chapter 10 in *Treatise on Solid State Chemistry Vol. 4: Reactivity of Solids*, N. B. Hannay (ed), Plenum, New York, 1976.

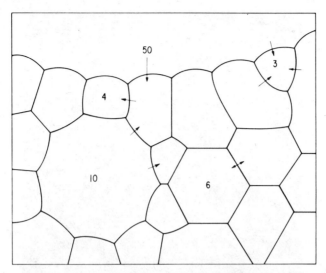

Fig. 26.9 Section through a polycrystalline solid showing grains with a different number of sides; arrows indicate direction of boundary displacement. (Reprinted with permission from J. E. Burke and J. H. Rosolowski, in *Treatise on Solid State Chemistry Vol. 4: Reactivity of Solids*, Plenum, New York, 1976.)

growth. For a log-normal distribution of grain sizes, the mean grain size d may be calculated from the mean intercept length $\bar{L}$ between boundaries using the relation $d = 1.56\bar{L}$, where $\bar{L}$ is determined using random test lines superimposed over a large number of grains in the plane of polish. An assumption in deriving Eq. 26.10 is that the form of the distribution of grain size is constant with time—i.e., regular grain growth.

Pores and solid inclusions smaller than the grains may intersect the grain boundaries. When these inclusions disappear from the boundaries, grains exaggerated in size are commonly observed to appear in the microstructure, creating a bimodal grain size distribution (Fig. 26.10). This mode of grain coarsening is called exaggerated grain growth or discontinuous grain growth; it is usually undesirable unless a very coarse-grained microstructure is desired such as in a refractory material (where creep resistance is most important). An inclusion intersecting a boundary exerts a drag force on the boundary and reduces its effective mobility. The limiting grain size when immobile inclusions of radius (r_i) are dispersed along the boundary is*

$$d = r_i\left(\frac{8}{f_i^v}\right)^{1/2}$$

(26.11)

where f_i^v is the volume fraction of inclusions in the solid. In the intermediate

---

* J. E. Burke and J. H. Rosolowski, Chapter 10 in *Treatise on Solid State Chemistry Vol. 4: Reactivity of Solids*, N. B. Hannay (ed), Plenum, New York, 1976.

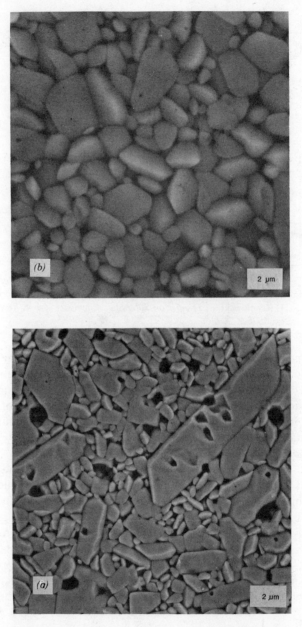

Fig. 26.10 Scanning electron micrographs of sintered alumina showing (a) regular and (b) exaggerated grain growth. (From D. Miller, M.S. thesis, Alfred University, Alfred, NY, 1986.)

stage and early part of the final stage, pores intersecting the grain boundary retard grain growth. Under the pull of the grain boundary, pores may migrate with a mobility M_p by means of mass transport across the pore. The condition for pore attachment depends on the ratio of M_p/M_b as well as f_p^v and the pore radius r_p. For pore migration by surface diffusion (surface diffusivity $= D_s$) the mobility has been expressed by Brook* as

$$M_p = \frac{KD_s}{Tr_p^4} \tag{26.12}$$

which indicates that M_p is larger for smaller pores. Pore growth can occur by

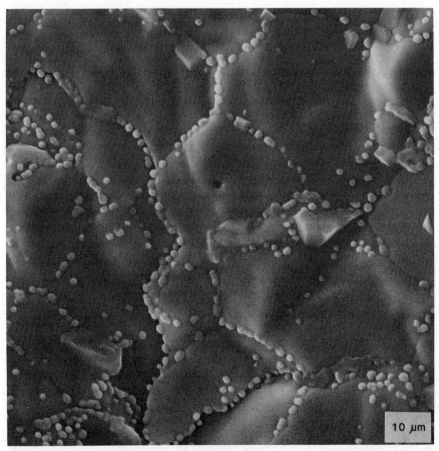

Fig. 26.11 Grain growth-inhibiting inclusions segregated in the grain boundaries of alumina doped strontium zirconate (thermally etched). (From C. Scott, M.S. thesis, Alfred University, Alfred, NY, 1974.)

* R. J. Brook, in *Treatise on Materials Science and Technology Vol. 9: Ceramic Fabrication Processes*, F. F. Y. Wang (ed), Academic Press, New York, 1976.

either the coalescence of pores on the boundary or the diffusion of vacancies from smaller pores to larger pores rather than to the grain boundaries, which is called Ostwald ripening.

In the final stage of sintering, exaggerated grains separated from pores commonly appear in some regions of the microstructure when the processing is not carefully controlled. The tendency for exaggerated grains is higher when powder aggregates or dense, extremely coarse particles are present and the pore distribution is inhomogeneous. A high sintered density with minimal grain growth occurs when the material contains fine particles packed very densely and when an additive called a grain growth inhibitor is dispersed and homogeneously distributed. A classic grain growth inhibitor is MgO in alumina, which apparently inhibits exaggerated grain growth by increasing D_s and consequently M_p;[*] when this system is sintered in an appropriate atmosphere, the density may exceed 99.9%. Small second-phase solid inclusions on the boundaries (Fig. 26.11) may also retard grain growth when the pores disappear, as has been demonstrated in yttria doped with La_2O_3 and ferrites and silicon nitride doped with Y_2O_3.[†] The general combinations of grain size and pore size for regular (pore-boundary attachment) and exaggerated (pore-boundary separation) grain growth and the effect of an inhibitor that reduces M_b are illustrated in Fig. 26.12.

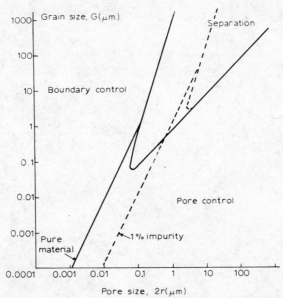

Fig. 26.12 Dependence of pore-boundary interaction on microstructural parameters in a system where the pores move by surface diffusion. (Reproduced with permission from R. J. Brook, *J. Am. Ceram. Soc.* **52**(1), 57 (1969).)

---

[*] K. A. Berry and M. P. Harmer, *J. Am. Ceram. Soc.* **69**(2), 143–149 (1986).
[†] M. F. Yan, Chapter 6 in *Advances on Power Technology*, G. Y. Chin (ed), American Society for Metals, Metals Park, OH, 1982.

A rapid heating schedule may produce densification with a smaller concomitant grain size. Surface diffusion, which predominates at low temperatures, can cause grain coarsening; when fast heating increases the vacancy diffusivity (D_v) and densification at a faster rate than diffusion causing grain coarsening, fast firing with a minimum isothermal hold at the maximum temperature should be beneficial.

The atmosphere is also important in sintering. Gas trapped in closed pores will limit pore shrinkage unless the gas is soluble in the grain boundary and can diffuse from the pore. Alumina doped with MgO can be sintered to essentially zero porosity in an atmosphere of H_2 or O_2, which are soluble, but not in air, which contains insoluble nitrogen. Insoluble gas evolved into pores, such as SO_2 or Cl_2 from anion impurities in powders, may also limit pore shrinkage. The density of oxides sintered in air is commonly less than 98% and often only 92–96%. The sintering atmosphere is also important in that it may influence the sublimation or the stoichiometry of the principal particles or dopant. The oxygen pressure and partial pressure of PbO and ZnO must be controlled when sintering compounds such as lead titanates and zinc ferrites, because incongruent vaporization may produce PbO or ZnO vapor. The sublimation rate is lower when the oxygen pressure is higher and in a closed sagger saturated with the vapor. However, chromic oxide (Cr_2O_3) can be densified only on sintering at a low oxygen pressure (10^{-10} to 10^{-11} atm), because in an oxidizing atmosphere it sublimates as CrO_2 or CrO_3.[*] In ceramic ferrites, the oxidation state of the transition metal ions and lattice vacancies depend on the sintering atmosphere, and the partial pressure of oxygen is controlled to obtain a high density and the proper magnetic phases. When sintering nonoxide ceramics

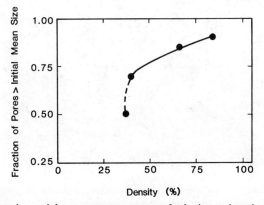

Fig. 26.13 On sintering an inhomogeneous compact of submicron zirconia powder, pores larger than the grain size shrink relatively slowly and persist, and the apparent mean pore size determined by mercury porosimetry increases on densification.

[*] H. U. Anderson, *J. Am. Ceram. Soc.* **57**(9), 34–37 (1974).

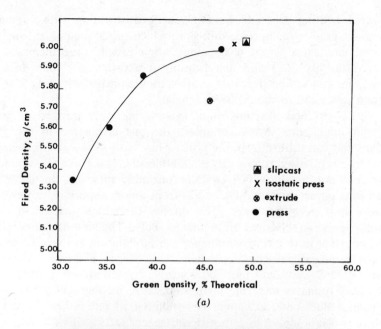

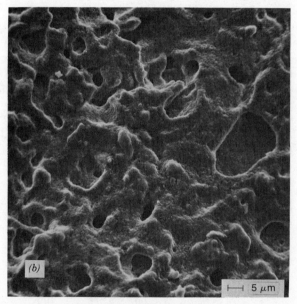

Fig. 26.14 The ultimate sintered density is higher for (a) compacts of a submicron zirconia powder having a more uniform pore size and a higher green density, but (b) large pores from air in extruded material persist. (From T. Carbone, M.S. thesis, Alfred University, Alfred, NY, 1975.)

such as silicon carbide at a temperature that may exceed 2000°C, a nitrogen atmosphere plays an important role in inhibiting the sublimation of SiC and the B_4C dopant. The infiltration of a powder compact with a reactive gas may produce reaction bonding and a significant increase in density, as in producing reaction-bonded Si_3N_4 from silicon powder.

The sintering of ceramic materials depends importantly on the distribution of pore sizes and the homogeneity of the porosity in the compact.[*] On sintering an ideal homogeneous compact of uniform particles, the uniform interstices smaller than the grains should shrink at a relatively fast, uniform rate. However, large pores in a compact shrink relatively slowly (Fig. 26.13). Compacts pressed at a higher pressure that have a more narrow pore size range (Fig. 26.14) and a higher green density may achieve a higher sintered density when fired under identical conditions.

Inhomogeneous sintering and shrinkage may occur in materials containing relatively densely packed regions or regions containing finer particles that are separated by more porous regions containing larger pores. Regions containing smaller pores or particles that densify earlier may precede the more porous regions into the final stage of sintering, and the coarser pores in the boundary regions tend to persist (Fig. 26.14). A similar effect may occur when a chemical densification aid is distributed inhomogeneously in the material.

26.4 SINTERING OF GLASS PARTICLES

A product composed only of glass particles may also be sintered to form a dense structure. In the initial stage, viscous flow produced by the driving force of surface tension causes neck growth. For two glass spheres of uniform diameter a, the initial shrinkage $\Delta L/L_0$ is given by the classic Frenkel equation[*]

$$\frac{\Delta L}{L_0} = \frac{3\gamma_s t}{2\eta a} \tag{26.13}$$

where t is the isothermal sintering time. The initial shrinkage behavior for a compact of a standard glass powder is of the form predicted by Eq. 26.13, as seen in Fig. 26.15.

In the intermediate stage, it can be assumed that gas diffuses rapidly through the interconnected pores and that the rate of pore shrinkage is proportional to γ_s/η, which is very dependent on temperature. Sintering occurs very rapidly when the glass is heated to its softening point of $10^{6.6}$ Pa s, but gravity-induced flow causes slumping at this temperature; the

[*] F. F. Lange, *J. Am. Ceram. Soc.* **67**(2), 83–89 (1984).
[*] J. Frenkel, *J. Physics (USSR)* **9**(5), 305 (1945).

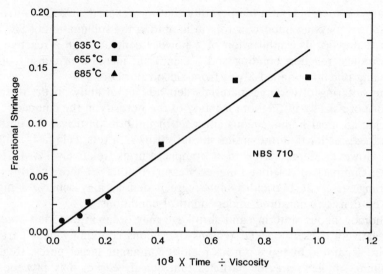

Fig. 26.15 The variation of (fractional shrinkage · time)/viscosity for a compact of NBS 710 glass powder. The curve corresponds to Eq. 26.13 with $\gamma = 300\ \text{mN/m}$ and $a = 10\ \mu\text{m}$. (After T. Clark, Ph.D. thesis, Alfred University, Alfred, NY, 1984.)

heating schedule must be controlled to obtain a moderate densification rate without slumping, which corresponds to an apparent viscosity of about $10^8\ \text{Pa s}$. Gels may be sintered at or below T_g.

Closed pores form on entering the final stage, and these should be spherical when the medium is isotropic. The densification rate dD/dt of uniform spherical pores dispersed in an isotropic, incompressible viscous medium is described by the Mackenzie-Shuttleworth equation*

$$\frac{dD}{dt} = \frac{K\gamma_s n^{1/3}(1 - D)^{2/3}(D)^{1/3}}{\eta} \tag{26.14}$$

where D is the bulk density fraction of the porous glass containing n pores per unit volume of glass and K is a constant that depends on geometry.

When sintering a crushed-glass powder, the sintering rate is higher than predicted using Eq. 26.13, because the effective radius of curvature is smaller than the equivalent spherical radius. Water vapor in the sintering atmosphere may reduce η and increase the apparent densification rate. In the final stage, gas solubility and its diffusion in the glass and pore coalescence which decrease n, affect the final pore shrinkage. Finer pores have a larger driving force for shrinkage. An increase in the gas pressure due to an increase in temperature or from gas evolved from within the glass, or a reduction of the driving force due to pore coalescence, can cause pore

---

* J. Mackenzie and R. Shuttleworth, *Proc. Phys. Soc.* (*Lond.*) *B* **62**, 833 (1949).

enlargement and a volume expansion, called bloating, which is usually undesirable. Some glasses may be crystallized after being densified by sintering, to produce a glass ceramic.

26.5 SINTERING IN THE PRESENCE OF A SMALL AMOUNT OF A WETTING LIQUID

A significant proportion of ceramic products used in low-temperature applications are predominantly crystalline materials containing a minor amount of a glassy phase distributed in the grain boundaries. Some examples are alumina substrates, grinding media, spark plug insulators containing a calcium magnesium aluminosilicate glass phase, titanates, and hard ferrites containing 2 vol% lead silicate glass. When the liquid coats each grain, the material may often be sintered to a higher density at a lower temperature (Fig. 26.16) with less of a tendency for exaggerated grain growth. Also, the glassy phase may be essential to improve the adherence of a printed thick film or a glaze. In systems where the glassy phase is inhomogeneously distributed, however, differential rates of densification and grain growth may produce an inhomogeneous microstructure and a lower average density.

Less than 1 vol% liquid phase is sufficient to coat the grains when the liquid is distributed uniformly in a material with a nominally 1-μm grain size, and at this concentration the viscosity of the liquid does not have to be high, to resist slumping during sintering. The liquid draws the particles together, and angular particles may rotate, enabling sliding and rearrangement into a denser configuration. Pores surrounded by liquid have a driving force for shrinkage that is opposed by solid particles in contact. Liquid-

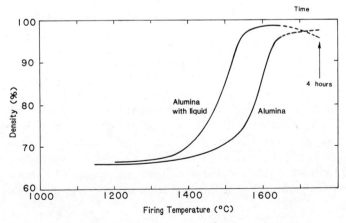

Fig. 26.16 Comparative densification behavior of a nominally 1-μm alumina powder and the alumina powder containing 5 wt% of an alkaline earth aluminosilicate glass phase.

phase-assisted sintering continues when there is some solubility of the solid, and a higher diffusivity in the liquid may increase the rate of mass transport and shrinkage. Sharp edges and small particles will preferentially dissolve, and diffusion through the liquid and crystallization in another region produce grain growth. A liquid phase that wets and dissolves the solid has been shown to rapidly penetrate between grains and disperse powder aggregates; this can reduce the tendency for exaggerated grain growth.* Also, viscous flow at boundaries may aid densification by providing a mechanism to accommodate strains between microscopic regions shrinking at slightly different rates. In some systems it may be desirable and possible to partially crystallize the glassy phase after densification. If the liquid becomes nonwetting or coalesces during sintering, a heterogeneous microstructure may be produced.

26.6 SINTERING OF WHITEWARE BODIES (VITRIFICATION)

The sintering of whitewares is complex, because densification occurs simultaneously with the reaction and dissolving of raw materials that produce new glass and crystalline phases. Fine clay coats the quartz and feldspar. The elimination of the water of crystallization in the clay minerals, the burnout of organic matter, and the decomposition of carbonate impurities occur prior to sintering. A characteristic shrinkage is observed when the metakaolin within the clay relics transforms into needle-shaped mullite crystals and silica glass in the range of 950–1000°C (Fig. 26.17).

Pure potassium and sodium feldspars melt at about 1150 and 1050°C, respectively. A liquid phase may form below 1000°C when potassium feldspar is in contact with silica and in a mixture of the feldspars when water vapor is present. The fluxing reaction of the feldspar with the kaolin above 1050°C produces a glass and needle-shaped (primary) mullite nearer the feldspar and platelike (secondary) mullite nearer the kaolin side. Shrinkage occurs by sintering in the presence of a liquid phase (Fig. 26.17).

Quartz in contact with the feldspar liquid dissolves slowly above 1250°C (Fig. 26.18). The rate of mullite formation and quartz dissolving is very dependent on the particle size of the materials and the type and concentration of impurities and secondary fluxing additives. Water absorbed from the kiln atmosphere reduces the viscosity of the glassy phase. Resistance to slumping is maintained by the low initial liquid content distributed somewhat heterogeneously and the increase in its effective viscosity due to the dispersed mullite crystals as the liquid content increases to above 50 vol%. At higher temperatures near the final stage of sintering, the mullite crystals

---

* W. J. Huppmann, S. Pejovnik, and S. M. Han, Rearrangement during Liquid Phase Sintering of Ceramics, in *Processing of Crystalline Ceramics*, Plenum, New York, 1978.

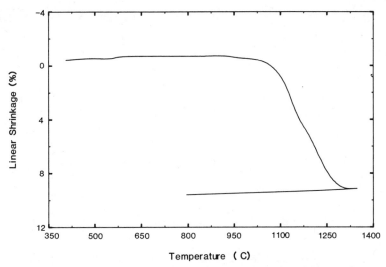

Fig. 26.17 Firing shrinkage behavior on firing an electrical porcelain body in a recording dilatometer.

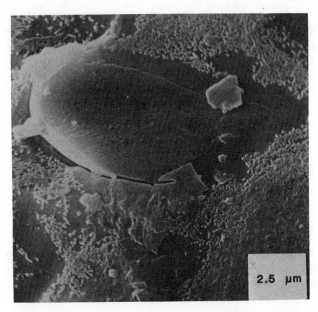

Fig. 26.18 Scanning electron micrograph of a fired electrical porcelain showing primary mullite and glass, secondary mullite, and a quartz grain with a silica-rich glassy rim and a crack at the quartz-glass interface. (Photo courtesy of G. Steere, M.S. thesis, Alfred University, Alfred, NY, 1977.)

465

become prismatic in shape, and some mullite may dissolve, but the dissolving of quartz increases the silica content and viscosity of the glass.

Coarse quartz may persist. The substitution of alumina for silica retards mullite formation but reduces the mullite dissolving at high temperatures. Because the reactions described above are controlled by chemical diffusion and viscous flow, the microstructure produced is dependent on the heating rate program and the maximum temperature. Sintering in the final stage after closed pores have formed is similar to that described in Section 26.5, except that in the glassy phase fine mullite and coarser quartz are present and the glassy phases differ in composition and viscosity. Bloating occurs when gas is evolved on heating. A high green density and control of the heating program and maturing temperature are required to achieve a high sintered density without slumping and bloating. A small amount of a secondary flux as calcium carbonate or talc may reduce the firing temperature.

Steatite porcelains are produced from bodies containing clay and at least 60% talc. Fired bodies contain a high content of glass and enstatite ($MgSiO_3$) and cristobalite (SiO_2) crystals. Cordierite materials of very low thermal expansion are produced by firing a mixture of clay, talc, and alumina. Complex reactions between the raw materials produce transient crystalline phases such as cristobalite, mullite, and protoenstatite and a minor amount of a transient liquid phase. Cordierite forms slowly at 1250–1400°C. A long firing schedule is required to produce the cordierite product, and the maximum sintering temperature is quite close to its incongruent melting temperature of 1460°C. The quantitative microstructural analysis of these fired ceramics is a complex topic, as discussed by Kingery et al.[*]

26.7 SINTERING OF GLAZES AND GLASSY THICK FILMS

A dried glaze is a relatively thin coating commonly <1 mm in thickness that contains raw materials that react to form a glassy phase, glass particles called frit, finely ground crystalline ceramic colors and opacifiers, and the processing additives. During firing, the thickness of the coating decreases as the particles fuse and the coating densifies. Reactions between the glaze and the near-surface substrate material and the glaze and the kiln gas have a significant effect on the bonding, appearance, and quality of the fired coating.

Presintering reactions described in Section 26.2 should be completed before the final stage of liquid phase sintering. Spreading of the vitreous phase is a function of the density and surface tension of the glaze, which

[*] W. D. Kingery, H. K. Bowen, and D. R. Uhlmann, *Introduction to Ceramics*, Wiley-Interscience, New York, 1976.

commonly decrease slowly with temperature, and the viscosity, which decreases more rapidly. When the surface tension and contact angle are high, a break in the glaze coating due to a drying crack or the incomplete merging of screen-printed flow units may appear to widen—a process called crawling. The surface tension and contact angle may be reduced by lead oxide in the glaze; this is commonly added in the form of a lead frit in which the lead oxide is less soluble, and much less volatile on firing. Dispersed and agglomerated particles of the refractory colors and opacifiers increase the effective viscosity of the glaze and the temperature for its flow.

Air contained in closed pores must diffuse to the surface where it can be eliminated when bubbles break. Bubbles are produced rapidly on heating when gas dissolved in the frit or molten glaze or gas that is a decomposition product produces bubbles. In fast firing, the bubble growth may temporarily increase the thickness of the glaze. The maximum bubble size increases rapidly and may exceed several hundred microns (Fig. 26.19) before decreasing to below 100 μm, as the concentration of bubbles decreases. In a relatively thin glaze, the concentration of bubbles decreases more rapidly, and there is a practical upper thickness for a quality glaze. Smoothing of the surface depends on viscous flow after surface bubbles break.

Dense pigment particles may settle in a fluid glaze but at a slower rate in a higher-density lead glaze. The molten glaze commonly penetrates into pores and the glassy boundary phase of the substrate and may dissolve some crystalline phases preferentially. As the penetration-reaction layer is formed, the glaze coating thins. Some reaction of the glaze is desirable to increase the glaze-body interface area and to produce a stronger reaction bond. Dissolving of some constituents of the body, such as quartz or foreign particles during relatively slow firing, may decrease the solubility of gas in the glaze and produce a second growth of bubbles; a pinhole defect may appear when the glaze coating is thin. In a two-fire process, the firing of the glaze (glost fire) is programmed to facilitate its sintering and maturation on the sintered (bisque-fired) substrate; however, in a one-fire process, the sintering of the glaze must be synchronized with the sintering of the underlying body.

Glaze defects produced during sintering include pinholes (described above), blisters, crawling, and gray coat. Blisters are caused by an organic particle, a spot of organic liquid or water, rust, dust, or the like that reacts with the glaze and produces gas. Crawling during sintering of the glaze is negative spreading in a region on the surface. It is produced by carbon produced from oil or grease on the substrate, which increases the interfacial tension there, or by liquid glaze with a relatively high surface tension near breaks in the applied glaze. Dissolved ceramic colors may sometimes increase the surface tension, and the dissolving may be higher when their specific surface area is extraordinarily high.

Thick film conductors, resistors, and dielectrics on electronic substrates and decorations contain glass frits that provide a vitreous fired bond. The

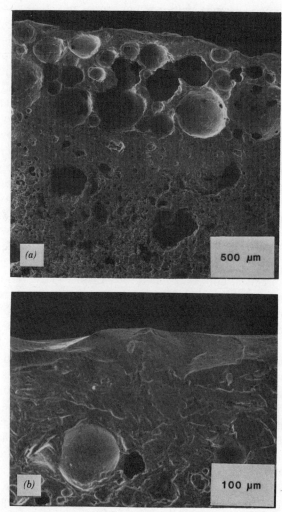

Fig. 26.19 Fracture surface of a glazed whiteware showing bubbles in glaze after (a) 10-min and (b) 60-min soaking.

frit must have a viscosity and surface tension dependence on temperature that is convenient for firing and a thermal contraction on cooling that is compatible with the substrate. For electronic applications, the frit should have a high resistivity and a low dielectric constant and loss. During sintering, the glassy phase must wet and bond the partially dispersed particles to the substrate so that shrinkage occurs only in the thickness. Some dissolving of the substrate commonly occurs, but the sintering time must be restricted. A discontinuous glass interface with particles in contact with the substrate decreases the tendency for a continuous crack separating the film from the substrate (Fig. 26.20).

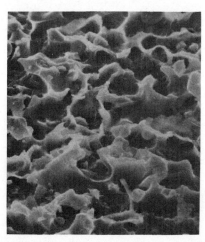

Fig. 26.20 Scanning electron micrograph of the interface of a frit-bonded Ag/Pd thick film fired on an alumina substrate (dispersed particles dissolved before photographing). (Reproduced from R. W. Vest, *Am. Ceram. Soc. Bull.* **65**(4), 631–636 (1986).)

26.8 COOLING

Fired ceramics that do not require either thermal or chemical annealing may be cooled quite rapidly, and the practical cooling rate is often controlled by the thermal shock resistance of the setters and the kiln furniture. Products with thicker cross sections and only average thermal stability are cooled more slowly. Some electrical ceramics are cooled in an atmosphere that is different from the sintering atmosphere to purposefully change the oxidation state of component ions, alter the stoichiometry, or alter the phase equilibria. In the industrial firing of manganese zinc ferrites (Fig. 26.21), a composition containing excess Fe_2O_3 is fired in an atmosphere with a relatively high partial pressure of oxygen to obtain a small grain size and to suppress the volatilization of zinc oxide. During cooling, the oxygen partial pressure is reduced to establish the content of ferrous iron[*] and produce a single ferrite phase.

When cooling products containing a glassy matrix, it is common practice to cool slowly through the glass transition temperature (viscosity of 10^{12}–10^{13} Pa s), to anneal stresses in the glass produced by thermal gradients. Slow cooling is also necessary when a crystalline phase undergoes a transformation causing a large change in volume, such as the thermal inversion of quartz at 573°C and cristobalite at 220–280°C. Stresses of sufficient magnitude caused by differential thermal contraction produce small cracks in the

---

[*] T. Reynolds, Firing, in *Treatise on Materials Science and Technology Vol. 9: Ceramic Fabrication Processes*, F. F. Y. Wang (ed), Academic Press, New York, 1976.

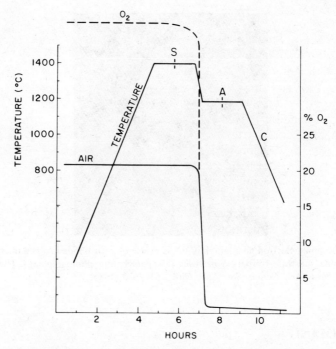

Fig. 26.21 Schematic of firing cycle for manganese zinc ferrites showing controlled atmospheres during sintering, annealing, and cooling cycles. (Courtesy T. Reynolds, p. 211 in *Treatise on Materials Science and Technology Vol. 9: Ceramic Fabrication Processes*, F. F. Y. Wang (ed), Academic Press, New York, 1976.)

product or complete failure; in whiteware, this failure is called dunting. Tensile stress is a glaze produced by a differential contraction between the glaze and substrate may produce fine cracks in the glaze, a glaze defect called crazing; a high compressive stress may cause peeling.

26.9 HOT PRESSING

Sintering during the application of an external pressure (P_e) is called hot pressing. A mechanical pressure can increase the driving pressure for densification (P_T) by acting against the internal pore pressure (P_i), without increasing the driving force for grain growth;*

$$P_T = P_e + \frac{2\gamma}{r_p} - P_i \qquad (26.15)$$

---

* D. S. Wilkinson and M. F. Ashby, p. 473 in *Sintering and Catalysis*, G. C. Kuczynski (ed), Plenum, New York, 1976.

A practical consequence of hot pressing is that a ceramic may often be produced with a comparable density but a finer grain size or a comparable grain size but a higher density. In uniaxial hot pressing, the material is pressed and sintered in a refractory die manufactured from graphite, alumina, stabilized zirconia, or a refractory metal. The industrial use of uniaxial hot pressing is limited, because its productivity is relatively low and the die maintenance is expensive. It is used to produce some commerical materials when a sintering aid or grain growth inhibitor is unknown or unacceptable, and for producing specimens for research. The hot forging of ceramics containing a liquid phase has been investigated.

Hot isostatic pressing (called HIP, or hipping), in which a product previously sintered to the final stage of sintering is subjected to a hot pressurized gas, is used industrially to reduce the size of closed pores in high-performance structural materials. Applied pressures used are in the range of 10–200 MPa. Porous materials or products may be hipped if first enclosed in a refractory, compressible enclosure or when the surface is sealed.

SUMMARY

Ceramic products are manufactured in a wide variety of sizes, shapes, material compositions, and production rates, and many types of kilns and furnaces are used for industrial firing. Initial heating volatilizes organic processing additives and decomposable constituents in raw materials and may cause the diffusion and reaction of chemical dopants with particles of the primary materials. In the first stage of sintering, the formation of grain boundaries or a glassy matrix causes the particles to join together, and an aggregate having some strength is formed. A chemical dopant acting as a densification aid may reduce the temperature for the onset of densification and in some materials is requisite to complete neck growth by chemical diffusion.

Significant densification with a decrease in the sizes of pores and a small increase in the grain size occurs during the intermediate stage of sintering. Grain growth reduces the boundary free energy but is retarded by pores that intersect the grain boundary. The transition from interconnected pores to closed pores marks the beginning of the final stage of sintering. The transition occurs rapidly in material containing pores, grains, and chemical dopants that are uniform and homogeneously distributed and more slowly when densification lags in some regions of the material.

Microstructural changes in the final stage of sintering may proceed in a desirable or an undesirable manner. To achieve a high final density and a small to moderate grain size, small pores should be associated with grain boundaries, and the gas in the pore should diffuse from the pore. The association of the pore and the boundary may be enhanced by doping the powder with an additive that either reduces the boundary mobility or increases the pore mobility and using a faster firing schedule.

Powder aggregates are a common source of chemical and/or physical nonuniformity in the green material and exaggerated grain growth (heterogeneous sintering). A few percent of an additive that creates a reactive liquid in the grain boundaries may aid in dispersing aggregates and reduce the sintering temperature, and this type of additive is often used in producing nonrefractory products. Shaped products, glazes, and thick films containing a glassy matrix may be sintered into a monolithic product or coating. The sintering of ceramic whitewares to a high density without slumping depends on a complex progression of reactions which alters the crystalline phases and increases the proportion of the viscous glass in the microstructure.

The atmosphere and cooling schedule following sintering are important to develop the proper oxidation states and to anneal differential strain in products containing a glassy phase. Pressure sintering called hot pressing may be used to increase the density without increasing the sintering temperature.

SUGGESTED READING

1. *Process Mineralogy of Ceramic Materials*, W. Baumjart, A. C. Dunham, and G. C. Amstutz (eds), Elsevier, New York, 1984.

2. M.F. Yan, Sintering of Ceramics and Metals, *Advances in Powder Technology*, Gilbert Y. Chin (ed), American Society for Metals, Metals Park, OH, 1982.

3. W. D. Kingery, H. K. Bowen, and D. R. Uhlmann, *Introduction to Ceramics*, Wiley-Interscience, New York, 1976.

4. J. E. Burke and J. H. Rosolowski, Sintering, in *Treatise on Solid State Chemistry Vol. 4: Reactivity of Solids*, N. B. Hannay (ed), Plenum, New York, 1976.

5. R. J. Brook, Controlled Grain Growth, in *Treatise on Materials Science and Technology Vol. 9: Ceramic Fabrication Processes*, Franklin F. Y. Wang (ed), Academic Press, New York, 1975.

6. F. H. Norton, *Fine Ceramics*, Robert E. Krieger, Malabar, FL, 1978.

7. G. C. Kuczynski, N. A. Hooten, and C. F. Gibson (eds), *Sintering and Related Phenomena*, Gordon and Breach, New York, 1967.

8. K. A. Berry and M. P. Harmer, Effect of MgO Solute on Microstructure Development in Alumina, *J. Am. Ceram. Soc.* **69**(2), 143–149 (1986).

9. T. M. Shaw, Liquid Redistribution during Liquid Phase Sintering, *J. Am. Ceram. Soc.* **69**(1), 27–34 (1986).

10. W. H. Rhodes, Agglomerate and Particle Size Effects on Sintering Yttria-Stabilized Zirconia, *J. Am. Ceram. Soc.* **64**(1), 19–22 (1981).

11. George W. Scherer, Sintering of Low Density Glasses: I, II, and III, *J. Am. Ceram. Soc.* **60**(5), 236–246 (1977).

12. J. H. Rosolowski and C. Greskovich, Theory of the Dependence of Densification on Grain Growth during Intermediate Stage Sintering, *J. Am. Ceram. Soc.* **58**(5–6), 177–182 (1975).

13. F. Thummler and W. Thomma, The Sintering Process, *Metal. Mater.* **12**(115), 69–108 (1967).

14. G. C. Kuczynski, Self Diffusion in Sintering of Metallic Particles, *Trans. Am. Inst. Min. Met. Eng.* **185**, 169–178 (1949).

PROBLEMS

26.1 Explain why the temperature for complete binder burnout is commonly higher when products of a larger size or finer particle size are fired along with the regular product.

26.2 What are the causes of shrinkage gradients and warping when firing ceramic products?

26.3 Explain how the densification behavior shown in Fig. 26.5 would be altered if (1) the particle size was larger, (2) a dopant that increases neck growth and the densification rate is added to the powder, (3) very large pores are present in the compact, and (4) the compact density is significantly higher.

26.4 The mean intercept length determined for a polished and etched section of a sintered material exhibiting regular grain growth is 1.3 μm. What is the mean grain size?

26.5 Derive an equation expressing the bulk density as a function of the isotropic shrinkage during sintering and the fired density. Assume the mass is constant during sintering and cooling.

26.6 Describe the microstructural defects that might be expected after sintering compacts containing (1) clusters of pores much larger than the grain size, (2) powder aggregates that densify more rapidly, and (3) powder aggregates that densify less rapidly.

26.7 Explain using an equation why faster firing to a higher temperature may sometimes produce a product with an equal density but a smaller grain size.

26.8 An unfired glaze coating is 0.4 mm thick and 45% porous. What is its thickness after firing when the porosity is 10% and 20% of the glaze has penetrated into the body?

26.9 What is the relative temperature for initial shrinkage in uniform compacts containing particles of 5.0 and 0.5 μm diameter, respectively? Would this ratio be altered if the 0.5-μm particles are 5.0-μm aggregates?

26.10 Explain why pore coalescence in the glass matrix of a material may cause bloating during isothermal sintering.

26.11 During sintering, a material forms closed pores when the bulk density is 92%. Estimate the grain size if the pore diameter is 0.3 μm. What are your assumptions?

26.12 Sketch the fractional grain size distribution curves for a sequence of times for regular grain growth and on the appearance of exaggerated grains.

26.13 A resistive film printed on a fired alumina substrate is 0.02 mm thick and 40% porous. During firing 0.01 mm of the substrate is dissolved. The fired density is 90%. What is the fired thickness of the film?

26.14 Compare the driving force for the closure of a pore 5 μm in diameter ($\gamma_s = 350$ mN/m) to a hot isostatic pressing pressure of 35 MPa.

26.15 The mean pore side should decrease when sintering a compact containing uniform pores smaller than the grain size, but the mean pore size may increase when a compact containing a wide range of pore sizes is sintered. Explain.

26.16 Binder burnout may produce considerable gas. Calculate the volume of gas produced from 1 cm^3 of an alumina compact containing 2 wt% polyvinyl alcohol binder. Assume that the compact is 60% dense and the gas is at 400°C and contains H_2O and CO_2. What is the volume of air that must infiltrate the compact?

Aperture Size of U.S. Standard Sieves

Sieve Number	Aperture (μm)
3.5[a]	5660
4	4760
5[a]	4000
6	3360
7[a]	2830
8	2380
10[a]	2000
12	1680
14[a]	1410
16	1190
18[a]	1000
20	841
25[a]	707
30	595
35[a]	500
40	420
45[a]	354
50	297
60[a]	250
70	210
80[a]	177
100[6]	149
120[a]	125
140	105
170[a]	88
200	74
230[a]	63
270	53
325[a]	44
400	37

[a] Recommended series.

475

Density of Ceramic Materials

Material	Composition	Density (Mg/m^3)
Albite	$NaAlSi_3O_8$	2.61
Aluminum nitride	AlN	2.8
Anorthite	$CaAl_2Si_2O_8$	2.77
Andalusite	Al_2SiO_5	3.15
Anorthoclase	$KNaAl_2Si_6O_{16}$	2.58
Barium ferrite	$BaFe_{12}O_{19}$	5.31
Barium titanate	$BaTiO_3$	6.01
Beryllium oxide	BeO	3.0
Boron carbide	B_4C	2.5
Calcite	$CaCO_3$	2.71
Cordierite (beta)	$Mg_2Al_4Si_5O_{18}$	2.6
Corundum	Al_2O_3	3.98
Forsterite	Mg_2SiO_4	3.22
Galena	PbS	7.5
Graphite	C	2.25
Halloysite	$Al_2Si_2O_5(OH)_4 \cdot 2H_2O$	2.62
Hematite	Fe_2O_3	5.25
Ilmenite	$FeTiO_3$	4.7
Kaolinite	$Al_2Si_2O_5(OH)_4$	2.61
Kyanite	Al_2SiO_5	3.60
Lime	CaO	3.3
Magnesite	$MgCO_3$	2.96
Microcline	$KAlSi_3O_8$	2.58
Montmorillonite	$(Al_{1.67}Na_{0.33}Mg_{0.33})(Si_2O_5)_2(OH)_2$	2.50
Mullite	$Al_6Si_2O_{13}$	3.23
Muscovite	$Al_2K(Si_{1.5}Al_{0.5}O_5)_2(OH)_2$	2.9
Orthoclase	$KAlSi_3O_8$	2.55
Periclase	MgO	3.58
Pyrite	FeS_2	5.02
Pyrophyllite	$Al_2(Si_2O_5)_2(OH)_2$	2.80

Material	Composition	Density (Mg/m^3)
Quartz	SiO$_2$	2.65
Rutile	TiO$_2$	4.26
Silica, vitreous	SiO$_2$	2.20
Silicon carbide	SiC	3.21
Silicon nitride	Si$_3$N$_4$	3.1
Sillimanite	Al$_2$SiO$_5$	3.23
Spinel	MgAl$_2$O$_4$	3.60
Spodumene (beta)	Li$_2$Al$_2$Si$_4$O$_{12}$	2.35
Spodumene (alpha)		3.20
Talc	Mg$_3$Si$_4$O$_{10}$(OH)$_2$	2.75
Titania (see rutile)		
Tungsten carbide	WC	15.7
Zinc oxide	ZnO	5.68
Zircon	SiZrO$_4$	4.7
Zirconia, stabilized	ZrO$_2$(8%Y$_2$O$_3$)	6.0

Viscosity
and Density Values
of Water and Air

Temperature (°C)	Viscosity of Water (mPa s)	Density of Water (Mg/m^3)	Density of Air (Mg/m^3)
15	1.1404	0.999099	0.001226
16	1.1111	0.998943	0.001221
17	1.0828	0.998774	0.001217
18	1.0559	0.998595	0.001213
19	1.0299	0.998405	0.001209
20	1.0050	0.998203	0.001205
21	0.9810	0.997992	0.001201
22	0.9579	0.997770	0.001196
23	0.9358	0.997538	0.001192
24	0.9142	0.997296	0.001188
25	0.8937	0.997044	0.001184
26	0.8737	0.996783	0.001180
27	0.8545	0.996512	0.001176
28	0.8360	0.996232	0.001173
29	0.8180	0.995944	0.001169
30	0.8007	0.995646	0.001165

Conversion from Metric to English/American Units

Parameter	From SI Units	Multiply by	To Get
Length	m	39.370	Inch
		3.2808	Foot
Area	m^2	1.5500×10^3	$Inch^2$
		10.764	$Foot^2$
Volume	m^3	6.1024×10^4	$Inch^3$
		35.315	$Foot^3$
		1.0567×10^3	Liquid quart (U.S.)
Mass	kg	2.2046	Pound
		1.1023×10^{-3}	Ton
Density	kg/m^3	3.6127×10^{-5}	$Pound/inch^3$
		6.2428×10^{-2}	$Pound/foot^3$
Speed	m/s	3.2808	Foot/second
Acceleration	m/s^2	3.2808	$Foot/second^2$
Force	N	0.22481	Pound force
		10^{5a}	Dyne
Pressure, stress	Pa or N/m^2	1.4504×10^{-4}	Psi
		10^a	$Dyne/cm^2$
		2.0885×10^{-2}	$Pound/foot^2$
		7.5006×10^{-3}	Torr
Energy	J	9.4845×10^{-4}	Btu
		0.23901	Calorie
		10^{7a}	Erg
		2.7778×10^{-7}	Kilowatt hour
Power	W	9.4845×10^{-4}	Btu/second
		3.4144	Btu/hour
		0.23901	Calorie/second
		1.3410×10^{-3}	Horsepower
Heat capacity	$J/(kg \cdot K)$	2.3901×10^{-4}	$Cal/(g\,°C)$
		2.3901×10^{-4}	$Btu/(lb\,°F)$
Thermal conductivity	$W/(m \cdot K)$ or	6.9380	$Btu\,in./(h\,ft^2\,°F)$
	$J/(s \cdot m \cdot K)$	2.3901×10^{-9}	$Cal/(s\,cm\,°C)$
		0.57816	$Btu/(h\,ft\,°F)$

[a] These factors are exact; others to five digits.

Index